D0445269

DISCARDED
PASADENA CITY COLLEGE

Empire of the Stars

BOOKS BY ARTHUR I. MILLER

IMAGERY IN SCIENTIFIC THOUGHT
Creating 20th-Century Physics

INSIGHTS OF GENIUS
Imagery and Creativity in Science and Art

EINSTEIN, PICASSO
Space, Time, and the Beauty That Causes Havoc

EMPIRE OF THE STARS
*Obsession, Friendship, and Betrayal
in the Quest for Black Holes*

DISCARDED
PASADENA CITY COLLEGE

EMPIRE
of the
STARS

Obsession, Friendship, and Betrayal
in the Quest for Black Holes

Arthur I. Miller

HOUGHTON MIFFLIN COMPANY
BOSTON • NEW YORK
2005

Copyright © 2005 by Arthur I. Miller

ALL RIGHTS RESERVED

For information about permission to reproduce selections from
this book, write to Permissions, Houghton Mifflin Company,
215 Park Avenue South, New York, New York 10003.

Visit our Web site: www.houghtonmifflinbooks.com.

Library of Congress Cataloging-in-Publication Data
Miller, Arthur I.
Empire of the stars : obsession, friendship, and betrayal
in the quest for black holes / Arthur I. Miller.
p.   cm.
Includes bibliographical references and index.
ISBN 0-618-34151-X
1. Chandrasekhar, S. (Subrahmanyan), 1910– 2. Eddington,
Arthur Stanley, Sir, 1882–1944. 3. Astrophysicists—India—
Biography. 4. Astrophysicists—Great Britain—Biography.
5. Astrophysics—History—20th century. 6. Black holes
(Astronomy) 7. Relativity (Physics) 8. Quantum theory.
9. Hydrogen bomb. I. Title.
QB35.M55 2005 520'.92'2—dc22
[B] 2004060909

Printed in the United States of America

Book design by Robert Overholtzer

QUM 10 9 8 7 6 5 4 3 2 1

To Lesley

SHATFORD LIBRARY

NOV − − 2006

1570 E. Colorado Blvd.
Pasadena, CA 91106

In my entire scientific life, extending over forty-five years, the most shattering experience has been the realisation that [Kerr's] exact solution of Einstein's equations of general relativity provides the *absolutely exact representation* of untold numbers of massive black holes that populate the universe. This "shuddering before the beautiful," this incredible fact that a discovery motivated by a search after the beautiful in mathematics should find its exact replica in Nature, persuades me to say that beauty is that to which the human mind responds at its deepest and most profound.

— *Subrahmanyan Chandrasekhar*

The conditions in the star are very extreme; but the ultimate things to be dealt with — electrons, atomic nuclei, x-rays — are the same in the star as in the laboratory, and we can apply our laboratory knowledge of them. Calculate "according to the laws of physics as known to terrestrial experiment"; and then turn to the man with the telescope and ask, "Is that anything like the stars *you* come across in the sky?"

— *Arthur Stanley Eddington*

The thought that the laws of the macrocosmos in the small reflect the terrestrial world obviously exercises a great magic on mankind's mind; indeed its form is rooted in the superstition (which is as old as the history of thought) that the destiny of men could be read from the stars.

— *Max Born*

Contents

PART III

What Happens When Stars Die

Acknowledgments

Researching the dramatic story of Chandra and Eddington, of Chandra's mathematical verification of black holes and his four decade wait until the scientific community accepted it, involved getting to grips with esoteric theoretical physics and state-of-the-art technology as well as meeting as many as possible of the major players in the story.

Above all, I am deeply grateful to Chandra's widow, Lalitha, who generously spent many hours sharing her memories of her late husband. I am also much indebted to James Cronin, Valeria Ferrari, Robert Geroch, Roger Hildebrand, Norman Lebovitz, T. D. Lee, Eugene Parker, Edwin Salpeter, Noel Swirdlow, Saul Teukolsky, Kip Thorne, Michael Turner, Peter Vandervoort, Robert Wald, and Don York, all very distinguished long-term colleagues of Chandra's who contributed personal reminiscences and details about his life and career.

Many thanks to Chandra's cousin, V. Radhakrishnan, of the Raman Institute in Bangalore, India, who filled in details of Chandra's life, provided superb and companionable hospitality, and showed me around his latest project, his catamaran. Dr. U. Ponnambalam, chairman of the Physics Department at Presidency College, Chennai, very kindly gave me an informative tour of Presidency College, where Chandra was an undergraduate.

At the University of California at Santa Cruz, Stan Woosley patiently took me through the present state of research on supernovae. Don Osterbrock kindly filled me in on day-to-day life during Chandra's years in residence at Yerkes Observatory and shared his historical knowledge and acumen as an astrophysicist. He also read large chunks of the manuscript and made invaluable comments.

I enjoyed an unforgettable weekend at Los Alamos in the company of Stirling Colgate, who gave me a fascinating account of the relationship between nuclear weapons and the birth of modern research into supernovae and black holes, among much else. His colleague back in the 1960s, Richard White, kindly provided more details of the flavor of the times. Sergey Blinnikov generously filled me in on his work with Yakov Zel'dovich on supernovae in the Soviet Union. I am grateful to Edward Frieman, Richard Garwin, Marvin Goldberger, and Conrad Longmuire, who told me much about Chandra's association with Los Alamos.

My work would have been quite impossible without access to the University of Chicago's magnificent Joseph Regenstein Library, where Chandra's correspondence and manuscripts are deposited. I was very fortunate to have Jay Satterfield's assistance there. I am also most grateful to Takeshi Oka for his splendid hospitality at the University of Chicago and generous assistance with archival matters.

Peter Hingley, the librarian at the Royal Astronomical Society, Burlington House, London, generously devoted much time to dealing with my frequent requests for information.

As in all my other historical research, it was invaluable to have access to interviews carried out over more than four decades by the American Institute of Physics' Center for the History of Physics in Maryland. Every historian of physics owes an enormous debt to the center and its director, Spencer Weart.

I would also like to offer my thanks to the following archivists and librarians who provided invaluable assistance: Judith Goodstein (California Institute of Technology), Shaun J. Hardy (Carnegie Institution of Washington), Mark Hurn (Institute of Astronomy, Cambridge University), Dan Lewis (Huntington Library, San Marino, California), Kathi Murphy (McGill University), Jonathan Smith (Trinity College, Cambridge University), Tony Hogarth Smith (Christ's Hospital, Horsham, England), and G. Madhaven (executive secretary, Indian Academy of Sciences, Bangalore, India). K. C. Wali's biography of Chandra, written with the benefit of extensive interviews with him, was extremely helpful.

Meg Weston Smith graciously shared with me memories of her father, Edward Arthur Milne. Marianne Douglas and Patrick Douglas, relatives of Eddington's biographer Allie Vibert Douglas, kindly provided further information about her. David Dewhirst provided information about Eddington's daily life at the Cambridge Observatory.

Many thanks as always to my close friend Mike Brady, who read an early version of this book from cover to cover and offered much sagacious advice as well as his usual support. One of the unexpected rewards of this book was that I renewed my old friendship with Gary Steigman, whose "speed of light" e-mail correspondence helped clarify many issues. And warm thanks to my pal Dave Scott for filling me in on his excursion to the Moon on *Apollo 15* as well as for good cheer and many companionable dinners.

Many colleagues generously took time from their busy schedules to respond to my queries and to read portions of the manuscript. I am very grateful to Mitch Berger, Jeremy Bernstein, Stirling Colgate, Andy Fabian, Chris Fryer, Roy Garstang, Jeremy Gray, Chryssa Kouveliotou, Keith Mason, Leon Mestel, Martin Rees, Virginia Trimble, Steve Weinberg, Martin Weisskopf, and Stan Woosley.

My appreciation to Ken Brecher, Steve Brush, David Cassidy, Emma Goddard, Andrew Gregory, Simon Singh, Steve Miller, Roger Stuewer, and Andrew Warwick for helpful and enlightening conversations.

I am hugely indebted to my agent, Peter Tallack of Conville & Walsh Literary Agency, for his encouragement, enthusiasm, and faith in this project. The agency has been immensely helpful in smoothing out the route from first to final draft and all the travail in between. Particular thanks too to Patrick Walsh. I owe an enormous debt to my editors, Tim Whiting at Little Brown, who provided much support and helpful criticisms, and Amanda Cook at Houghton Mifflin, whose insightful comments and suggestions were invaluable in shaping the final version. This book would not be what it is without their help.

I shifted from physics research to the history of scientific ideas in order to explore the way in which scientists think as they seek out deep truths about the cosmos, in both the large and the small, and to examine what happens when scientist meets nature in a one-on-one confrontation. Looking into the story told in this book, I could not avoid becoming aware of the dauntingly hard and very solitary work involved in research at the cutting edge of science. At this level, scientists often find themselves in thrall to their intellectual endeavors with a degree of intensity and competitiveness that sometimes exacts a high psychological toll. Playing for such high stakes, they sometimes act and react in ways that may seem inexplicable to lesser human beings. That was true for the major players in my story. Working out why has involved spinning a scenario based on what can be inferred from the information available. For this interpretation I bear full responsibility, as well as for any scientific errors.

A little way into the research for this book I happened to go to an art gallery opening in London where I met a beautiful and intriguing woman. Lesley became my inspiration, my muse, my wife, and my soulmate. She has greatly improved this book, and my life.

Prologue

Or, if there were a sympathy in choice,
War, death, or sickness did lay siege to it,
Making it momentary as a sound,
Swift as a shadow, short as any dream,
Brief as the lightning in the collied night,
That, in a spleen, unfolds both heaven and earth,
And ere a man hath power to say, "Behold!"
The jaws of darkness do devour it up:
So quick bright things come to confusion.

— William Shakespeare,
A Midsummer Night's Dream (Act I, Scene 1)

EVER SINCE the evocative term "black hole" was coined in 1967, these mysterious voids in the universe have assumed an almost mystical appeal. The attraction of a vast emptiness that imprisons not only matter but also light is quite literally inescapable.

Imagine that you are an astronaut, seduced by the grandeur of a black hole into straying too close. Trapped by its immense gravitational field and the tornado-like swirling of space around it, you sweep feet first over the horizon. As you fall, what you see is truly awesome. Just before you slip over the edge, the whole universe of stars and galaxies appears to rush together into one bright spot. The intense gravity of the black hole funnels the light from distant objects into a tighter and tighter cone, like tunnel vision. All the while you are entranced by the fireworks display of atoms snared by the black hole's immense gravity. They dance in a cosmic traffic jam, bumping into each together, becoming hotter and hotter until they blaze with the brightness of a million Suns as they stray too close and plunge with you into nowhere. Then you start to feel the irresistible attraction of the collapsed star deep inside, at once unimaginably small and infinitely dense. As the collapsed star sucks you deeper into the black hole, the gravitational pull grows stronger and stronger. You stretch like a piece of toffee, longer and longer and thinner and thinner, until you are torn apart. The potent gravitational force around the black hole means that light takes longer and longer to reach distant observers. They see you poised on the edge of the black hole, frozen in space and time forever.

A black hole is a well in space, the final resting place of a collapsed star. For decades, scientists resisted the very idea as a theoretical freak that couldn't actually exist, an ugly solution to the most beautiful theory ever created, Albert Einstein's general theory of relativity. But astronomers now know that the universe is littered with these monsters and that a giant black hole sits at the center of our own galaxy. What's more, we can actually observe black holes by detecting the x-rays emitted by particles as they spiral in toward the event horizon before plummeting in.

Having taken their place in our picture of the fabric of nature, black holes have opened our minds to staggering and sometimes frightening speculation: are they spawning baby universes, of which ours may be but one? Might they open a shortcut to a distant part of the universe or even be portals for time travel? How might we devise an experiment to create a black hole in a laboratory here on Earth?

Many scientists now believe that black holes hold the key to understanding how our universe has evolved and how nature behaves at its most extreme. At the very edge of space and time black holes are the engines that power quasars, the brightest objects in the universe, brighter than a trillion Suns. Black holes have pushed our knowledge of the cosmos to its limits.

Black holes may ultimately reveal the microstructure of matter and the fate of the universe itself. For in these gravitational plugholes, atoms are crushed into their basic building blocks and possibly even crushed right out of existence. Here the two great scientific theories of the twentieth century — general relativity, which describes the world of the very large, and quantum mechanics, which describes the very small — meet head on, and the laws of physics as we know them may break down completely.

*

In recounting the epic story of the discovery of black holes, *Empire of the Stars* is a window through which we can glimpse humankind's remarkable quest to understand how stars are born, live, and die, and the way in which this knowledge has profoundly altered our scientific and cultural views of the world. It is the story of one man's fight with the scientific establishment for recognition of his idea — an episode that sheds light on what science is, how it works, and where it can go wrong.

Fascinated by the stars from time immemorial, we have come to believe that their fate is ultimately ours. What is so extraordinary about the early research into the life history of stars is that it consisted almost purely of theoretical speculation, the imaginings of scientists who had the confidence to make grandiose assumptions about some of the largest objects ever known. With only meager experimental data on stellar evolution, scientists were forced to play God. Their biggest assumption was that the science discovered by humans, who are the accidental products of the stars and a tiny fraction of the age of the universe, could be used to explore the evolution of stars thousands of trillions of miles distant, from their birth some billions of years ago to their deaths untold billions of years in the future.

What breathtaking chutzpah that was! Tracing the story of those pioneering astrophysicists is an intellectual adventure, with twists and turns that will take us into the deepest realms of theoretical science, the designs for manufacturing hydrogen bombs, and the effect of the arms race on astrophysical research. The story also sheds light on the epic conflict between the concepts of classical and modern science embodied in the vastly different outlooks of physicists and astrophysicists and the clash of cultures at the twilight of the British Empire in the 1930s. All this led to a forty-year delay before the notion of black holes was finally accepted, by which time just about everyone had forgotten the role played by the man who had originally provided clear-cut evidence for their existence.

Subrahmanyan Chandrasekhar's flash of inspiration came when he was an unknown nineteen-year-old in the hot summer of 1930. In ten minutes, sitting in a deck chair overlooking the Arabian Sea, Chandra (as he was universally known) carried out some calculations that augured a disturbing fate for the small, dense stars known as white dwarfs. At the time scientists assumed that white dwarfs were dead stars in their final resting state. Those that had been found had more or less the mass of the Sun but were no bigger than Earth. Chandra's calculations showed that there was an upper limit to the mass of these white dwarfs. Any star more massive than that when it burned out would not end its life as an inert rock but would begin an endless process of collapse, crunched by its own gravity into a singularity—a minuscule point of infinite density and zero volume, many trillions of times smaller than the period at the end of this sentence and many trillions of times denser than Earth.

Only one person understood the full implications of Chandra's discovery: Sir Arthur Stanley Eddington, the greatest astrophysicist in the world at that time. Eddington himself had flirted with the idea that a dead star might collapse indefinitely in this manner, so he should have been delighted with Chandra's mathematical verification. Instead, without any warning, he used a meeting of the Royal Astronomical Society to savage Chandra's result cynically and unmercifully. The encounter cast a shadow over the lives of both men and hindered progress in astrophysics for nearly half a century.

As a teenager I read Eddington and was sufficiently taken by his writings to want to become a scientist. Although I didn't understand much of what he described, it all sounded exciting. The sheer sweep of the subject matter — from atoms to the life and death of stars — was breathtaking, the language vivid and gripping.

Chandra's writings were inspiring in another way. They exemplified how a superbly gifted scientist could use mathematics to study the nature of the stars. The more I discovered about Chandra's story, the more intriguing it became. For all his brilliance, his life was tinged with tragedy. After Eddington refused to take his idea seriously and subjected him to public ridicule, Chandra never really regained his confidence. Despite a long and incredibly productive scientific career, no amount of recognition could ever satisfy him. I wondered what other great discoveries he might have made, had his early life not been blighted by disappointment. By all accounts he was a reserved, deeply private, and highly serious man. Who was the real man behind that stern façade?

And what of Eddington? Why did the greatest astrophysicist in the world choose to demolish the youthful Indian in such a vicious way? Eddington, famous for his sharp tongue, was harsh and cynical with other scientists, too. But with Chandra, Eddington's criticisms took on a different and darker tinge.

By the time I met Chandra, he was eighty-three. The occasion was a conference on creativity at the Chicago Academy of Sciences, where Chandra was to give the keynote lecture. The ballroom-sized auditorium was filled to capacity. There was a buzz of excitement as a distinguished-looking Indian man walked through the massive double doors. Here was a Nobel laureate, one of the most important scientists of the age, who had come to address the audience on matters of high scientific creativity. Despite his slight figure, he was a man of great presence. Elegantly dressed and no more than five feet six in height, he carried himself with dignity, though his shoulders were beginning to stoop. His sparse white hair was carefully combed across the top of his still handsome dark face, with its prominent forehead, piercing eyes, and full mouth clamped tight in an expression of iron determination.

As he spoke, he looked up from his notes from time to time to reminisce with obvious delight about great scientists of the past whom he had known. The audience was spellbound. Afterward I was able to exchange a few words with him and shake his hand. It was a thrilling moment. This man, who had transformed our understanding of the heavens, had been a great inspiration to me throughout my life. He spoke of the book he was writing on his intellectual hero, Isaac Newton, and of the recent exciting discoveries that had been made researching black holes. Thoughtlessly, I mentioned the Eddington episode—and his face clouded. He graciously shook my hand, and we agreed to speak again.

Later reports of the altercation with Eddington claim that Chandra rapidly put it behind him and that the two men were actually firm friends. But even a cursory glance at Eddington's scientific articles and the correspondence between the two reveals a very different story. Again and again Chandra expressed deep anger, frustration, and resentment, while Eddington clung stubbornly to his own view of the universe, ridiculing Chandra's discovery as "stellar buffoonery."[1] Chandra never missed an opportunity to recount the events of that fateful day at the Royal Astronomical Society, emphasizing that he had been right and Eddington wrong, even though Eddington refused ever to admit it. He was always careful to speak in glowing terms about Eddington the man. But in interviews he revealed how profoundly he had suffered from Eddington's treatment of him.

After that memorable meeting in 1993, I continued to puzzle over Chandra's complex, tragic story. A couple of years ago I decided to explore it more deeply. Sadly, by that time Chandra was dead. But I was fortunate enough to meet Lalitha, his devoted wife of over fifty years. I also interviewed many of his colleagues, some of whom had studied under him. In Bangalore I met his cousin, the eminent astrophysicist V. Radhakrishnan, son of Chandra's uncle, C. V. Raman, India's first Nobel laureate in physics. From their recollections and from letters and documents that Chandra and, later, Lalitha deposited at the University of Chicago's Joseph Regenstein Library, little by little the real Chandra — the man behind the persona he built up around himself — began to emerge.

An important step in getting to know Chandra was to experience the tropical heat and dust of South India, where he was born and lived until he finished his undergraduate schooling at Presidency College in Madras (now Chennai), at the age of nineteen. There I walked across the burning sands of the Marina, the long beach that edges Chennai. In Chandra's time it was elegant and fashionable. He would cycle there in the evenings with his brothers to escape the stifling heat and gaze in wonder at the brilliant stars. As he recalled in a letter to his brother Balakrishnan, he sometimes went alone to the silent beach to fling himself on the sand and pray to God that he could be another Einstein.

Chandra's work lay dormant for more than three decades, to be looked at anew only when an entirely different development in world affairs — the race to design the hydrogen bomb — sparked renewed interest in the possibility of black holes. The explosive power that the scientists were looking for turned out to be the very same power that had created these massive holes in space, and the universe itself. Chandra and his discovery returned to their proper place at the forefront of scientific endeavor. Toward the end of his life, he received a Nobel Prize for his achievement. But he did not feel vindicated, for his achievement had been virtually ignored for almost forty years.

This book is the biography of an idea rather than of a man. Nevertheless, science is a human endeavor, driven by hopes, dreams, and aspirations. At the highest level, scientists are playing for the highest stakes. They may be brilliant, even geniuses. But as human beings they may also be seriously flawed, as both Chandra and Eddington were.

PART I

The Mystery of White Dwarfs

1. Fatal Collision

IT HAD BEEN a momentous meeting of the Royal Astronomical So ciety at Burlington House, just off Piccadilly, that Friday, January 11, 1935.[1] Yet as the hundred distinguished members filed out of the hall, chatting excitedly, no one could put their finger on exactly what had happened. What is certain is that at 6:15 P.M. a shy, boyish, twenty-four-year-old astrophysicist from Cambridge, Dr. Subrahmanyan Chandrasekhar, rose to present a dramatic discovery that had been ignored for nearly five years.

What Chandra had to say was entirely new and went against all the accepted tenets of science. He was aware that it might be met with frowns and criticisms, even downright opposition. A great deal was at stake, both scientifically and personally. But he was not worried. He was sure he could hold his own in any scientific give and take. Young though he was, he had a maturity and a command of astrophysics far beyond his years. He had given his first lectures when he was just eighteen, at Presidency College in Madras. There he regularly held forth at the physics colloquia normally dominated by mature research scientists. In India his contemporaries spoke of him as a genius. Now Chandra had a full thirty minutes to argue the case for his groundbreaking discovery; most speakers were given only half that time. Sir Arthur Eddington would deliver the next paper.

The two men could not have been more different. The diminutive Chandra, fresh-faced and immaculately dressed, was cherubically handsome and so dark he was practically black. With a broad nose, full mouth, and glistening black hair that swept back from his forehead, he carried himself with the aristocratic bearing of a Brahmin. He radiated youthful innocence, in stark contrast to Eddington's air of worldly superciliousness.

Eddington, then fifty-two, held the prestigious Plumian Professorship of Astronomy and Experimental Philosophy and had received practically every award science could bestow. A tall, etiolated figure with a lugubrious pale face, thin-lipped and sardonic, he kept a pince-nez carefully balanced on his narrow nose and stood ramrod straight. In his rumpled three-piece tweed suit with a fob watch in his breast pocket and an air of carefully cultivated dishevelment, he was the archetypal Oxbridge don. He carried himself like a representative of the British Empire in its grandest days. Eddington was renowned for his ruthlessness and rapier wit. Many of those present had come simply to see him perform. He never disappointed.

Chandra's great discovery concerned nothing less than the ultimate fate of the universe. What became of stars at the end of their lives, when they had finally burned up all their fuel? Astrophysicists assumed that they shrank and shrank until they became white dwarfs. But what happened then? Chandra had published two papers on white dwarfs a few years earlier. The problem they examined was one that Eddington had raised in his influential 1926 book, *The Internal Constitution of the Stars*. Nevertheless, the young Indian's work had been totally ignored.

Chandra was just twenty when his first brief paper on white dwarfs was published, in 1931. He had written it just as he was embarking on his postgraduate career at Cambridge, with no time for corrections or improvements. In his next paper, in 1932, he laid out his arguments in a more polished form. But still the scientific establishment refused to pay any attention. There was no encouragement from anyone — not from Eddington, or from Chandra's thesis supervisor, Ralph Fowler, or from his new friend Edward Milne, the Rouse Ball Professor of Mathematics at Oxford. Were they trying to protect a young innocent from foolishly publishing an outrageous idea? Was it professional jealousy — were they trying to safeguard their own positions in the scientific cosmos? Or was there some other, more sinister factor, to do with Chandra's color and racial origins? Such a thought would be taboo today, but in those days the Raj still ruled India, and Englishmen still basked in the belief in their innate superiority. Could it be that these stalwarts of Empire found it unacceptable to be overtaken by a young man from one of the colonies and refused to accept that he might have anything to teach them?

The train ride from Cambridge to London seemed interminable that freezing January Friday. It was not that Chandra was nervous about delivering a paper before such an august body as the RAS; he had already done so several times. All those presentations had been well received and duly published in the society's proceedings, the *Monthly Notices of the Royal Astronomical Society*. But his new paper was different. In his previous work he had simply elaborated on other scientists' research, mostly filling in details — no mean task, the sort of work that novices undertake to prove their technical worth. But this latest paper contained his own discovery. Here was his chance to prove himself, the day when everyone would sit up and take notice. He was full of excitement and anticipation.

Nevertheless, he sensed that something was not quite right. The previous day the assistant secretary of the RAS had told him in confidence that Eddington would speak after him. From the title of Eddington's paper, Chandra knew that it concerned his discovery. He wondered uneasily whether Eddington would be critical. When the two had met over dinner at high table the previous evening, Eddington had been evasive, refusing to reveal anything about his own talk. He had arranged, he said, for Chandra to have extra time to speak because he understood that the young man had so much to say. Chandra later recalled that he was "really annoyed because here was Eddington coming to see me, week after week, about my work while he was writing a paper himself and he never told me about it."[2] But perhaps some criticism at the well-attended RAS meeting by a scientist of Eddington's stature would be taken as a mark of the work's importance. Chandra had not the slightest inkling of what would actually happen.

Huddled in his dark overcoat, collar turned up against the chill January winds, Chandra took a taxi from Liverpool Street Station to South Kensington, as he did every Friday before an RAS meeting, to have lunch with his friend William McCrea. Six years older than Chandra, McCrea was a cheery young man with dark hair and a trim mustache. A Reader in mathematics at Imperial College, he too had studied under Fowler. Having completed his Ph.D. in 1929 on the constitution of the Sun's outer layers, he was a little more experienced than Chandra in the ways of academia, so Chandra naturally confided in him. McCrea tried to ease his friend's anxiety. He gently reminded Chandra that the RAS meeting was a wonderful opportunity to air his important and intriguing result and that he would undoubtedly learn a great deal from the ensuing discussion.

After lunch they walked briskly over to Burlington House, arriving at four, in time for tea. Chandra was eager to discover what research others were doing and to pick up the gossip. Suddenly, there was a commotion. Eddington appeared, moving through the crowded anteroom like Moses parting the Red Sea, giving an imperious nod here and there. McCrea knew Eddington slightly from his Cambridge days. Boldly, he stepped into the great man's path and asked him about his lecture. Without breaking his stride and without so much as a "How are you?" to Chandra, Eddington replied, archly, "That's a surprise for you!"[3] Eddington's brusqueness shook Chandra. It reminded him of something that had happened the first time he had met him, in 1931. Discussing a comment by his chief rival, Sir James Jeans, on his correlation between the mass of a star and its brightness, Eddington suddenly sat bolt upright and barked that Jeans was "a damned liar."[4]

The high-ceilinged, white-walled meeting room was intimidatingly large. At 4:25 P.M. the audience began to filter in. The rows of seats were arranged in tiers, as in a theater, with a narrow aisle down one side. On the stage at the front was a podium and a long table where the president and other leading figures sat, looking out over the audience. The walls were hung with portraits of past presidents of the society and great astronomers. A portrait of Newton hung behind the president's chair, and one of the English astronomer Sir William Herschel, the discoverer of Uranus, was on the wall to the right. There was not a single window.[5] As the sessions progressed, the room became increasingly stuffy. At a lecture on whether there was air on Venus, someone was heard to murmur, "Never mind Venus, is there air in here?" The first row was reserved for eminent figures such as Eddington, Fowler, and Jeans. Milne, on whose theory of stellar structure Chandra had written several papers, was in the second row. Junior figures such as Chandra and McCrea were relegated to the seats high up at the back.

The room was packed. Just about every major player in British astronomy and astrophysics was there. Not only were these monthly meetings of great academic interest, they were enjoyable social occasions too. The ferocious debates were a popular spectacle, making it well worth the trip. On the surface all seemed playful. Although the protagonists walked the boards like thespians, making grand gestures and turning out exquisite literary flourishes, English understatement and indirectness prevailed. Each speaker used his razor-sharp wit to snare his opponents or score points off them. In private, however, everyone was assumed to be on excellent terms, though that was not always the case. Some people were particularly hurt by Eddington's acid wit. A great deal of animosity seethed beneath the veneer of high manners. A reviewer of G. H. Hardy's wonderful little book *A Mathematician's Apology* wrote that it "is a reminder of a way of life where the participants did their best to hurt each other by day and dined together at night."[6]

At 4:30 the president, F. J. M. Stratton, called the meeting to order. Stratton, holder of the chair of astrophysics at Cambridge and director of the solar physics laboratory there, was an imposing figure both as a scientist and in person. He had an exemplary war record and in public life was known as Colonel Stratton. But to friends, this short, solid man with his rumpled suit buttoned to the neck, happiest in the company of soldiers, was known simply as Chubby. Many of his former students held prestigious positions in astronomy.[7]

Chandra remembered the first time he met Stratton, at a tea party in Eddington's garden in May 1933. Eddington had invited him specially to meet his close friend the famous American astrophysicist Henry Norris Russell, who was visiting Cambridge from Princeton University. To Chandra's relief, Russell was complimentary about his recent calculations on the shape of stars, work that drew on the research of their mutual friend Milne. Chandra wrote proudly to his father that he had "had a very pleasant time discussing with them 'big guns.'"[8] At the time, Chandra felt he had Eddington's support, for why else would the great man have gone so far out of his way to be kind to him? Eddington and Russell were the leading figures in British and American astronomy. They controlled appointments; they could make or break a man's career. After the party, Chandra had begun to feel more at home in the Cambridge world of academia.

One of the first formalities at the meeting was to award Milne the society's gold medal for his research achievements. Chandra was delighted. Milne had been one of those most deeply offended by Eddington's cutting wit, to the point that it had affected his career and personal life. Chandra recalled Milne telling him that he had no confidence in his own mathematical ability, because his university training had been interrupted by the Great War.[9] Eddington had sniffed out this weakness. In every showdown they had, at the RAS or elsewhere, he homed in unerringly on the mathematical elements in Milne's ideas. Only recently, during one of Chandra's visits to Milne at Oxford, Milne had blurted out that his run-ins with Eddington had "sapped his energy."[10]

Six fifteen-minute papers preceded Chandra's, followed by comments from the floor. Finally, at 6:15, Stratton called Chandra to the podium. By then the air in the room was so dank and heavy it was almost tropical.

Nineteenth-century science held that stars form when the inward crush of gravity is balanced by the outward pressure of the star's own gas particles and the radiation pressure they emit. But as a star grows older and burns up its fuel, it becomes dimmer, and gravity begins to take over, crushing the star into a dense ball. Could it be that stars eventually collapsed entirely? It was a tantalizing mystery, and one that Eddington made every effort to play down. After all, such a scenario seemed to suggest that some stars would not end up as inert rocks, and that was a troublesome notion indeed. Surely it was impossible for something as big as a star to disappear into nothingness. What would become of it if that was the case?

In 1926 Eddington's colleague Ralph Fowler had proposed using quantum physics to resolve the situation. Fowler, the first theoretical physicist in England to grasp the deeper meaning of the new atomic physics, was unique in walking the thin line between astrophysics and physics. At that time, physicists looked upon astrophysics as a backwater. While physicists were concocting mind-boggling theories on the unpredictable world of quanta, astrophysicists remained resolutely classical. For physicists the most interesting question about stars was what made them shine — in other words, what were they burning? Where did they get their fuel? Everyone assumed that stars generated their energy in the nuclei of their atoms, but no one knew exactly how. Nuclear physics was a subject that would remain murky until the discovery of the neutron in 1932. Astrophysicists, on the other hand, set this problem aside, attempting instead to explore the structure of stars by creating models that avoided looking too deeply into the problem of what exactly stars are made of. Fowler, applying one of the remarkable new laws of quantum theory, showed that a star's core could stop the gravitational contraction. The star would cease to shrink and then die in peace, in keeping with the vision people have always had of the universe.

But Chandra saw that Fowler's solution was incomplete. What he had left out — remarkably — was Einstein's special theory of relativity, which says that the speed of light (not space or time) is constant. When Chandra applied relativity to Fowler's results on his voyage from Madras to England, he came up with something extraordinary, something no one had done. He calculated that there was an upper limit to the mass of a white dwarf star. It took him just ten minutes to deduce this, but he spent the rest of the journey puzzling over its implications. He knew he was onto something but "didn't understand what this limit meant, and . . . didn't know how it would end."[11] It was beginning to dawn on him that the inescapable, though almost unimaginable, conclusion was that white dwarf stars with a mass greater than the maximum he had just discovered simply could not exist. Their own gravity would crush them until they disappeared into nothingness.

Apart from his calculation of the limiting mass, Chandra's discovery confirmed a scenario that Eddington had raised in one of his books but had dismissed as absurd: that white dwarf stars might not end up as inert rocks but could collapse completely. As Chandra saw it, he had solved Eddington's problem.

Standing at the podium, dwarfed by the white domed ceiling, the dark-skinned young Indian gazed around at the sea of starched shirts, dark jackets, and European faces. At last he had the audience he wanted. Everything rode on this moment. He now fully understood the magnitude of what he had achieved as a young man of nineteen on that sunny summer day, in transit between two worlds, two civilizations, two cultures that could not be more different.

Chandra must have felt rather as Einstein had in 1905, the year the young patent clerk discovered special relativity while working at the Swiss Federal Patent Office in the intellectual backwater of Berne. Like Einstein, he was convinced that the great scientists of the day were "theorizing out of their depth."[12] Only he could see the truth. There he was, Subrahmanyan Chandrasekhar, a nonentity, just as Einstein had been in that magic year. Like Einstein, he had lifted a corner of a great veil, revealing a majestic yet terrifying picture of the fate of stars and of humanity.

The young man loosened his collar a little and brushed a drop of sweat from his forehead. The windowless room was hot and still, sealed against the January winds that whistled outside the venerable building. Glancing at his watch, he turned to the last page of his paper and read out his conclusion in confident tones:

> The life of a star of small mass must be essentially different from that of a star of large mass . . . For a star of small mass the natural white dwarf stage is an initial step towards complete extinction. A star of large mass cannot pass into the white dwarf stage and one is left speculating on other possibilities.[13]

He gathered up his papers and returned to his seat near the back of the hall. The walk must have seemed an eternity. Perhaps he daydreamed that this might be the moment that would launch him on a glorious career. He had given his all. He had done the best he could.

The discussion was thrown open to the floor. Milne compared his own research program on white dwarfs with Chandra's, insisting that Chandra's was but a small part of his own. Chandra listened with half an ear. He was waiting to hear what Eddington had to say.

Ever theatrical, Eddington remained seated for a measured fraction of a minute, building up the dramatic tension. Then he rose from the front row and in one step mounted the podium. He turned, cocked his head, and began with an arrogant flourish: "I do not know whether I shall escape from this meeting alive!"[14] The point of his paper was that the very basis of Chandra's theory was downright absurd. There was no such thing as an upper limit to the mass of a white dwarf. The audience gasped. Chandra was shocked. Could he have heard Eddington correctly? If so, Eddington was claiming that Chandra was utterly wrong and all his work was for nothing.

In essence Eddington was reverting to Fowler's theory, which had no restrictions on the mass of a white dwarf; such a star would inevitably die a peaceful death as an inert rock, not disappear into nothingness. This theory was more to Eddington's taste; he applauded Fowler for getting astrophysics out of a mess and derided Chandra for reopening a can of worms. His response was not simply English game-playing: it was a direct slap in Chandra's face, and he felt it in just that way.

Nevertheless, Chandra had essentially solved the deeply important problem Eddington had proposed. Eddington would have had nothing to lose by saying "Yes, I was only joking in 1926 when I proposed that stars might collapse to nothingness. But now this brilliant young man has shown that this possibility actually exists in nature. He and I are going to explore this unexpected result." The great scientists such as Einstein and Bohr would surely have done so. Eddington's reputation ranked alongside theirs. Had he not been of great help to Chandra the previous year? To ease the burden of Chandra's extensive numerical calculations, he had pulled strings to procure for him an advanced mechanical calculator. So what had happened? Why had Eddington not revealed his disagreements with Chandra — disagreements so strongly felt that he presented them with an acidity and cynicism that went far beyond even the limits of the RAS's monthly skirmishes?

Chandra recalled the hours upon hours of conversation he had had with Eddington in Chandra's own rooms: "I was telling him, 'How can a star evolve? Massive stars must behave differently,' and so on — all this was being talked about."[15] Eddington had always appeared interested and thoughtful. Now Chandra knew that Eddington's sphinxlike face hid a duplicity of shocking proportions.

Eddington's true agenda was revealed when he said, "I think there should be a law of Nature to prevent a star from behaving in this absurd way!"[16] In other words, "The hell with physics!" With an awful feeling in the pit of his stomach, Chandra realized that "despite this man's incredible physical insight, he has always operated with preconceived ideas."[17] While physicists were at that very moment happily taking on board quantum theory, with all its weirdness, contradictions, ambiguities, and apparent impossibilities, astrophysics was, it seemed, bogged down in the past. Astrophysicists refused to follow the mathematics of their theories, which led inexorably into ever more implausible domains of physical reality. Instead they insisted on ignoring results about the fate of the stars that upset their preconceived view of the universe. It is an unwritten rule of science that when something potentially observable is predicted to become infinite, it is a sure sign that the theory is breaking down. As Albert Einstein was fond of saying, "Only two things are infinite, the universe and human stupidity, and I'm not sure about the former." Astrophysicists like Eddington simply refused to believe that something as big as a star could ever become infinitely tiny.

Chandra must have been deeply frustrated. Looking around at the nodding heads and mesmerized expressions, he saw clearly that everyone in the audience believed Eddington, even his own friend McCrea. How could this be? Just a few hours earlier, McCrea had been agreeing with him. But now McCrea was whispering to Chandra that he "thought Eddington sounded plausible."[18]

What was happening? Why didn't anybody speak up and tell Eddington that he was plain wrong? Chandra stood up to reply in his own defense. He was dumbfounded when President Chubby Stratton refused to let him respond; instead, he brought down the curtain on Act I of the Chandra-Eddington confrontation with the curt words "The arguments of this paper will need to be very carefully weighed before we can discuss it." And without further ado, he introduced the next presentation. Eddington's "authority was so great, that people accepted him," Chandra recalled ruefully. "He made jokes about [my theory] and made me look like a fool."[19] "At the end of it, everybody came by and said, 'Too bad. Too bad.'"[20]

Although minutes were taken of the meeting and each man's presentation was duly published, there was no consensus on Eddington's reason for launching into his ferocious attack without any warning. The basis of his argument was not at all clear, yet no one in the audience questioned it. According to McCrea, Eddington was guided by instinct. If anyone other than Eddington had offered such an argument, McCrea and others would have had real doubts about its validity. But despite the flimsiness of his reasoning, the academic community chose to support him rather than an outsider. Such was the force of Eddington's reputation and personality. In later years Chandra always said bitterly that there was only one case in which Eddington was wrong and never admitted it: his own.[21] He never forgot the laughter and guffaws that rose on cue during Eddington's carefully orchestrated lecture.

The attack was all the more incredible to Chandra because Eddington understood his discovery better than anyone else, including Chandra himself. Eddington's insight and intuition were enormous, but he utterly refused to accept the physics. He stood by his preconceived conceptions of nature. Soon after the meeting, at Trinity, Eddington commented to Chandra, "You look at it from the point of view of the star. I look at it from the point of view of nature." Bewildered, Chandra asked, "Aren't they both the same?" Eddington replied firmly, "No."

"Well, you see," Chandra recalled, "that sort of shows his attitude. Somehow, he felt that nature must conform to what he thought was right."[22] So did most other astrophysicists.

On the train back to Cambridge that Friday evening, Chandra recalled Milne saying, "I feel in my bones that Eddington is right."[23] Chandra snapped back angrily, "I wish you felt it elsewhere." Privately

Milne, who also had preconceived notions, was elated because his own theory of stellar structure was at odds with the concept of stars disappearing into nothingness; something had to prevent complete collapse. When Milne visited Chandra in 1934, he had begun to accept Chandra's results. "But when Eddington said that the formula [for the upper limit of a white dwarf's mass] was wrong, Milne was all aglow."[24] Soon after, Milne wrote that even if Chandra's theory was right, he would ignore it.[25]

Arriving back at Trinity well past midnight, Chandra could not face going into his room. Instead, he went to the senior common room and stood in front of the great fireplace, gazing into the flames. Alone with his thoughts, he reviewed the day's events. Until 6.45 that evening, he had been full of elation, bursting with self-confidence. He had revealed in all its rigor a momentous scientific discovery. In the red glow of the coals he saw Eddington's face. The attack had been "a totally unexpected occurrence, that came close to destroying my scientific confidence."[26] Had it all been for nothing — the sacrifices his family had made in sending him from India to Cambridge, the tearful farewells at the dockside in Bombay?

Less than a year after arriving in England he had heard that his beloved mother had died. Perhaps, he thought, he was reading too much into it, but just two days after hearing the news he had met Eddington for the first time. Despite his deep grief, which he could share with no one, he could not postpone that meeting.[27] He had been shown into the study in Eddington's house next to the Cambridge observatory. The great man sat at his desk with his back to the French windows that looked out onto a beautiful garden. The room was wreathed with pipe smoke and laced with the sweet smell of apples. Eddington was always eating apples, which he chewed up entirely, core and all.[28] Though Chandra felt empty inside, he performed well in that meeting.

Now he stood in front of the fire, repeating to himself T. S. Eliot's famous lines as he remembered them, "This is how the world ends. This is how the world ends: Not with a bang but with a whimper."[29] He had hit rock bottom, but he would not give up. Sometimes Chandra and Milne sarcastically referred to Eddington as the devil incarnate.[30] What happened on January 11, 1935, was Chandra's first experience of that manifestation of him. Chandra never forgot that moment of humiliation. Convinced that he was right, he could not understand why

Eddington had attacked him in such a vicious and unbridled manner.

It was indeed extraordinary that a nineteen-year-old Indian youth had managed to make a discovery that had eluded the great minds of European astrophysics. To understand how that came about, we have to step back in time to India not long after the end of the Victorian age, when the Raj was beginning to show signs of weakness and young Indians were daring to step forward and claim their heritage.

2. A Journey Between Two Worlds

FOR CHANDRA and for many other ambitious young Indians of his generation, science offered the best way to break through the seemingly insuperable barriers erected by the British Raj. Chandra wanted to express his patriotism by showing "that Indians could be accomplished in a way which the outside world [could] recognise." "Certainly one of the earliest motives that I had," he said later, "was to show the world what an Indian could do."[1] To the end of his life, he never forgot the day — April 27, 1920, when he was just nine — his mother told him "that a famous Indian mathematician, Ramanujan," had died the day before, at the age of thirty-two.[2]

Srinivasa Ramanujan's story was like a fairy tale. A lowly clerk in the Madras Accountant's General Office, he had sent off a letter in January 1913 to G. H. Hardy, Eddington's colleague at Cambridge, outlining some mathematical ideas and suggesting that he would send more. Hardy was intrigued and impressed. Six months later, Ramanujan was lifted from his hand-to-mouth existence and transported to Trinity College, Cambridge, where his only commitment was to do mathematics. In March 1918 he became the first Indian to be elected a Fellow of the Royal Society, a singular honor in the British scientific world. That autumn he became a Fellow of Trinity College, able to devote himself entirely to research for the next three years. Later Ramanujan would be recognized as one of the century's most original mathematicians. It was Hardy who had brought all this about. Ramanujan's rags-to-riches tale inspired Chandra and gave him the hope that he too could "break [his] bonds of intellectual confinement and perhaps soar the way Ramanujan had."[3]

Chandra was born on October 19, 1910, in Lahore. Now in Pakistan, Lahore in those days was the capital of the Punjab province of British India. His father had been posted there as the assistant auditor general for the North West Railways. As the first son, Chandra was given his grandfather's name, Chandrasekhar, which means "moon" in Sanskrit. The family moved on to Lucknow, and then, when Chandra was eight, returned to their roots in South India, settling in Madras.

In contrast to Ramanujan's early poverty, Chandra grew up in a free-thinking, upper-class Tamil Brahmin household. There were always servants around. The only responsibility for him and his three brothers and six sisters was to study.[4] Vidya, one of his sisters, recalls that "our home was really more like an educational institution. People seemed to be reading or debating or conversing about interesting subjects all the time."[5] The emphasis on the importance of learning was typical of Brahmins, for whom knowledge counted far more than wealth.

Ever since the 1850s, when India was established as the jewel in Britain's colonial crown, the establishment of a British education system had been a cornerstone of colonial policy. As Lalitha, Chandra's widow, points out, the ultimate aim was "not so much to educate Indians as to have an army of cheap clerks."[6] Nevertheless, the result was the development of an elite cadre of young Indian men, mostly Brahmins. For Indians, says Lalitha, education provided "a window to the West." In fact, it was this very "exposure to Western culture that led to the [Indian] renaissance," a national awakening in all spheres of creativity but most notably in science, sparked by the political movement to bring about India's independence.

Lalitha is the custodian of Chandra's memory. Now in her nineties, she is a tiny, birdlike woman with an oval face, wavy gray hair, and round-rimmed glasses perched firmly on her delicate nose. In her youth, when she wore her hair smoothed into a long plait, she was a beauty. Her exuberance, ready smile, and assertive speech still belie her years. Chandra's story, she insists, is inextricably entangled with what was happening in India when he was growing up. "The reason for the emergence of scientists like Ramanujan, [Satyendra Nath] Bose, and Chandra was the renaissance in India that commenced in 1910," she says emphatically. "You must start here. The country was ready for something to happen. It was a crucial time. Mohandas K. Gandhi was about to return from South Africa."

In 1910, the year that Chandra was born, Gandhi had been effectively in exile for more than two decades. After a conventional childhood, a friend had persuaded him that he should go to England to study law and become a barrister. For Indians, traveling "over the water" meant becoming a virtual outcast. In later years Gandhi wrote of his terrible loneliness abroad in words that Chandra would echo three decades later: "I would continually think of my home and country. My mother's love always haunted me. At night the tears would stream down my cheeks, and home memories of all sorts made sleep out of the question. Everything was strange."[7]

Gandhi mastered several European languages, learned to dance, and studied the violin. He read widely, probing Christianity, Hinduism, Theosophy, and the emerging ideological movement of pacifism. But when he returned to India in 1891, he found his chances for employment almost nonexistent. He decided to try life in South Africa. He spent the next two decades in Durban and Johannesburg, working as a lawyer, fighting the discrimination suffered by Indian settlers. There he discovered his life's work: to bring about radical change in India by the methods of nonviolent civil disobedience he had used so effectively in South Africa.

The movement toward Indian independence had begun on December 28, 1885, with the first meeting of the Indian National Congress. But it was not until after the Great War (World War I) that the struggle began in earnest. India's role in the war inspired a new sense of pride and self-confidence. Half a million Indians traveled overseas to fight. What they saw in Europe fired their imaginations.

When Gandhi returned from South Africa in 1915, he exchanged the life of a successful middle-class lawyer for that of an ascetic. He gave up his possessions and wore the simple homespun garments of an Indian peasant. Soon people were calling him the Mahatma, the Great Soul or Holy One. After the war, along with the inspirational poet Lokamanya Tilak, Gandhi led a movement that combined Indian nationalism with the rediscovery of traditional mores and the first moves toward breaking down the caste system. His method was pacifist noncooperation. Initially the representatives of the British Raj dismissed him. In 1917 the then secretary of state, Edwin Samuel Montagu, described Gandhi contemptuously as a fringe figure who "dresses like a coolie, forswears all personal advancement, lives practically on air and is a pure visionary."[8] It was an accurate description but a fantastic underestimate.

The British responded to the growing clamor for independence with a series of increasingly desperate laws, culminating in the detested Rowlatt Act which made wartime restrictions on civil rights permanent. In the infamous massacre at the Sikh capital of Amritsar in the Punjab, a British Army contingent opened fire on an unarmed crowd of protesters, killing hundreds. That was the last straw. Multitudes who had patiently endured the yoke of imperialism flocked to the nationalist cause.

Growing up in the southern city of Madras, young Chandra was at first insulated from the growing foment of political unrest around him. In his family the emphasis was very much on the importance of education and, specifically, of science. The focus on learning was the legacy of Chandra's grandfather, Ramanathan Chandrasekhar, who had been a professor of mathematics at Ankitham Venkata College in Vizagapatam (now Visakhapatnam), on the coast about 360 miles north of Madras. He died the year Chandra was born. But he left a library full of books on mathematics, including some he had written himself. Chandra treasured them throughout his life. Ramanathan also fathered ten children, two of whom were exceptionally brilliant.

Chandra's father, Chandrasekhara Subrahmanyan Ayyar, was the eldest of Ramanathan's children. An exemplary student at his father's college and then at Presidency College, he took the All India Examination and entered British government service in the Indian Audits and Accounts Office. There he worked in the railways section, traveling extensively to different parts of India. In those days it was a tremendous achievement to enter the Indian civil service, which offered great opportunities for advancement. But his children looked down on him. "The standards by which Chandrasekhar's family measured success were by intellectual and cultural achievements," Chandra's younger brother, Balakrishnan, wrote.[9] They considered their hard-working father "a relative failure" because he had chosen the civil service route. It was a harsh judgment.

Nevertheless, it was he who had built up the family's high standard of living. Ayyar, as he was known, was a strikingly tall, powerfully built man with a handsome round face, piercing eyes, and a powerful, rather authoritarian personality. At home he dressed in the South Indian style, in a shirt and *veshti* (a garment rather like a sarong). But for work he wore a Western shirt, tie, jacket, and trousers. Like most Indians in British government service, he always wore a tightly wrapped turban to emphasize his reluctance to abandon the old ways completely. He was a brilliant exponent of Karnatic music, which he played expertly on the violin and for which he also developed a notation system.

But for all his achievements, from a very young age Ayyar found himself in the shadow of his younger brother, Chandrasekhara Venkata Raman, who was the star of the family. In photographs, while Chandra's father has a gentle cast to his face, Raman has a look of ferocious intelligence and no-nonsense determination. He raced through his education at an extraordinary speed, graduating from Presidency College at the age of seventeen with a degree in physics and finishing his M.A. when he was nineteen. Dogged by ill health, he gave up his studies in physics and, like Ayyar, joined the Indian Audits and Accounts Service. The brothers were posted to Calcutta, where they and their wives shared a flat. In his spare time Raman continued his physics research at the University of Calcutta. In 1917 he was offered a chair or professorship there and left his lucrative civil service posting to take up research full-time. In 1930 he would receive the Nobel Prize in physics. According to Chandra's cousin Sivaraj Ramaseshan, there was bitter sibling rivalry between Ayyar and Raman. Not only did Raman seem to do everything better than Ayyar, but he had left the security of the civil service and eventually achieved world-wide fame. When the two brothers lived together in Calcutta, the atmosphere between them was brittle. Ramaseshan has studied their correspondence, and "there is no doubt that C. S. Ayyar disliked C. V. Raman," he says.[10]

Ayyar was a typical Indian father of his day, showing little affection toward his children. He assigned professions to his sons and husbands to his daughters. But beneath his crusty exterior, he cared a great deal about Chandra's emotional well-being, as is clear from their correspondence. Yet their relationship was far from smooth. Lalitha recalled that the "children were disappointed in Ayyar and, as grown-ups, did little to hide their frustrations with him. They were closer to their mother."

Sitalakshmi Balakrishna, Chandra's mother, was an extraordinary woman. She came from a typical Indian household with an overbearing mother-in-law and an unhappy aunt; her marriage to Ayyar had been arranged when she was fourteen. Over the years she bore ten children, yet she always continued to study. With Ayyar's help she learned English, and she went on to translate Ibsen's *A Doll's House* into Tamil. It became the standard text for high school students.

In those days, middle-class families in India began their children's education at home. Early in the morning, before he went to the office, Ayyar gave Chandra instruction on most subjects. His mother taught him Tamil. Ayyar was particularly concerned that Chandra learn to speak English perfectly; English was the key to everything, from entering the Indian civil service to attaining every bright young Indian's dream — study in England. Chandra's parents were aware from early on that he was an extraordinarily brilliant child. He made lightning progress in mathematics and was often found sitting alone in the library, devouring his grandfather's mathematics texts. By the time he finished high school, at the age of fifteen, he had read them all, including George Salmon's *Conic Sections,* G. H. Hardy's *Pure Mathematics,* George Boole's *Differential Equations,* and W. Burnside and W. Paton's *Theory of Equations.* In later years, all ten children were to look back on their individualistic and free-ranging early education with gratitude. As Chandra's brother Balakrishnan put it, it prepared them for school "without being cramped in the early years by a load of educational lumber in the head."[11]

The family did not really put down roots in Madras until 1923, when Chandra was thirteen. That December, Ayyar laid the foundation stone for Chandra Vilas, which was to be the family home. It was at 46 Edward Elliots Road in Mylapore, an affluent suburb of Madras, a world away from the squalor of neighboring Triplicane, where Srinivasa Ramanujan had grown up and lived. In Chandra's day the house was an elegant, whitewashed two-story mansion, built in the neocolonial style traditional in Madras at that time, with long shady verandas, roofed wooden balconies, and decorative eaves over the windows. The garden was planted with broad-leafed mango and coconut trees. It was a cool and inspiring place to read and talk, a haven from the intense light and heat of the day.

Ayyar had taken on the expense of providing a home for his family, but he was all too aware of the consequences. His family would no longer be with him when he was posted away from Madras. "A lonely life for earning a livelihood faced me," he recalled in his autobiography.[12] By 1924, when Chandra Vilas was completed, Chandra was already in his last year at the Hindu High School in Triplicane. Those were carefree days. The three eldest brothers — Chandra, Visvanathan, two years his junior, and Balakrishnan, four years younger — were inseparable. At first they took the tram to school together; later they cycled. Today Edward Elliots Road has become Dr. Radhakrishnan Salai, a noisy, traffic-filled thoroughfare — a far cry from the idyllic setting in which Chandra grew up. Chandra Vilas still stands, though it is much reduced and rundown, tucked along an alleyway off the main road. Still, one can imagine Chandra and his brothers riding their bicycles along Edward Elliots Road to school nearly eighty years ago.

From 1925 onward, Chandra's bicycle ride took him along the Marina beach to the splendid pink-domed halls of Presidency College. He was fifteen, two years younger than everyone else, but he had already acquired a reputation for brilliance. "Chandrasekhar was born great and lucky — lucky at least to start with. From his early years he was recognised by all the people he came into contact with, including all the members of his family, as a mathematical prodigy or even genius,"[13] recalled his brother Balakrishnan. At Presidency he became famous for answering every question correctly in examinations, even those that were not required. He usually scored more than 100 percent. He read voraciously, devouring mathematics and physics texts in a single reading. His tastes in literature ranged from Robert Louis Stevenson to Thomas Hardy and Shakespeare; he also learned German. One of his college friends, S. R. Kaiwar, recalled that he was "lightning-quick in comprehending, [and] could read a hundred pages in an hour quite easily. [He] was interested in everything around him." He had time to enjoy a normal boyhood. "We laughed and joked a lot," remembered Kaiwar. "He wasn't by any means a prude."[14] But he was prone to fits of gloom, sometimes withdrawing to his room to sit for hours in the dark.

Family life helped nurture the young man's brilliance. Conversations over dinner were not merely intellectual but focused on science. As Chandra later wrote, "The atmosphere of science was always at home."[15] By now his uncle Raman had become one of India's most eminent scientists. One day Chandra overheard Raman talking to his father about Ramanujan. Chandra was fired up. He yearned to follow in Ramanujan's footsteps and become a mathematician.

After completing his first two years at Presidency College, Chandra had to choose what to study for his honors degree. He immediately opted for mathematics. But his father had other ideas; he could see no future in mathematics. He wanted Chandra to play it safe, as he had, and pressured him to join the civil service. Knowing that his son was determined to be a scientist, Ayyar insisted that he study physics, which at least had practical applications. For Chandra physics was acceptable, but the civil service was not. Unusually for Indian society, his mother supported him. "You should do what you like. Don't listen to him; don't be intimidated," she told him.[16] In the end, Chandra took his father's advice and in later years he was grateful. He wrote to Balakrishnan from Cambridge, "I am ever so thankful to Babuji for having guided me properly in making the proper choice."*[17] After all, Chandra continued, his physics research was so highly mathematical that "my 'first love' is not killed."

Chandra often went to the Marina beach to think and dream. His hero Ramanujan had also strolled there to escape the hot, dusty streets of Triplicane. Chandra was consumed with the urge to make his mark on the world as a scientist, to discover something that would transform our understanding. "When I was in 5th form and later, I used to go to the beach and pray God — actually prostrate on the ground (I blush to tell you so) — to mould my life like that of an Einstein or a Riemann," he later admitted to Balakrishnan.[18]

Chandra, Balakrishnan, and Visvanathan would sprawl on the sand, joke as boys do, and listen to Chandra hold forth on mathematics, science, or literature. In those days the Marina was a highly fashionable resort. In the evening, automobiles sputtered along side by side with the increasingly rare horse-drawn shuttered carriages in which Muslim ladies could take the night air without being seen. The beach, one of the longest and broadest in the world, stretches south for several miles, from Fort St. George to the sixteenth-century Portuguese town of San Thomé. From the venerable pink-domed clock tower of Presidency College, it is a long walk across the burning sand to the shimmering blue waters of the Bay of Bengal.

Established in 1840, Presidency College became the nucleus of the University of Madras, founded seventeen years later. In Chandra's day, as now, it was the best college in South India. The lecturers, most of whom were British, encouraged students to study rather than learn by rote, which was the norm at other Indian colleges. Chandra still complained that the system was exam-oriented and stressed the solving of problems instead of the understanding of science. One of the teachers, H. Parameswaran, had done his Ph.D. at Cambridge. He was, Chandra recalled, "quite a competent experimental physicist [who] understood what research was."[19] Chandra himself certainly did. His career at Presidency College was enormously successful. "*Everybody* knows

---

* *Babu* is an affectionate Tamil term for "father"; *ji* indicates respect.

Chandrasekhar, he is a very famous man," Balakrishnan wrote excitedly to him in November 1930. Today photographs of Chandra are proudly displayed in the hallways of the physics department, alongside Bohr, Einstein, Parameswaran, and Raman. In his study, the present head of the department has four photographs on his desk: two of Chandra and one each of Raman and Einstein.

Most physicists and physics students struggle through the texts equation by equation. Chandra could read them almost like novels, effortlessly grasping the essence and the details. But his studies did not exclude other interests. As a student in India in the late 1920s, Chandra could not fail to be drawn into the struggle for Indian independence, even though his contribution was primarily to make his academic mark, to show that an Indian could succeed in the Westerners' world of science. He did not go so far as to join the Indian National Congress. But when Jawaharlal Nehru, its president, came to Madras in 1928 to speak at a mass rally on the Marina, Chandra was among the students who joined the crowd to listen, in defiance of the college principal's injunction against it. In later years, Chandra remembered Nehru in a way many others had, as "a complex product of intellectual brilliance, physical attractiveness and stamina, human sensitivity and charm, which made him at once the hero to India's youth."[20]

That summer of 1928, Raman offered his precocious nephew a priceless opportunity, giving him a job in his laboratory in Calcutta. There Chandra had firsthand experience of real scientific research. It was an extraordinarily exciting time. The previous February, Raman had discovered what quickly became known as the Raman effect, a way to glimpse a molecule's structure from its interaction with light. Chandra was swept along in the euphoria over this groundbreaking discovery. No doubt he hoped that he would achieve something similar someday. But his own experiments were disastrous. Broken equipment began to pile up. It became clear to everyone that his true calling was theoretical physics, especially when he began to give brilliant seminars to the others in the laboratory, lucidly explaining his uncle's discovery.

In the spring Raman had paid a visit to Chandra's home in Mylapore, bringing with him a book with the tantalizing title *The Internal Constitution of the Stars,* by Arthur Stanley Eddington, which took the reader to the very frontier of astrophysics. Eddington laid out the key issues and ended with one of his unforgettable flourishes: "It is reasonable to hope that in a not too distant future we shall be competent to understand so simple a thing as a star."[21] Chandra was spellbound.

A titan of science, Eddington had single-handedly established the field of modern astrophysics. By clarifying the relationship between a star's mass, brightness, and temperature, he had begun to make sense of its life journey from its birth, amid a vast cloud of interstellar gas and dust, to its death as a cold, inert rock; he had worked out methods of analyzing the movement and distribution of the stars; and he had made vital contributions to our ideas about stellar composition and energy generation, presciently suggesting that stars shine by converting hydrogen into helium. Most important, however, Eddington was the first to apply general relativity to astronomy, having introduced the theory to the English-speaking world in popular books still read to this day.

Eddington's research papers were masterpieces of elegant prose. With minimal use of mathematics, of which he was a master, he derived results from first principles, linking the basic laws of physics to experimental data, always with a bon mot or superb metaphor. He proceeded to show how, although at first sight understanding the stars seems hopeless, it is not. Stars have a gravitational field and emit light that "manages to struggle to the surface and begin its journey across space. From these two clues alone a chain of deduction can start," based on universal laws of nature.[22] Eddington had a charmingly vivid and utterly unforgettable way of bringing astrophysics alive. Describing a photon of light, he wrote:

> You may picture a photon of radiation, barging first one way, then another, like a man in a rioting mob — absorbed by an atom and flung out again in a new direction. In this way a photon in the Sun will wander aimlessly round in the interior for a million years or more until, just by accident, it finds itself at the exit of the maze — shoots through — and makes a bee-line across space to the Oakridge reflector, where Professor Shapley photographs it.[23]

The American astrophysicist Henry Norris Russell wrote of *The Internal Constitution of the Stars*, "The book itself is a work of art . . . the product of a great teacher as well as a great investigator."[24] It was not only a journey through almost everything that was known about the stars; it also set out the key problems that remained to be solved. Eddington was an inspiration to everyone, from professional scientists to blazingly ambitious and idealistic young men like Chandra.

Working in Raman's Calcutta laboratory, Chandra leaped at the opportunity to borrow his uncle's copy of the book and read it closely. He was fascinated by Eddington's wide-ranging discussion of how atoms emit and absorb radiation according to quantum theory and how stars can be described in mathematical terms as spheres of gas. Eddington's discussion of quantum theory helped Chandra understand Raman's discovery.

But then he came across the classic *Atomic Structure and Spectral Lines*, by the great German physicist Arnold Sommerfeld, which had first appeared in 1919. It was regularly updated and became the bible for researchers; the version Chandra read was the English translation of the 1924 edition. Born in 1868, Sommerfeld was involved in virtually every phase of theoretical physics right up to his death in 1951. He was immensely talented both as a researcher and as a teacher. His famous institute in Munich produced many of the pioneers in quantum physics, from Hans Bethe and Wolfgang Pauli to Werner Heisenberg, his prize student. As Heisenberg often said, if anyone deserved a Nobel Prize it was Sommerfeld — though inexplicably he was passed over by the Nobel committee, one of its few serious omissions in physics.

By pure chance, that September of 1928 Sommerfeld showed up in Madras to give a lecture. Full of youthful bravado and confidence, Chandra found out where he was staying and knocked on his door. What followed changed the course of his life.

Chandra was totally unaware of the discoveries that had been made in atomic physics over the previous two years, which had utterly changed scientists' views of matter at the microscopic level. In the extraordinary new quantum world that was unfolding, electrons and light behaved in astonishing ways. The old images of electrons as particles and light as waves no longer held; both turned out to be both waves and particles at the same time, a bizarre and unimaginable conclusion known as the wave-particle duality.[25] From this, scientists extrapolated all manner of quantum weirdness. Electrons, they concluded, were spread across space and time, capable of being in several places at once. Stranger still, electrons appeared sometimes to be able to "sense" and react to an experiment carried out on another electron far away, as if by ESP. The old certainties of classical physics — that scientists could calculate exactly the position and velocity of an electron and plot its path through time and space — were overthrown. There was now a completely new way of looking at these properties in this volatile new world, and Sommerfeld, in his clipped, precise English, gently explained it all to Chandra. He gave Chandra copies of two papers he had published earlier that year in which he showed the results of applying quantum theory to a gas of trillions upon trillions upon trillions of electrons squashed tightly together, as in a metal.*[26] It seemed that the quantum nature of electrons unexpectedly produced an outward pressure that held the metal together; this pressure helped explain why metals were not crushed by the massive inward pressure of gravity.

Eager to learn everything he could, Chandra sought out the seminal articles that Sommerfeld had recommended. Chief among these were papers by Pauli, at Hamburg University, and Heisenberg, at Leipzig, both of whom had gone on to become the instigators of the new quantum physics. No more than a decade older than Chandra, both men were already legendary names. Chandra was particularly inspired by the research that underlay Sommerfeld's work. He pored over articles by P. A. M. (Paul) Dirac, in Cambridge, and Enrico Fermi, in Rome, who had shown the dramatic effects produced by the wave-particle duality when electrons are cramped together in a gas.

Physicists try to understand a concept by using it to solve a problem, and that is what Chandra proceeded to do. Although only seventeen, he had already homed in on the area of quantum physics he wanted to pursue — the quantum properties of a gas of electrons — which would lead him to his discovery about white dwarf stars. From his reading of Eddington's book, he explored how light, electrons, and atoms interact in a star. Combining this with what he had learned from Sommerfeld, within a week or two he wrote his first paper. It was published in the *Indian Journal of Physics* in 1928.[27]

The precocious youth was convinced that his results were important enough to merit publication in the *Proceedings of the Royal Society*, but a paper for the *Proceedings* had to be submitted by a Fellow of the society. As it happened, he had recently read Ralph Fowler's newly published *Statistical Mechanics* and noticed that Fowler was a Fellow. Boldly, Chandra decided to take a chance and send him a copy of a second paper he had written. It was January 1929.

* The sheer size of stars, together with their densities, the number of atoms they contain, and their distances from Earth and the Sun can be expressed only in huge numbers. As shorthand I will use powers of 10: 10^2 is one hundred (100); 10^3, one thousand (1,000); and so on. Similarly, 10^{-1} is one tenth (0.1); 10^{-2}, one one-hundredth (0.01); and so on. I use the standard American notation, in which 10^9 is one billion and 10^{12} one trillion.

That same month there was a meeting of the Indian Science Congress (modeled on the British Association for the Advancement of Science) in Madras. Raman, who had by now achieved worldwide fame, presided. The meeting was packed to the rafters. Chandra, a fresh-faced eighteen-year-old, rose to address the venerable gathering and presented a summary of his paper. It was his first professional presentation at a scientific meeting. Professor Parameswaran announced that the young man was only in his second year at college and added proudly that "he had written the paper without any guidance or advice from anyone."[28] There was thunderous applause.

Chandra's brother-in-law A. S. Ganesan, who was a physicist, had left his copies of the *Monthly Notices of the Royal Astronomical Society*, running from 1925 to 1929, in the well-stocked library of Chandra Vilas. Glancing through them in the early spring of 1929, Chandra came across a paper in which Fowler proposed that quantum physics might offer a solution to Eddington's paradox that a white dwarf could collapse completely instead of ending up as an inert rock. Chandra was excited to see that the paper used the new way of studying a gas of electrons that he had learned from Sommerfeld and had already applied successfully in his papers. "Right there, there was something which I could do. So that is how I started, you see," he later recalled.[29] Chandra was now convinced that Cambridge was the place for him: that was where the action was.

In June he sent his father a letter outlining his work in progress; he was working on five papers at the same time.[30] In one of them Chandra elaborated on the paper he had sent to Fowler and had submitted to the *Indian Journal of Physics*.[31] He waited impatiently, hoping desperately that Fowler would reply. Finally he heard from him.

When the paper from the unknown Indian lad landed on Fowler's desk, he read it with growing interest. He made a few suggestions, to which Chandra agreed. Chandra immediately withdrew the paper he had just submitted to the *Indian Journal* and incorporated its results into the one he had first sent to Fowler. Fowler was very impressed. To Chandra's great excitement and pride, his work came out "as a fairly big paper" in the October issue of the *Proceedings of the Royal Society*.[32] He then wrote yet another paper which, he told his father, he passed to a professor at Presidency College "for correction. But he indefinitely postponed it. So I sent the paper to the Phil. Mag. [*Philosophical Magazine*]." In Chandra's view, the paper needed no corrections. He was right; it was published the following year.[33] Chandra also sent a letter to Eddington, but sadly he lost the reply. "After loosing Eddington's letter, I am very careful about my letters from Fowler," he wrote to his father.[34]

In October 1929, Presidency College was abuzz. Werner Heisenberg had arrived to give a lecture, as Sommerfeld had the previous year. Everyone was poring over Heisenberg's amazing research results, particularly the new atomic physics — dubbed quantum mechanics — which he had discovered in 1925, at the age of twenty-four. Two years later he had discovered his famous uncertainty principle.

Léon Rosenfeld, Bohr's close colleague, wrote, "A wonderful combination of profound intuition and formal virtuosity inspired Heisenberg to conceptions of striking brilliance."[35] Heisenberg was fond of recalling that he learned physics backward — first quantum physics, then classical physics; he had, he wrote, "a very irregular education and training in physics."[36] This hybrid education was no doubt the source of his daring, no-holds-barred approach to research. He had joined Sommerfeld's class in theoretical physics in the autumn of 1920, when Sommerfeld was discussing atomic physics, an area then in turmoil, and immediately began writing groundbreaking papers. By 1927 he was a professor at Leipzig University and was already being spoken of as a future Nobel Prize contender.

The staff at Presidency College suggested that their foremost physics student, Chandra, give a lecture on Heisenberg's work and introduce the great man. Chandra's family "had our own way of looking at the matter," wrote Balakrishnan. "We told one another that the professors realised their inadequacy to talk on the work of Heisenberg and left it to Anna (as we used to call him, the word in Tamil meaning 'elder brother')."[37]

Chandra was thrilled to be able to spend time in the company of the virtuoso physicist. The two men were about equal in height, but there the physical resemblance ended. While Chandra was slight and dark-skinned, Heisenberg had short, fair hair, fair skin, and the robust build of someone who regularly hiked and skied. People who met him in later life recalled his rock-hard handshake, youthful looks, and purposeful walk. Chandra borrowed a car and showed Heisenberg around Madras. The two young men spent the entire day together. "On one day by merely talking to him I could learn a world of physics. At night, we drove along the Marina, and he was telling me about America. On the whole his stay in Madras was extremely profitable. He has suggested one or two problems also," Chandra wrote with enormous pride to his father.[38]

A month later Chandra sent Fowler a paper titled "On the Probability Method in the New Statistics," based on a problem Heisenberg had suggested to him. But the paper was never published. Chandra himself may have withdrawn it after Heisenberg pointed out an error to him. "What days they were!" he recalled "wistfully," writing to Balakrishnan many years later.[39]

In January 1930, a few months after Heisenberg's visit, Chandra was invited to attend a meeting of the Indian Science Congress Association in Allahabad. There he met the distinguished Indian scientist Meghnad Saha, who congratulated Chandra on the paper he had recently had published in the *Physical Review,* describing it as "very suggestive."[40] Chandra reported excitedly in a letter that Saha had also invited him to tea, and then to a small luncheon, so that Chandra could meet Saha's group. The young man felt as if he had truly arrived.

But on the train journey back to Madras, he was rudely reminded that he was still just an Indian under the Raj. Because his father worked for the railway, Chandra always enjoyed free first-class travel. That day he found himself sharing a carriage with an English couple. The train had barely left the station when the woman began complaining loudly at having to share the compartment with an Indian. The only consolation, she added, was that at least he was wearing Western clothes. Infuriated, Chandra stormed out of the carriage and returned in a *veshti* and shirt. When the conductor came through, the woman demanded that Chandra be moved to a second-class compartment. Chandra suggested that she move to a second-class compartment instead. With supreme arrogance, she pulled the emergency cord, then threatened to pull it again when the train restarted. In the end, Chandra won the battle. "After a while they moved to another compartment — second class or something, I don't know," he wrote triumphantly.[41]

That month Chandra wrote to tell his father the strictly confidential news that the government was planning to offer him "a scholarship *straight without advertising for it . . .* as an exception they do it for me as a very special case." The money would enable him to study in England. In exchange, Chandra was to return to Presidency College and take up a professorship there in theoretical physics. The staff at Presidency had high hopes for his future. The principal, P. F. Fyson, had fined Chandra not long before for attending the rally on the beach where Nehru had spoken. Now he asked Chandra if he thought he could become a Fellow of the Royal Society by 1934. "I told him that if I could succeed [by] *1940*, I shall consider myself lucky. I explained to him how Dirac and G. P. Thomson both of them are not yet Fellows," Chandra wrote to his father.

Meanwhile, Chandra's final examinations were looming, and he had to put his calculations aside. From working at the cutting edge of physics, he reverted to undergraduate status. Around this time, Chandra's mother became very ill. He was devoted to her and contemplated abandoning his plans to study abroad. "You must go," she insisted. "You must pursue your own ideals to the utmost."[42] She advised him strongly not to accept advice or help from his uncle Raman. There had been a rift in the family ever since the two brothers and their wives had shared the flat in Calcutta years before. Raman had impudently criticized Sitalakshmi's looks and lack of formal education and had contrasted her with his own wife, whom he had married after a courtship rather than in an arranged marriage like Chandra's parents'.[43] Apart from such thoughtless cruelty, Chandra's family also disliked Raman's arrogance and self-importance.

To complicate matters further, Chandra had fallen in love. Lalitha Doraiswamy was vivacious, outgoing, and extremely determined — essential qualities for a woman who had set her sights on entering the male-dominated world of physics.[44] Because of the burgeoning women's movement that was part of the Indian renaissance, Lalitha received an education and thus, as she proudly explains, met her husband. Like Chandra, she came from a well-known family. Her mother's sister, Subbalakshmi Iyer, had become a widow at the very young age of twelve. This was almost invariably a disaster for a Brahmin girl, who would be hidden away, shunted off to do menial work, and denied a second chance for marriage and motherhood.[45] Subbalakshmi was lucky. Instead of punishing their daughter, her liberal-minded parents sent her to college. She graduated in 1911, outshining all the men in her class. A Brahmin widow's winning such plaudits made headlines across India. Subbalakshmi went on to found a home for child widows, and she lobbied the government to build a college for women. Queen Mary's College stands opposite the Marina, very near Presidency College.

Subbalakshmi's story affected Lalitha's family greatly. Instead of being married off at an early age, the daughters were encouraged to complete high school and university, then marry husbands of their choice.[46] To this day, love marriages are rare in Indian culture.

Lalitha was born just four days before Chandra, on October 15, 1910, in Triplicane. Her mother, Savitri Doraiswamy, was as formidable as Subbalakshmi. When Lalitha was ten, her father, Captain Doraiswamy, died suddenly, probably as a result of trauma suffered in the Great War. Left alone, Savitri continued her education in Hindu tradition and culture and, with the help of her husband's wartime pension, educated her children. She even saved enough money to buy a house in Mylapore, just behind Chandra Vilas. A wall separated the two properties.

In the honors course in physics at Presidency College, Lalitha was one year behind Chandra. The main lecture room there is steeply banked, with twenty rows of fixed wooden tables with benches. As one of the few women, Lalitha was permitted to sit in the front row. She always sat on the far right-hand side. Chandra shyly followed her around. It was love at first sight, although, as Lalitha remembers, "Indian society makes it difficult to actually begin conversations." The lecture room had only one entrance, so Lalitha must have seen Chandra every day as he came in and quietly slipped into the seat behind hers.

Eventually Lalitha took the initiative and asked Chandra if she could borrow his laboratory notebooks — "and that's the way we formally met." They had other chances to speak on school trips. At colloquia they sat as close as possible but did not converse. Lalitha remembers how impressed she was when Chandra demonstrated his deep understanding of physics in lectures he gave to much older and more experienced scientists. He seemed to know everything. Shyly and protectively, he would cycle home along Edward Elliots Road behind Lalitha, and the two would turn off into their respective driveways. Smiling, Lalitha recalls one of Chandra's first approaches, when he gave her a rose to pin on her sari at an annual party.

One of his first gifts to her was an inscribed copy of the 1924 English edition of Sommerfeld's *Atomic Structure and Spectral Lines,* which had been such an inspiration for him. He remembered that he gave it as "a gift to my girl friend, who later became my wife."[47] They were soulmates who shared a passionate interest in science. For a long time nobody suspected that there was anything romantic between them. Just before he was due to leave on his momentous voyage to England, Chandra called on Lalitha at home to give her a list of books she had asked for. Suddenly they found themselves alone and unchaperoned. Speechless, they stood and stared at each other. The spell was broken when a family member brought tea.

By the time he left India, Chandra had read widely and deeply in mathematics and physics. He had developed a sophisticated scientific taste and method of learning. The day before he boarded ship for Europe, he wrote a letter to his brother Balakrishnan from Bombay. Balakrishnan had asked him what he ought to read and whether he should take up a career in physics. Chandra advised him to begin with the 1930 edition of Sommerfeld's book. "Of course you may not fully understand it on first reading," he wrote. "Never mind that. Read say the 1st chapter through — simply as if it were prose. Along with the book you can take E. N. C. Andrade's *Structure of the Atom.*" He also recommended Arthur Hass's *Wave Mechanics and Quantum Theory,* "the one which is at present at home," and Richard C. Tolman's *Statistical Mechanics.* "But before all this," he continued, "you will have to know calculus very well." For this he suggested Horace Lamb's *Infinitesimal Calculus* — "(it is at home)" — and also advised Balakrishnan to "work through Daniel Alexander Murray's *Elementary Course on Differential Equations.*"[48] (By "working through" a text, Chandra meant not only reading it but doing all the practice problems, a challenging task indeed.) Balakrishnan replied that he would "begin with the physics books" and mentioned the subject no more. The mathematics books were dauntingly detailed, while the physics texts took the reader into the strange new world of the atom. For Chandra to be so well acquainted with these very difficult texts at such a young age was quite extraordinary.

Chandra spent his last week in India taking care of last-minute details and giving a lecture at the Royal Institute of Science in Bombay. Then came the tearful farewells at the dock. His mother was too ill to make the trip to see him off. His uncle Raman wrote that great things were expected of him: "We are all looking forward to your doing for Indian physics what Ramanujan did for Indian mathematics."[49] This was high praise, especially from his famous uncle. The family stayed on board the *Lloyd Triestino* until the last minute. No doubt Chandra breathed a sigh of relief when, in the late afternoon of July 31, the ocean liner slipped out of her moorings into the Arabian Sea. Bursting with youthful optimism, he was looking forward to studying with Fowler in England and spending time in the company of Eddington. It would be an incredible adventure.

3. Rival Giants of Astrophysics

RTHUR STANLEY EDDINGTON once wrote, in reference to himself, "Human personalities are not measurable by symbols any more than you can extract the square root of a sonnet."[1] Eddington's portraits reveal little of the man. His stiff pose, impenetrable gaze, domed forehead, long nose, and unsmiling lips are reminiscent of the prismlike expression of a previous Cambridge man, Sir Isaac Newton. Chandra perceived Eddington as the quintessential product of an Edwardian England in which everyone knew his place in the grand scheme of things and was confident of his right to the life of privilege he enjoyed:

> He was a man who was very distinguished, in the sense that one felt when one talked to him that one was talking to someone really substantial. The British, particularly in earlier times, can be very nice and kind, but at the same time, an element in their behaviour makes it very clear that [they feel] they're on a different level. There's no snobbery involved in it. It sort of comes naturally to them. Eddington was that sort of man.[2]

We know little of Eddington's private life. He usually destroyed his correspondence as soon as the business at hand was completed. In the autumn of 1944, as he lay dying, he began to dispose of his personal letters. His spinster sister, Winifred, four years older, with whom he had lived for most of his life, destroyed much of what remained. In his will he bequeathed to the Royal Astronomical Society a large standing cabinet filled with papers. The venerable Colonel Chubby Stratton examined them, pronounced them "merely of biographical interest" and threw them away.[3] The only biography of Eddington, commissioned by Winifred, is more like a hagiography. Written by a former student of his, Allie Vibert Douglas, it contains no personal information of any sort.

Eddington was a study in contrasts. Students recall him as painfully shy but also so arrogant as to be almost unapproachable; he expected great reverence. As a teacher he was appallingly dull, in the worst Cambridge style. Yet among his peers, particularly when dining at high table, he could be the life of the party. In a notebook full of random thoughts and memories, Chandra listed twenty-five stories he had heard Eddington tell at high table. One story reveals Eddington's distinctly offbeat sense of humor. At a gathering one night, a solemn discussion was under way on major technological discoveries. "Eddington sat back," recalls Chandra,

> puffed on his trademark pipe, and said that "as far as he was concerned, zips [zippers] were the only genuine discovery." [He then] commented that the use of zips in women's dresses might lead to awkward situations. For example at an evening party one may inadvertently unzip a woman's evening dress and the dress may fall to the ground to every one's embarrassment . . . Eddington was amused at the spectacle he imagined but the staid prudes next to him did not even smile.[4]

Eddington was a highly cultivated man. The books he wrote, both technical and popular, bristle with quotations from French, German, and Italian literature, cited in the original languages. He was an expert chess player and a whiz at the *Times* crossword puzzle, which he could complete in the time it took to move his pen across the grid. Despite his dull classroom lecturing style, he was a well-regarded public speaker.

The American physicist W. H. Williams, with whom he shared an office in Berkeley, California, for a couple of months in 1924, wrote that "he suffered from British uncommunicativeness in most severe form." Eventually, however, the two discovered a mutual love of golf. Thereafter they played together twice a week at the Claremont Club, "perhaps the worst golf ever seen on that course." He also remembered that Eddington was an admirer of *Alice in Wonderland* and a composer of entertaining doggerel.[5]

The eminent American astronomer Harlow Shapley recalled that when Eddington visited Harvard University for its tercentenary celebrations in 1936, he "divided his interests between galaxies and the Red Sox," who were in town at the time.[6] He dominated the world of astrophysics. Shapley told Chandra that he had sent a circular to the leading American astronomers that same year, asking them to rank their colleagues to decide who should be awarded an honorary degree. "Eddington was the first in every single list received!" he told him.[7]

Eddington, indisputably the greatest astrophysicist in the world, had developed the field virtually single-handed. All other scientists were in awe of him. He wielded enormous power, less because of his position than because of the sheer brilliance of his work and a network of highly influential colleagues around the world. He was an intimidating figure who cast a spell (as Milne was later to write of him) purely through force of intellect. "Eddington had the gift, characteristic of genius, of being able to reach correct conclusions by arguments at first sight dubious," wrote Thomas Cowling, a student of Milne's who many times witnessed Eddington in action at the RAS.[8] Questioning any aspect of his work was a risky business that involved crossing not only Eddington himself but devoted colleagues such as the powerful American astrophysicist Henry Norris Russell, who wholeheartedly supported Eddington's wildest theories and also controlled appointments in the United States.

Eddington was born on December 20, 1882, in the picturesque town of Kendal among the hills of the Lake District. Contrary to Chandra's assessment of him, his background was far from privileged. His family were Quakers. His father, the headmaster of a famous old Quaker school, died in a typhoid epidemic when Eddington was just two years old. His mother, who must have been a formidable woman, moved to Weston-super-Mare, on the coast near Bristol, taking her two young children with her. There he and Winifred grew up in some poverty. Even as a child, Eddington was interested in astronomy and would survey the night sky through a three-inch telescope that had been lent to him. He had a great fascination for large numbers and a prodigious memory. One day his mother went out for the afternoon, leaving him alone. When she returned she asked how he had spent his time. "I've been counting all the words in the Bible," the precocious child replied. He actually succeeded in getting through the Book of Genesis.[9] At night he counted the stars in the sky.

Whereas Chandra came from an affluent background, Eddington had to apply for every available scholarship. He went to a small school in Weston-super-Mare as a day pupil, not a boarder, to save money. He was popular at school. Besides mathematics, his other great love was

English literature. In later years he was fond of recounting that he had once won a contest for the best "jabberwocky" poem — "Stand by the hedge and sound like a turnip" was his impenetrable winning phrase, according to Chandra's recollection of the story.[10] He also loved cycling. A brilliant youth, he won a scholarship to Owens College (shortly to become Manchester University) when he was not yet sixteen, officially too young to enter. A series of scholarships enabled him to study physics as well as mathematics, mechanics, English history, and Latin there. He graduated with first-class honors in 1902, at the age of nineteen. Another scholarship took him to Trinity College, Cambridge, where he studied mathematics. In his second year he came first in the fiercely competitive mathematics examinations and was awarded the rank of Senior Wrangler, effectively guaranteeing him a research career. He was the first person ever to achieve the rank in such a short time.[11]

By 1906 he had decided to focus on his childhood love — astronomy. That year the Astronomer Royal offered him the position of chief assistant at the Royal Observatory in Greenwich, where he quickly realized how extraordinarily little was known about the stars. Astronomers and astrophysicists put forward a ragbag of confused hypotheses but seemed to have little clear idea about what stars were, how they moved, and what their life cycle was. Here was a challenge — to order and advance the field of astrophysics. He began by looking into the movement of stars. He won a fellowship at Trinity College in 1907 and produced such distinguished work that in 1913, when George Darwin, one of Charles Darwin's sons, died, Eddington was offered the prestigious position left vacant, that of Plumian Professor of Astronomy and Experimental Philosophy at Cambridge. The following year he was appointed director of the Cambridge Observatory. The post provided him with a comfortable home in the observatory house, and he brought his mother and sister to live there with him in 1914, the year the Great War broke out.

Eddington lived in the sort of world for which the term "dreaming spires" was coined. Oxbridge dons formed almost monkish communities, living in college for their entire adult lives and devoting themselves single-mindedly to their studies. G. H. Hardy and John Littlewood, Eddington's colleagues at Trinity, were of that sort. The most famous team in twentieth-century mathematics, they produced superb papers on number theory. One would dash off a letter to the other with the words "just had an idea," and shortly afterward a paper would be in the mail.[12] Such men played their cards close to the chest. In middle age Littlewood was often seen in the company of an attractive, much younger woman, whom he always introduced as his niece. A few eyebrows were raised, but in proper Cambridge tradition no one uttered a word. In his eighties, in the Combination Room, the staff common room at Trinity, Littlewood finally let out the news that the woman was his daughter, from a long and secretive liaison with a married woman. "Next day he was oppressed that nobody had blinked an eyelid!" wrote a former colleague.[13] Aggrieved though he may have been, such sangfroid was totally in keeping with the Cambridge ethos.

In Eddington's day, the dons who lived at college spent most of their time in the company of other dons. The late Victorian ideal was still in force, of "romantic friendship within exclusively male communities, a phenomenon so normal and respected throughout the period   that the greatest care must be taken to avoid slick and dismissive judgements."[14] Decades were to pass before a man could safely confess that his relationship with another man was anything more than friendship.

G. H. Hardy embodied that ethos. With his high cheekbones, ice-clear eyes, narrow nose, austere demeanor, and look of perpetual youth, colleagues routinely described him as the quintessential "beautiful" man.[15] He was a member of the Cambridge Conversazione Society, an elite secret society better known as the Apostles, because when it was established in 1820 it had twelve members. Its luminaries had included Alfred, Lord Tennyson, Bertrand Russell, and James Clerk Maxwell, some of the most brilliant minds that Cambridge had ever produced. By Hardy's time the society had such a strongly homosexual atmosphere that one member wryly observed, "Even the womanisers pretend to be sods, lest they shouldn't be thought respectable."[16] In the late 1890s its members included the nucleus of what would become the Bloomsbury literary circle, as well as the economist John Maynard Keynes. In the 1930s the society became notorious as a breeding ground for Communist agents, many of whom were seduced and then recruited by Guy Burgess. Homosexuality was taken for granted among its members. Littlewood, speaking at a time when homosexuality was not only taboo but illegal, referred to Hardy as "a non-practising homosexual."[17] The pioneer of digital computing, Alan Turing, was more straightforward. He spoke of Hardy as "just another English intellectual homosexual atheist."[18]

And what of Eddington? Women found him charming. Unlike most of his Cambridge colleagues, he went out of his way to be courteous to them. One of the few female students of science, Cecilia Payne, recalled how rude the eminent physicist Ernest Rutherford was to her. He often began his lectures, for which she was usually the sole woman in the audience, by looking her straight in the eye and uttering the word "Ladies," followed by a long pause, " . . . and gentlemen." Rutherford's daughter, Eileen, a friend of Payne's, told her that Rutherford, prickling with macho arrogance, had said, "She isn't interested in you, my dear. She's just interested in me."[19] Payne was so offended by his attitude that she gave up physics and switched to astronomy, a field that had attracted her ever since she heard Eddington lecture. Eddington immediately set her a problem on the structure of stars and was very helpful to her.

Many years later, Payne told Chandra that "my interest in astronomy originated with a lecture by Eddington on his results verifying relativity. At that time I was doing the natural science Tripos, but because of Eddington I changed to the mathematics Tripos in order that I could become Eddington's student. . . . I do not mind telling you that, in fact, I fell in love with him."[20]

But although women found Eddington attractive, there is no evidence that he found them so. His biographer, Allie Vibert Douglas, noted that Eddington never wanted to get married. Apart from his mother and sister, "his interest in women was simply and solely as acquaintances or, in the case of the very few women astronomers in various countries, as friendly colleagues."[21] There was, she records, only one person with whom Eddington "formed a lasting friendship [and with whom he] could throw off all the hesitant diffidence which formed an almost impenetrable barrier to intimacy with others."[22] This was a man whom Douglas refers to as Eddington's constant companion and "most intimate friend," Charles Trimble.

Trimble's acquaintances from Trinity College recalled that Eddington was the "strongest, certainly a most rewarding influence on his life."[23] The two men, both from working-class backgrounds, met at Trinity, where working-class students were thin on the ground; perhaps that was part of the initial attraction between them. Born in Bath in 1883, Trimble studied at Christ's Hospital, a private school that offered an excellent education to boys from impoverished backgrounds. At Trinity College, he graduated fourth in his class (Eddington was first), then entered the civil service. In 1910 he took up a post as a mathematics tutor at Christ's Hospital. But he was plagued by mental problems and suffered several breakdowns.

Former students remembered that Trimble tried to instill in them a love of literature as well as mathematics. A photograph, taken in 1923, at the height of his friendship with Eddington, shows a pensive man with gentle features and immaculately combed dark hair. By that time Eddington, in contrast, had acquired the supercilious, worldly look, his hair already graying, of a man who lived amid the rough-and-tumble of international science.

Eddington usually cycled alone, but when he did not, he made a note of his company. In three separate entries in his cycling diary, he notes that he made trips of several days "with Trimble."[24] The two went on regular walking trips together and enjoyed glissading down icy and snowy slopes. On these jaunts Eddington abandoned his Quaker sobriety and cheerfully indulged in alcohol, tobacco, and visits to the theater and the cinema. Trimble recalled one glissading episode in which the two men tumbled through the snow and ended up with their clothes torn and wet. Eddington refused to take his shirt off and insisted that Trimble take a needle and thread and mend it right there while he was wearing it.[25]

It could well be that the two were more than simply friends. But in those days it was downright dangerous even to be suspected of homosexuality. Everyone was aware of the penalties of getting caught. The Oscar Wilde affair of 1895 still cast its shadow. If Eddington was a practicing homosexual he would have had to be extremely careful in his liaisons; had word got out, he would have been ruined. Robert Kanigel, Ramanujan's biographer, wrote of the fragile balance that Hardy maintained in his life: "He was a friend of many in Cambridge, an intimate of few."[20] In Eddington's case, he had just two intimates, his sister, Winifred, and Trimble. If indeed he was homosexual, the strain of maintaining secrecy must have been horrendous.

In his working mode, Eddington was a loner. Few students attended his classroom lectures, which were usually stilted and dull. They were more likely to glimpse the real Eddington in his public lectures, which were stylish as well as highly informative. Entering his study, visitors were struck by the papers and books lying all over the floor, littering the sofa and tables, and covering his desk. He sat right in front of French windows opening onto the Cambridge Observatory's beautifully manicured lawn. If students interrupted him in midthought, their reception was usually off-putting, verging on the comical. Several re-called walking into his study for an appointment at the precise time that he had arranged to see them. He would look up in surprise, usher the student in, and listen politely, all the time with a quizzical look on his face, as if he was wondering who on earth this person was and why he was there. Meanwhile he chewed up apple after apple. Once the Norwegian astrophysicist Svein Rosseland dropped by to see him for a prearranged appointment. At the time, Eddington was obsessed with the idea that he could deduce the existence of elementary particles from mathematics alone. Rosseland knocked again and again. There was no response. Then Eddington shouted, "Oh, it's you. Come in. I have just found the neutron."[27]

When Eddington took up his position at the Royal Observatory in Greenwich in 1906 and began his research into the stars, astrophysics was in its infancy. Astrophysicists assumed that stars — the Sun, for example — were essentially balls of gas, made up of atoms.[28] But very little was known about their structure until the Danish physicist Niels Bohr proposed a theory of the atom in 1913. He depicted it as a minuscule solar system with a positively charged nucleus in the middle, like the Sun, surrounded by enough electrons to give it a net charge of zero. The simplest and lightest atom is hydrogen, which has a nucleus and one electron circling it like a planet.[29] Helium, the next lightest, has two electrons.

The temperature inside stars is in the millions of degrees kelvin.* The intense heat energizes the electrons orbiting the nuclei of the atoms so much that they tear themselves free. As a result the inside of a star is made up of a huge number of swirling electrons with heavier nuclei lumbering along more slowly. In the 1920s astrophysicists investigating a star's structure focused on the electrons.[30] At the time, no one understood much at all about the nucleus, although Eddington, for one, was already beginning to speculate whether it might be the fuel that a star burned to make it shine.

To study stellar structure, astrophysicists made the assumption that a star was an idealized gas known as a "perfect gas." The concept of a

---

* In science, temperatures are usually measured in degrees kelvin, a temperature scale named after the nineteenth-century British scientist William Thomson, Lord Kelvin. Kelvin temperatures can be transformed into Celsius by subtracting 273. The temperatures of stars are in the millions of degrees kelvin, and at that level the difference between the two scales is negligible. In this book all temperatures are in kelvin.

perfect gas was first developed in the nineteenth century, when scientists discovered the simple relationship connecting the pressure, volume, and temperature of a gas of about the same low density as the air we breathe. No matter what chemicals the gas is made up of, if the temperature remains the same and the pressure on it increases, it will decrease in volume. If it is kept at a constant pressure, in a pressure cooker, for example, as the temperature rises, the gas will increase in volume; in other words, it expands as it gets hotter. The mathematical relationship among the pressure, volume, and temperature of a perfect gas is called the perfect gas law.[31]

Through his research, Eddington came across Henry Norris Russell around 1913. Then in his thirties, Russell was a professor at Princeton University. As an "Old Princetonian," he sported high-laced boots, starched collars, and expensively tailored suits. He even cultivated a British accent, acquired during a brief period at Cambridge. A prudish man with little sense of humor, he loathed the oft-repeated quip that he was a world authority on Eros, the asteroid he had studied for his Ph.D. thesis. He wielded great power on the American scene. Up-and-coming astrophysicists feared running up against him. Many remember him as selfish, overbearing, and opinionated. Russell bore a lifelong grudge against Eddington's colleague and archrival, the astrophysicist James Jeans, because Jeans had been recruited to Princeton in a more senior position and with a much higher salary. But despite his personal failings and eccentricities, Russell was a superb astrophysicist. His students populated nearly every major astronomical institution.

Russell was looking into the life cycle of stars: how they are born, how they live, and how they die. One of Eddington's first projects was to take up one aspect of Russell's work by investigating Cepheid variable stars. The brightness of Cepheids fluctuates from maximum to minimum, then back to maximum; the period of fluctuation can range from a few hours to many days. By 1908, more than 1,700 of these variable stars had been found, and a definite relationship had been established between a Cepheid's period of fluctuation and its luminosity.[32] Eddington was able to deduce a relationship between fluctuation and density and was pleased to find that his result, obtained mathematically, more or less fitted astronomers' observations.[33]

He then began to wonder how stars became Cepheids and what happened to them when they exhausted their energy, ceased to shine, and stopped pulsating. He found his first clue in a lecture that Russell gave

| Gamma rays | | X-rays | Ultraviolet | | Visible | Infrared | | Microwaves | | Radio, TV |

The full electromagnetic or radiation spectrum. To the left end, beyond
ultraviolet, are x-rays and gamma rays, which have short wavelengths, high
frequencies, and high energy. At the opposite end, beyond infrared, are
radio waves, with long wavelengths, low frequencies, and low energy. Only
a small part of the spectrum is visible to the naked eye.

at the Royal Astronomical Society in 1913, the year Eddington became
Plumian Professor. That day Russell was the last speaker in a long day of
presentations. By the time he presented his last slide, everyone was doz-
ing off. Eddington, however, snapped to attention.

In the nineteenth century, scientists had discovered that the universe
is crisscrossed by different sorts of light, some visible to the human
eye, others not.[34] At one end of the radiation spectrum are radio waves,
with very long wavelengths — several meters long — and very short frequency. At the other end is very short wavelength radiation — ultraviolet light, x-rays, and gamma rays, the shortest of all. The only part of the
spectrum visible to the eye is the small "window" between ultraviolet
and infrared. From Earth the universe looks quiescent and unchang-
ing. But telescopes mounted on orbiting satellites high above the atmosphere blanketing the planet use instruments sensitive to parts of the
spectrum that we cannot see. They reveal a startlingly violent universe,
with gamma rays exploding and x-rays shooting out.

Every atom in a star emits light of a particular frequency, much
the way a tuning fork produces sound of a particular frequency when
struck. Starlight is a hodgepodge of frequencies, made up of all the
light emitted by a huge number of atoms of many different elements.
If millions of tuning forks of different frequencies were all struck at
once, the sound produced would be continuous, and it would be impossible to hear any individual frequency. In the same way, starlight appears as a continuous spectrum of light. Astrophysicists unravel the different light frequencies using a spectroscope set in the eyepiece of a
telescope. As a single beam of starlight enters, a fan of light of different
frequencies emerges. When astrophysicists shine this spectrum onto a
photographic plate, each frequency appears as a separate "spectral

line." Somewhat like DNA or fingerprints, each chemical element has a unique set of spectral lines.

In the late nineteenth century, astronomers at the Harvard Observatory, backed by an army of underpaid female workers, took on the epic task of classifying the light from 500,000 stars. This involved teasing out similarities in the characteristics of light from each star as displayed on stacks of glass photographic plates. They then divided the stars into spectral classes by grouping together those with similar spectral lines. They classified the classes with the letters O, B, A, F, G, K, and M. The brightest stars were O, and the dimmest M; O-type stars were the hottest, M-types the coolest. Stars in spectral class O had a surface temperature of about 50,000 degrees kelvin, and those in M about 3,000 degrees. These temperatures were worked out by measuring the wavelengths of the light emitted by the star.[35]

Russell's brilliant idea was to take certain of these stars and plot them on a graph, with brightness along one axis and temperature on the other. The result was a fascinating visual representation of the heavens.[36] By chance, a Danish photochemist turned astronomer, Ejnar Hertzsprung, had been doing similar work, so the diagram became known as the Hertzsprung-Russell (H-R) diagram. This was the diagram that grabbed Eddington's attention during Russell's 1913 talk.

Russell's diagram showed 300 stars whose distances from Earth were accurately known. To his own amazement, he found that they were not randomly scattered. Most were in a diagonal strip, from hot, bright O type stars at the upper left to cool, dimmer M-type stars at the lower right, and most were about the same size as the Sun. He called this strip the main sequence. The other large grouping of stars lay on a horizontal line jutting out from the main sequence; all were very big and more or less of the same degree of brightness. Following a suggestion of Hertzsprung's, he classified stars on the main sequence as "dwarfs" and those on the horizontal strip as "giants." From this he proposed that there were "two great classes of stars" — giants (bright stars ten to a hundred times bigger than the Sun) and dwarfs (about the same size as the Sun and of decreasing brightness).[37]

Some years later Russell recalled that he invented the H-R diagram "because it represented the phenomena I wanted to describe in such a way that I could get them on a page."[38] But the diagram turned out to be much more than an economical way to present a lot of information at a lecture. It soon became apparent that studying stellar structure and

The original Hertzsprung-Russell (H-R) diagram, with spectral types listed along the top and absolute magnitudes on the left. (From Henry Norris Russell, *Nature* 93, 1914)

the life and death of stars involved understanding why stars populate certain regions of the H-R diagram and not others, and how they "trek," or evolve from one area to another. Eddington was one of the first astrophysicists to understand the diagram's importance and try to fathom its cosmic message. His research was at the heart of the debate that led eventually to firm knowledge of the chemical constitution of stars and to an understanding of what makes them shine.

One star presented a particularly tantalizing puzzle. It was completely outside Russell's neat layout. This was o^2 Eridani B, the dim companion of o^2 Eridani A.[39] The two stars revolve around each other in an endless dance known as a binary system. o^2 Eridani B did not fit any pattern; it was neither a giant nor a dwarf. It was quite hot — about 10,000 degrees kelvin — but very dim, so it lay well below the main sequence. At first Russell classified it as a dwarf star, while Hertzsprung called it a "dark white star." It soon became known as a white dwarf. It was the first white dwarf to be identified.

Hertzsprung and Russell assumed that o^2 Eridani B was a freak occurrence because it fell so awkwardly outside the scheme of things and tried to ignore it. But then Russell remembered a conversation he had had in 1910 with Edward C. Pickering, at the time America's most influential astronomer and Director of the Harvard Observatory. "It is just such discrepancies which lead to the increase of our knowledge," Pickering had told him.[40] It was a salutary reminder. Then in 1914, an astronomer made some momentous new observations of another white dwarf, Sirius B, the star that was later to change Chandra's and Eddington's lives and the course of astrophysics.

Sirius B is the companion of Sirius A, the brightest star in the night sky, twenty-six times as luminous as the Sun. Sirius A had been studied extensively since 1844, when the German astronomer and mathematician Friedrich Wilhelm Bessel worked out that it was 54 trillion miles from Earth. To astronomers of the mid-nineteenth century, this was an unimaginably great distance. Earth is a mere 7,926 miles across, while the Sun is 93 million miles from us and a billion miles from Pluto, the most distant major planet of our solar system. Sirius A is many trillions of miles farther away, yet it blazes away so brightly that it all but obscures its companion star. The fact that it can be observed with a telescope shows how extraordinarily bright it is.

Nineteenth-century astronomers began to suspect that Sirius A had a companion star when they noticed that it wobbled instead of following a smooth path. In 1862 an American amateur astronomer and lens maker named Alvan Clark spotted Sirius B for the first time. Astronomers worked out that the next time the two stars would be far enough apart to allow observations of Sirius B would be fifty years later. In 1914, the year after Eddington encountered the H-R diagram, Walter S. Adams, America's leading observational astronomer, was ready. Working at Mount Wilson Observatory in California, he measured Sirius B's brightness and spectral class and discovered that it too was in spectral class A, like o^2 Eridani B, and slightly cooler, about 8,000 degrees kelvin. o^2 Eridani B was not a freak after all; Sirius B was equally strange.[41] How could these stars be so hot yet radiate so little light? Surely there was some mistake. Or, as Pickering had cautioned, maybe not.

Using the time it took Sirius A and Sirius B to orbit each other, the degree of irregularity of the orbit, and the distance of the two from Earth, astronomers were able to calculate the mass of Sirius B.[42] It turned out to be about the same as that of the Sun, some 2 billion trillion trillion grams or about 10^{33} grams.[43]

They also calculated the radius to be some 11,280 miles, making it about thirty times smaller than the Sun — a little more than three times the size of our planet. In other words, Sirius B packed the whole mass of the Sun into a volume not a great deal larger than Earth's. That made its average density a stunning 61,000 grams per cubic centimeter. Transferred to Earth, a single teaspoonful of this white dwarf matter would weigh nearly 6 tons, as much as a good-sized elephant. White dwarfs are so extremely dense that astrophysicists suspected they would have to find some new version of the perfect gas law to describe them, though hopefully not drastically new. Eddington's comment on the situation as it stood in 1914 was "absurd."[44] He decided to set aside the issue of white dwarfs and turn his mind to the physical structure of dwarfs and giants. What puzzled him was why stars should cluster in certain areas of the H-R diagram and not others.

The issue of clustering bore on one of the most important questions that had yet to be answered: what makes stars shine? Eddington was certain that the theory formulated late in the nineteenth century by the German polymath Hermann von Helmholtz and the British scientist William Thomson, Lord Kelvin, was completely wrong. According to that theory, stars shone as a result of the particles in their gas being squeezed together by gravity, increasing their temperature to the point that they emitted radiation. But if that theory was right, the Sun was only 20 million years old.[45] In 1917 radioactive dating estimated that Earth was around 2 billion years old. How could Earth possibly be older than the Sun?

Eddington proposed an alternative theory. Perhaps the protons and electrons in a star annihilated each other in a slow process that created light. The number of electrons making up a star was so immense that the energy supply thus created would be "almost inexhaustible."[46] The problem with the theory was that if the electrons and protons kept on destroying each other, the star itself would eventually disappear.

Moreover, one would expect very dense stars, in which the protons and electrons were packed together and jostling furiously, to be extremely bright. But in that case, why were very dense stars like white dwarfs so dim? Could some unknown processes be affecting the ongoing annihilation?

Eddington was sure that it had to be the nuclei of the atoms in a star that produced its energy, as opposed to the electrons, and that there had to be vast reserves of energy, enough to keep a star shining for billions of years. Then in 1920, Francis Aston, a forty-three-year-old chemist at the Cavendish Laboratory in Cambridge, made a startling discovery. When he measured the mass of the nucleus of the helium atom (also called an alpha particle) and compared it with the sum of the masses of the particles that made it up, the two figures were very slightly different; the mass of the helium nucleus was smaller than the mass of the particles by 0.8 percent.

He reasoned that if a helium nucleus was forged from its separate particles, the lost mass would reappear as a huge quantity of energy, as predicted by Einstein's famous equation, $E = mc^2$.[47] In this equation E is energy, m is mass, and c is the velocity of light (186,000 miles per second). The velocity of light multiplied by itself — c^2 — is a vast number, and thus the equation shows that a minuscule amount of mass will translate into an enormous amount of energy. Aston described his conclusion vividly: "To change the hydrogen in a glass of water into helium would release enough energy to drive the 'Queen Mary' across the Atlantic at full speed."[48]

Eddington immediately realized that Aston's discovery could explain why stars continued to burn for so long. Perhaps what made them shine was the fusion of hydrogen to form helium, using the processes implied in Aston's measurements. "What is possible in the Cavendish Laboratory may not be too difficult in the Sun," he pointed out.[49] It was an inspired hypothesis, but it could be no more than that, for at the time scientists knew very little about the nucleus. But they did know that for fusion to happen, the temperature inside stars would have to be extraordinarily high — high enough to give protons sufficient energy to overcome their immense mutual electrical repulsion and slam into each other with enough force to stick together. Some astrophysicists argued that the temperature inside stars might not be hot enough. To this Eddington replied fiercely, "We tell them to go and find a *hotter* place."[50]

Eddington had by now embarked on an ambitious quest to work out, entirely through mathematical means, what went on in the interior of a star. He called this his standard model, and it embodied his developing ideas on the constitution of the stars. As he said, "It is as though nature had a standard model before her in forming the stars, and (except for occasional lapses of vigilance) would not tolerate much deviation."[51]

In 1917, when he first put forward his standard model, it applied only to giant stars, which have a density low enough that they can be studied using the perfect gas law. The temperature inside a star is upward of 10 million degrees kelvin, which means that the radiation there is in the form of very high-energy and high-frequency x-rays. As the x-rays collide with the atoms in the star, they strip off their electrons, beginning with those in the outer electron rings, which are less tightly bound than those closer to the nucleus. Knocking out an electron depletes the amount and strength of the light inside the star. The detached ("free") electrons swirl around until they are captured by another atom, only to be knocked out again, further depleting the star's store of light.

Eddington described the process:

> The inside of a star is a hurly-burly of atoms, electrons and [radiation]. . . . Try to picture the tumult! Dishevelled atoms tear along at 50 miles a second with only a few tatters left of their elaborate cloaks of electrons torn from them in the scrimmage. The lost electrons are speeding a hundred times faster to find new resting places. Look out! there is nearly a collision as an electron approaches an atomic nucleus; but putting on speed it sweeps round it in a sharp curve. . . . Then comes a worse slip than usual; the electron is fairly caught and attached to the atom, and its career of freedom is at an end. But only for an instant. Barely has the atom arranged the new scalp on its girdle when a quantum of [light] runs into it. With a great explosion the electron is off again for further adventure.[52]

In order to explore a star's temperature and the amount of radiation it emits, Eddington had to specify the average number of free electrons per nucleus, the "mean molecular weight." At the time, astronomers made the perfectly reasonable assumption that a star is made up of the same elements as Earth — oxygen, iron, sodium, silicon, potassium, magnesium, aluminum, and calcium — and almost no hydrogen or helium.[53] Taking this composition into account, and assuming that not all atoms would be stripped of their electrons, Eddington adopted a mean molecular weight of 2.1. He proposed that the mixture of chemical elements was the same in all stars and that the mean molecular weight therefore applied in all cases.

Another issue that Eddington had to consider in modeling a star was opacity, the extent to which a star's chemical composition prevents radiation from escaping. If a star is not very opaque, more radiation will escape and it will cool down quickly. But if it is highly opaque, then radiation cannot easily escape and it may explode prematurely. Eddington began by taking the mathematical formula for opacity used in the laboratory, which he called the "physical value." He then altered it to fit the conditions inside a star, calling this version the "astronomical value." Taking a mean molecular weight of 2.1, he calculated a figure for the astronomical value of opacity, which he assumed was valid for every star.[54]

From this he made the important discovery of what he called the mass-luminosity relationship: the more massive a giant star, the brighter it is. Amazingly, this relationship turned out to be valid not only for low-density giant stars, which could be studied using the perfect gas law, but also for dwarf stars, even though they were much denser. This meant that if astronomers were able to measure a star's luminosity, they could work out its mass. It also provided firm theoretical backing for astronomers' observations that more massive stars are brighter than less massive ones and explained why, for example, Sirius A was so extraordinarily bright. The one sort of star that it was not valid for was the strangely anomalous white dwarfs.

Eddington was amazed at his own result. "In making the comparisons for these dense stars [that is, dwarfs, but not white dwarfs], the writer had no anticipation that they would agree to the curve [that is, to his theory]," he wrote.[55] His results completely disproved Russell's theory that dwarf stars were too dense to obey the perfect gas law.

But how was that possible? Eddington's inspired guess was that since the atoms inside stars are almost completely stripped of their electrons, they are some billion times smaller in volume than terrestrial atoms and occupy much less space. That was why the matter in the interior of a star would behave like a perfect gas even if it was very dense. In ordinary dwarf stars, matter may be as dense as platinum yet still behave like a perfect gas; even platinum is still a long way from the maximum density possible.

Although Eddington's mass-luminosity equation agreed perfectly with astronomers' observations, the fit was achieved with great difficulty. To make his calculations work out, Eddington had to assume that the astronomical value for opacity was ten times the physical value. This "opacity discrepancy" was a fundamental flaw in Eddington's standard model, as he readily admitted.[56] The only way to make the astronomical and physical values match was to postulate a huge amount of hydrogen inside stars.[57] Comparing the mathematical equations for the physical and astronomical values made this crystal clear. But increasing the amount of hydrogen would make the standard model untenable because, according to his mathematics, it would decrease a star's radiation pressure, a key ingredient of the model.

As it happened, the Indian physicist Meghnad Saha had already made a discovery that threatened Eddington's standard model. In 1920 he had discovered a way to link the temperature at the surface of a star with the chemical elements in its surface layers. From this he deduced that the Sun's atmosphere had a million times more hydrogen than any other chemical element. Could this mean there was hydrogen in the interior of stars as well as on the surface? And if so, why should there be so much hydrogen in stars when there was so little on Earth? Eddington fervently hoped that further refinements in Saha's theory would explain away his finding, but none ever did.

Then in 1925, Cecilia Payne, Eddington's former student, who was by then a twenty-five-year-old Ph.D. student at the Harvard Observatory, used an improved version of Saha's theory to reexamine spectra from the Sun. She proved conclusively that there was a huge amount of hydrogen in the Sun's atmosphere. The irascible Russell, an ardent supporter of Eddington and in awe of him, wrote to Payne that this was utterly "impossible."[58] Fearing Russell's wrath, Payne followed his line and toned down her remarks, declaring that the vast quantity of hydrogen "is probably not real."[59] When she visited Cambridge in September 1925 and told Eddington her result, he replied curtly, "Well, that is on the stars, but you don't know that it is in the stars."[60]

Eventually, independent estimates forced Russell to accept Payne's conclusion.[61] By 1932 the weight of evidence was overwhelming. The twenty-seven-year-old Danish astrophysicist Bengt Strömgren, who was to become a friend and colleague of Chandra's, proposed that at least a third of a star was made up of hydrogen. This hypothesis brought the astronomical and physical values for opacity into agreement, thus removing the opacity discrepancy.[62] Eddington gave in.[63]

Along with his work on stars, Eddington was also doing pioneering work on relativity theory. Einstein had discovered his general theory of relativity in 1915. But the world was at war and Einstein's revolutionary theory did not immediately cross the Channel. Through his astronomical studies, Eddington came into contact with the distinguished Dutch astronomer Willem de Sitter, a professor of astronomy at Leiden University in neutral Holland, who was examining the astronomical implications of Einstein's theory. In 1917 de Sitter sent Eddington a copy of Einstein's latest paper on general relativity. It was almost certainly the first copy to reach Britain. Eddington immediately understood the significance of Einstein's work. The following year the Physical Society commissioned him to write the first report on it. His masterly "Report on the Relativity Theory of Gravitation" brought the theory to the attention of many British scientists and established Eddington as the leading exponent of relativity in the English language.

Meanwhile the war intensified. Conscription had been introduced in 1916. As a Quaker, Eddington could have declared himself a conscientious objector, though his colleagues would have considered it an unpardonable disgrace to abstain from participating in a war when youthful patriots all over the country were volunteering to fight. Several major scientists at Cambridge prevailed on the government to exempt Eddington on the basis of his scientific achievements. The Home Office later appealed against the exemption. At the first hearing, on June 14, 1918, Eddington declared himself a conscientious objector. He saw it as no shame, he said, to join his Quaker friends in internment camps in Northern Ireland, peeling potatoes. In the end the Astronomer Royal intervened, proposing that Eddington be exempted on condition that he lead a British scientific team to observe the total eclipse of the sun on May 29, 1919, to verify Einstein's general theory of relativity. The board granted him a year's reprieve to do so.

Soon afterward the armistice was signed. But Eddington prepared for the expedition nonetheless. The aim was to test a prediction of the general theory of relativity: that light would bend near a massive star like the Sun. At that time observations of stars close to the Sun could be made only during a total eclipse.

It was a great adventure. In early March 1919 two groups sailed from England to Lisbon, where they parted company. One went on to Sobral, Brazil. The other, led by Eddington, set off for the tiny Portuguese island of Principe, off the west coast of Africa. By mid-May Eddington and his colleagues had their instruments in place and had carried out some test photographs. But heavy rain on the day of the eclipse, due to occur at two P.M., threatened to ruin all their plans. "The rain stopped about noon," Eddington recorded, "and about 1.30 . . . we began to get a glimpse of the sun. We had to carry out our photographs in faith. I did not see the eclipse, being too busy changing plates, except for one glance to make sure it had begun and another half-way through to see how much cloud there was."[64]

He and his colleagues photographed the stars that appeared in the sky close to the edge of the eclipsed Sun. Later they carefully measured the positions of the stars and compared them with their positions when the Sun was in another part of the sky. Eddington took a leading part in the observations and also made the critical micrometer measurements on the best photographs. To his and everyone's delight, the positions of the stars differed more or less by the amount predicted by the general theory of relativity. He never forgot the moment when he measured the first plate that "gave a result agreeing with Einstein."[65] In later years he referred to this as the greatest moment of his life. *The New York Times* headlined the results of the expedition in jocular fashion: "STARS NOT WHERE THEY SEEM OR WERE CALCULATED TO BE, BUT NOBODY NEED WORRY," for Einstein knew where they were.[66] The public particularly appreciated the irony that a British scientist had verified a "German theory" not long after the end of the Great War.

In celebration, Eddington penned a small epic describing the dramatic events on Principe in a parody of Edward Fitzgerald's popular *Rubaiyat of Omar Khayyam*. One verse reads:

> And this I know; whether Einstein is right
> Or all his theories are exploded quite,
> One glimpse of stars amid the Darkness caught
> Better than hours of toil by Candle-light.[67]

The expedition made Eddington famous as a world expert on relativity theory. In later years he enjoyed regaling high table with his tale of bumping into a renowned physicist (whom he chose not to name) at the Royal Society. "Well! Professor Eddington," the physicist exclaimed in jocular tones, "you must be one of the three people in the world who understands 'Relativity.'" Eddington savored the amusing exchange that followed: "I said, 'Oh! I don't know,' to which he said, 'Don't be modest, Eddington.' To which I replied, 'On the contrary! I am wondering who the third person is!'"[68]

The one topic that brought together all the different strands of Eddington's work was the puzzling matter of those mysterious hot yet dim white dwarfs.

Ever since scientists had begun studying the stars, astrophysicists had surmised that a star was probably formed in the following way. Deep in interstellar space, in regions rich in gas, particles of gas begin to coalesce through mutual gravitational attraction. Some of the particles continue to compress, changing from a diffuse, spread-out mixture to one that is more compact and well defined. As the particles jostle about, following the perfect gas law, they exert an outward pressure that partly counterbalances the inward tug of gravity. But gravity, still more powerful, presses the gas particles more and more tightly together, making them hotter and hotter, until eventually they begin to glow. The light they radiate creates another outward pressure, called radiation pressure.

In everyday life, radiation pressure is minute: the light of a flashlight or a car's headlight does not exert any pressure that we can feel. But a star is so vast and powerful that radiation streams out of it with the force of a hurricane, eventually creating a stable glowing mass. But what is the role of radiation pressure in creating a star that will not collapse? In 1917 Eddington made a brilliant suggestion. The key, he suggested, was balance. A stable star is formed when the gravitational pressure, squeezing inward, is in perfect balance with the outward pressure of the gas plus the radiation pressure of the light emitted by the gas particles. Eddington's hypothesis marked a major step forward in our understanding of the stars and was the cornerstone of modern astrophysics.

But what happens at the end of a star's life? White dwarfs, Eddington suspected, must be superannuated stars. Solving the puzzle of why they were both hot and dim would surely cast light on the final destiny of stars. He worked out the probable chain of events. As a star ages, it grows dimmer. Its radiation pressure decreases and the inward crush of gravity, now counterbalanced predominantly by the pressure of the gas itself, squeezes the star, making it smaller. But this scenario led to a paradox. As Eddington pointed out, when extremely dense, compact stars — like white dwarfs — are crushed by gravity, they become far denser than even the hardest rock on Earth. But in order to cool down, he reasoned, they would need to become less dense, until they were no denser than terrestrial rocks. Instead they seemed to be stuck, eternally shrinking and compressing. Although they grew dimmer they could not cool down. But could they carry on in this state forever? Could it be that they collapsed to nothing? As Eddington put it:

> The star seemed to have got itself into an awkward fix. Ultimately its store of subatomic energy would give out and the star would then want to cool down. But could it? The enormous density was made possible by the high temperature which shattered the atoms. If it cooled it would presumably revert to terrestrial density. But that meant that the star must expand to say 5,000 times its present bulk. But the expansion requires energy — doing work against gravity; and the star appeared to have no more energy available. What on earth was the star to do if it was continually losing heat, but had not enough energy to get cold?[69]

It was a tantalizing mystery. In *The Internal Constitution of the Stars* he played down the issue. Nevertheless, white dwarfs seemed an awkward anomaly in the astounding progress that was being made — largely by Eddington — in understanding the structure of the stars.

Another mystery that Eddington wanted to crack was how a white dwarf could be so small yet so dense. In 1914 he had described the incredibly high density of Sirius as "absurd." But ten years later, when he developed his mass-luminosity relationship, he discovered that the perfect gas law was valid even at the high densities inside dwarf stars such as the Sun. This led him to wonder whether, at the far higher temperatures inside white dwarfs, every atom might be completely stripped of its electrons, permitting even more atoms to be packed in.[70] In January 1924 he wrote to his friend the astronomer Walter Adams, who had carried out the key observations of Sirius B in 1914, that he had "recently been entertaining the wild idea" that the "incredible" density of Sirius B "may just be possible."[71] He went on to suggest an ingenious way to check whether white dwarfs really could have such high density by using Einstein's general theory of relativity to verify the small size of the radius.

General relativity predicts that the intense gravitational field of a white dwarf will affect the frequency and wavelength of the light it emits. Gravity on the surface of Sirius B is 100,000 times stronger than on the surface of Earth — so strong that it stretches the wavelengths of the light emitted by the atoms. From Earth the wavelengths appear longer. Long-wavelength light is toward the red end of the spectrum, so the light has a reddish tinge. This lengthening is called gravitational redshift.

The difference in the wavelength of light emitted from an atom on Sirius B and the same light from an atom on Earth turns out to depend on the mass of Sirius B divided by its radius. Sirius B's mass and radius had already been calculated; on the basis that Sirius B was 0.85 times the mass of the Sun, astrophysicists had worked out its radius, based on its luminosity and temperature. From these results Eddington was able to calculate the star's redshift.

He then suggested that Adams make a direct measurement of the redshift of Sirius B and compare it with the figure he had calculated.[72] The measurement was exceptionally delicate. Sirius A was so extremely bright that it strongly impinged on the spectra of Sirius B, making it difficult to determine which spectral lines belonged to Sirius A and which to Sirius B. Adams, however, felt up to the task and produced a result very close to Eddington's figure. Confirming the gravitational redshift also verified his figure for the unfeasible smallness of the radius which in turn confirmed the supposedly absurd density. Astronomers were sure that the mass was right. Eddington thus provided firm evidence that Sirius B really was extraordinarily dense, unlikely as that seemed. He, for one, was convinced by his own figures. His experiment also provided further proof of the general theory of relativity, in that Sirius B's gravitational field really did alter the light the star emitted, just as the theory stated. Eddington was doubly delighted.

As he was writing *The Internal Constitution of the Stars*, Eddington realized that general relativity revealed something extraordinary about the structure of space around extremely dense objects like white dwarfs. The supergiant Betelgeuse has a radius of about a hundred million miles, twice the size of Earth's orbit around the Sun, but it is a million times less dense than the Sun. What would happen if such a vast object were as dense as the Sun or even as dense as a white dwarf? "The force of gravitation would be so great that light would be unable to escape from it, the rays falling back to the star like a stone to the Earth. . . . The mass would produce so much curvature of [space] that space would close up round the star, leaving us outside (i.e. nowhere)," he wrote.[73]

The idea that there might be "dark stars" with a gravitational attraction so powerful that none of their light could escape was first proposed by the peripatetic English natural philosopher John Michell in 1784, some one hundred years after Isaac Newton discovered his theory of gravity. In 1796 the French mathematician and scientist Pierre-Simon de Laplace popularized the concept, though without mentioning Michell. Eddington, however, was not aware of any of their theories. Colleagues recall that he would contemptuously dismiss the history of science as irrelevant to working scientists.[74]

According to the perfect gas law, white dwarfs lacked the energy necessary to expand and were therefore unable to die as stars were supposed to, by turning into inert rocks. But instead of picking up on his own fanciful prediction — that they might be crushed to a point of infinite density and disappear into some niche in the fabric of the universe — Eddington suggested bypassing this "curious problem," adding that it was "not necessarily fatal."[75] Someone would have to return to it, however, because "white dwarfs are probably very abundant."[76]

They are indeed. In 1926, when Eddington was writing, only four were known with certainty. By 1938 about twenty had been found. Today hundreds of white dwarfs have been located in a small observable region close to the Sun, the only place where it is possible to see these dim stars. In this region white dwarfs account for about 9 percent of all observable stars, and those are only the ones accessible with our telescopes. Astronomers suspect that white dwarfs are the most abundant species of star in the universe.

As early as 1926, Eddington and others suspected that the main way stars die is by becoming white dwarfs. The problem was, as he wrote in his preface to *The Internal Constitution of the Stars*, that according to the perfect gas law, white dwarfs ought to collapse, a conclusion that he found simply unacceptable. It was an impossible conundrum, a stumbling block to the progress he was making in reasoning out what went on in the interior of stars. But that same year Eddington's colleague Ralph Fowler had the brilliant idea of applying the intriguing new quantum physics — on which Fowler was an expert — to stars. His aim was to clear up once and for all the knotty paradox that Eddington had uncovered — to prove that stars do not collapse but expire peacefully as inert rocks, thus upholding the harmony and beauty of the universe.

4. Stellar Buffoonery

RALPH FOWLER was "a big man, both physically and in the world of science; he was a hard hitter, in every sense of that phrase; he was a man of winning charm, who inspired friendships," as Arthur Milne described him.[1] Born in 1889, Fowler came from a privileged background and grew up fully expecting success in life. At Winchester School, he was particularly renowned as an athlete. Throughout his life he continued to be a keen sportsman; he played golf, cricket, and rugby, all with distinction, and was a formidable rock climber and bridge player. An affable, easygoing man with a famously hearty laugh and open smile, he was much liked.

Quite apart from his own stellar achievements in physics and astrophysics, he was at the heart of a circle of brilliant and influential scientists. It was he who introduced the brilliant and eccentric Paul Dirac, then his student, to quantum theory and later put Dirac, Heisenberg, and Bohr in touch with one another. In 1921 he married Eileen Rutherford, the only child of his close friend and colleague the legendary Nobel prize–winning physicist Ernest Rutherford, by then Lord Rutherford, who had discovered the nucleus of the atom. The Fowlers had four children.

Milne described Fowler with affectionate humor as "the kind of man you can still remain friendly with, even when he has sold you a motorbike; it is not possible to say more" and added that he was a "prince amongst men."[2] After his first meeting with him, in September 1930, Chandra wrote home excitedly, "Mr. Fowler — strong (big) healthy, middle-aged man, quite happy, full of the joy of life. He says 'Splendid' very often."[3] At that first meeting, Chandra was so nervous that he "fell down a whole flight of stairs . . . I can still hear Fowler's admonishing tone: 'steady, steady!'"[4]

In 1908, after Winchester, Fowler entered Trinity College, Cambridge, where he studied pure mathematics. He happened to specialize in the solution of certain equations that dealt with the properties of gaseous spheres and were thus of great interest to astrophysicists. He was awarded a fellowship at Trinity in 1914. When the war broke out, Fowler was commissioned in the Royal Marine Artillery. He was often seen at Cambridge wearing his gown over his military uniform. He took part in the disastrous Gallipoli campaign, where he was wounded so severely in the shoulder that he was removed from combat. This turned out to be fortuitous, for as a result he met the eminent physiologist Archibald Vivian Hill, then a captain in the Cambridgeshire Regiment.

With his sterling leadership qualities, broad scientific background, and extensive contacts with key Oxbridge players, Hill was the kind of scientist to whom governments naturally turn in times of war.[5] He was asked to form an Anti-Aircraft Experimental Section within the Munitions Inventions Department. The main task was to improve the effectiveness of antiaircraft gunnery against German zeppelins and later against the powerful twin-engine Gotha bombers by taking into account conditions at high altitude, such as wind, temperature, and pressure.[6] Hill set about assembling a group of mathematicians and physicists, and turned to Hardy, another Trinity Fellow, for advice. Hardy responded that "although he was ready to have his body shot at he was not prepared to prostitute his brains for purposes of war."[7] But he had no such compunctions regarding Fowler, whom he heartily recommended, as well as Milne, a student of Hardy's, then in his second year at Trinity. This was how Fowler, then twenty-seven, met Milne, seven years his junior. The two quickly became firm friends as well as colleagues.

Hill's group was based on HMS *Excellent,* docked at Whale Island, off Portsmouth. Unofficially the men called themselves Hill's Brigands, referring to their unorthodox methods in matters both social and scientific. It was a heady time. Fowler and Milne flew out from Farnborough airdrome in the powerful new F.E. 2D biplanes, powered by 250-horsepower Rolls-Royce engines. They persuaded the Royal Flying Corps pilots to fly as high as possible so they could check the atmospheric conditions, which they did by hanging over the wing to take readings of temperature and pressure. In an open cockpit and with no parachute or radio, this was not only thrilling but downright dangerous. The highly experienced combat flyers tried to scare their passengers by horsing around, performing daredevil nosedives and stalling turns. Fowler and Milne loved every minute of it.

Milne began with an equation for low-level trajectories of antiaircraft shells and expanded it to apply to much higher altitudes. His result is still in use. Fowler's contributions included meticulously recording and writing up data from aircraft runs and shell-burst tests, building good working relationships with naval officers, inspecting coastal antiaircraft facilities, and making many visits to the combat lines in France to encourage scientists in the field doing meteorological work and plotting artillery trajectories.

Hill's effect on both men was enormous. Milne always remembered Hill for teaching him how to do research. As a result of their wartime work, both Fowler and Milne changed their research interests away from mathematics, first to upper-atmosphere physics and then to astrophysics.

Fowler's forte was not so much in thinking up original ideas as in applying someone else's to solve a particular problem; his weakness, however, was that he tended to stop short of pushing an argument to its logical conclusion. The prime example is his attempt to solve Eddington's paradox by applying quantum statistics to a gas of electrons.

A few months after Eddington published *The Internal Constitution of the Stars* in 1926, Fowler proposed that quantum physics might resolve the apparent paradox that a white dwarf should logically collapse completely instead of ending up as an inert rock. Fowler was well aware that a gas of electrons millions of times denser than any terrestrial matter — as in a star — no longer behaves like a perfect gas. Most astrophysicists assumed that the perfect gas law would have to be modified to incorporate this discrepancy. Fowler, however, saw that the situation called for more sweeping measures, that such a dense gas of electrons would have to be studied using the laws of quantum physics, which state that when electrons congregate, they try to stay as far apart from one another as possible.

The need for a new law of nature emerged with Bohr's model of the atom as a solar system in which electrons circle the nucleus like planets. Crucially, they can occupy only certain orbits.[8] Bohr went on to show that electrons are distributed like beads on what can be visualized as rings. The simplest atom, hydrogen, has one electron in one ring. Helium consists of two electrons that complete this ring, making helium inert, that is, unable to react with any other chemical element. Then comes lithium, with three electrons threaded on two rings, followed by a succession of chemical elements, each of which has an added electron in the second ring, culminating in neon, in which the second ring is completed with eight electrons; neon is also inert. Bohr's model gave the satisfying result that the number of electrons in an element determines its chemical activity, thus making sense of the periodic table. Einstein declared his model "an enormous achievement."[9]

But Bohr could not explain why each ring had a specific number of electrons. Then in 1925, the flamboyant twenty-four-year-old Viennese physicist Wolfgang Pauli, a former student of Sommerfeld's, came up with the answer. The missing ingredient was another mysterious ingredient of the quantum world: electron "spin."

Spin is a quantum property of an electron and quite impossible to imagine, just as it is impossible to imagine an electron as a wave and a particle at the same time. Physicists coined the term "spin" because its mathematical representation is the same as if the electron were a spinning top. Spin turned out to be the missing quantity needed to explain why the rings filled up as they did. The full explanation required a new law of nature: that no two electrons in an atom can have the same properties. This is known as the Pauli exclusion principle.[10]

Scientists can study a very low density gas of electrons using the theory of the perfect gas, based on Newtonian physics. But for the highly dense gas of electrons in a star, laws of quantum physics come into play. Pauli's exclusion principle rules that each quantum state can contain at most only one electron with a specific position, velocity, and spin, akin to the restriction on the number of electrons that can occupy each ring in an atom. Given that no two electrons can have the same properties, one by one their various possible quantum states — their positions, velocities, and spins — are occupied. These quantum states begin to fill up rapidly, starting with the low-velocity states, which can accommodate electrons that are barely moving.

When a star begins to run out of fuel and starts becoming dark, the radiation pressure pushing outward decreases and the inward thrust of gravity begins to prevail. The star starts to contract. Deep inside, the many trillions of electrons are squashed tighter and tighter, like people trapped in a crowd. The quantum states fill up, beginning at the center of the crowd, where the electrons are milling slowly, and moving out to the fringes, where they swirl about more quickly. The leftover electrons scurry around trying to find a place to settle. Eventually the collection of electrons cannot be squeezed any tighter because they cannot infringe on each other's "territory," or state. Collectively they resist any further compression by exerting a powerful outward pressure. When the electrons begin to resist being squeezed together, they are in what is called a degenerate state; the outward pressure they exert is "degeneracy pressure." This is far stronger than their mutual electrical repulsion, which is almost nonexistent, having been neutralized by the positive charges of the nuclei of the atoms from which they were stripped.

Eventually the star goes completely dark and produces no more radiation pressure. The outward pressure exerted by the electrons is all that is left to counter the enormous inward push of gravity. The result of this degeneracy pressure is that the star becomes so hard that it cannot be crushed by gravity. It can now die quietly as a cold, dark rock.[11] Thus, incorporating quantum physics into astrophysics appeared to return the subject to its classical purity. It resolved Eddington's paradox, showing that stars expired gracefully and supporting the age-old vision of a quiescent universe.

Fowler had demonstrated that a theory which had emerged from studying the microcosmos could also be applied to some of the largest objects known — stars. He framed his work in daunting mathematical detail. To allow for the possibility that white dwarfs were far denser than anyone had thought, he assumed a density of 100,000 grams per cubic centimeter, double the density used by Eddington. But at that very high density, electrons within a white dwarf still do not move at velocities approaching that of light, so it was not necessary to incorporate relativity theory. Fowler was aware that even that figure for the density might be too low and that it might be necessary to incorporate relativity theory, but he chose not to pursue the matter. His goal had been to resolve Eddington's paradox, and he felt that he had succeeded in doing so. In his usual forceful and authoritative tone, the ex–Royal Marine declared that applying quantum theory "clears up Eddington's question in a convincing manner, and I am content to leave the matter so."[12] Eddington, too, was content.

Thanks to their wartime adventures and to Fowler's affable personality, Milne and Fowler remained close for the rest of their careers. Milne's relationship with Eddington, however, was far more troubled.

Small and slight, Milne was a warm, likable man with a round, boyish face and thick glasses. His poor eyesight had excluded him from active military service, so he eagerly welcomed the invitation to join Hill's Brigands. He came from a more modest background than Fowler. His parents were schoolteachers, and he had won a scholarship to Trinity entirely by his own brilliance; he scored the highest grade ever awarded in the examination. A colleague of his wrote, shortly after his death, "He could crack open an intractable ballistic problem, explain a forgotten theorem of Euclid, recite an ode from Horace or a speech from Shakespeare, and even crack a bawdy joke, all in the same luncheon hour."[13]

Milne was extremely popular, very charming, and completely dedicated to his work. Chandra's friend William McCrea wrote that Milne "could be said positively to exude vitality. His features were often set in intense concentration but in conversation they became wonderfully mobile and expressive. They became overspread with impish delight at any witticism that appealed to him." McCrea added that Milne "was essentially a religious man; faith and courage gained the victory in each succeeding struggle of his life."[14]

At the end of the war, Milne was twenty-two and reluctant to return to Cambridge to complete his undergraduate studies. So Fowler and Hill arranged for him to submit original work that would hopefully qualify him for a fellowship at Trinity. Amazingly, in a single year Milne completed no less than three theses: one on a mathematics problem, one on the propagation of sound waves in the atmosphere, and one on the properties of Earth's atmosphere at high altitudes. The last two were based on his wartime research, which Fowler had convinced him was outstanding. He was duly elected a Fellow.

But that did not satisfy Milne. According to Chandra, "The fact that he never went through the 'Tripos mill' produced in him an 'inferiority complex' (those were the words he used), and a feeling that he had never acquired that foundation which is the 'stuff of science.'"[15] This was extraordinary, for if anything, Milne's mathematical rigor was excessive. Eddington's papers, in contrast, were a model mixture of a great deal of physics with just enough mathematics to back up his conclusions. He once criticized Milne for taking "thirty-six pages of technicalities" to make a single point.[16] Milne was determined to prove at every possible opportunity that he really did know some mathematics.

He always retained "traces of the undergraduate attitude" toward most senior scientists.[17]

One night at high table in Trinity, Hugh Newall, director of the solar laboratory, mentioned that he was looking for an assistant director. Suddenly he realized that Milne, sitting right next to him, had studied atmospheric problems during the war. Why not work on the solar atmosphere? he suggested.[18] As a result, Milne spent ten years, from 1919 to 1929, researching the Sun's outer layers. It was the most successful period of his career; he was later — on January 11, 1935 — awarded the Gold Medal of the Royal Astronomical Society for his achievements during the 1920s.

But tragedies accumulated in Milne's life. In 1921, when his father died, he seriously considered leaving academia to support his mother and brother. Then in April 1924, he had an attack of epidemic encephalitis, which was prevalent in England at the time. Cambridge friends looked after him, and by July he had recovered, but he was never quite the same afterward. Milne then left Cambridge to become the Beyer Professor of Applied Mathematics at the University of Manchester, where he had a vigorous and fruitful academic career. When he met and married Margaret Scott, the adopted daughter of a professor of history, his life seemed to be back on track. Then in 1928, he was offered the position of Rouse Ball Professor in Mathematics and Fellow of Wadham College at Oxford. At that time, compared to Cambridge, Oxford was a desert for mathematics. Several others had turned the position down. Milne accepted. But establishing a mathematics presence there took up a great deal of his time, and he lost contact with colleagues and friends in Cambridge.

At Oxford, Milne's personal life suffered. Margaret had been happy in Manchester, taking part in academic activities along with her husband. But Oxford was different; for one thing, women were not permitted to dine in college.[19] Milne felt isolated and out of touch with developments in atomic physics and quantum mechanics, subjects that were beginning to have a bearing on research into the atmospheres of stars. So he decided to change fields and to work on creating models of the interiors of stars, which did not require such knowledge.

Milne had been invited to present the Royal Society's prestigious Bakerian Lecture on June 6, 1929. It seemed the ideal opportunity to introduce his new ideas, which he considered nothing less than revolutionary. Everyone present expected him to speak on his groundbreaking work on stellar atmospheres, but he had set his sights much higher. Milne's great idea was to look into the interiors of stars by extending his work on their atmospheres rather than by working from within the star to its surface, as in Eddington's standard model. But before presenting his lecture, Milne decided to visit Eddington and discuss his research with him. He should have guessed that he would encounter ferocious opposition, for Eddington was totally convinced that a star's surface temperature could be understood only by examining its internal structure.

When the two met, on a fine May day just a few weeks before the June 6 lecture, Eddington totally demolished Milne's theories. With an almost superhuman effort, Milne wrote another lecture from scratch and delivered it as planned. The lecture, a tour de force that not only consolidated his research on stellar atmospheres but also contained original discoveries, would be referred to by researchers in the field for years to come.[20] Yet Milne, dissatisfied, told friends that he did not consider his talk worthy of the occasion, and certainly no match for Eddington's lecture three years before, in which he had presented new work that brought the study of interstellar matter — gas and dust in regions of space far from any star — into the mainstream of astrophysics research.

A few months later, at the beginning of November, Milne had another chat with Eddington. He insisted that an important result such as Eddington's mass-luminosity relation could not be deduced just from balancing the outward push of the radiation and gas pressures with the inward tug of gravity. It would also be necessary to look into the structure of a star, and in particular at the source of its energy. The life of a star was far more complex than Eddington made it out to be, he declared excitedly.

We can imagine Eddington listening, puffing on his pipe, intermittently chewing up apples, trying to restrain himself. And then responding: Was not the mass-luminosity relation astounding? Did it not reveal that, amazingly, the perfect gas law, which everyone had thought applied only to less dense stars such as giants, extends to the incredible densities inside dwarf stars? Had it not put paid to Russell's giant–dwarf theory, except in the one case of white dwarfs? Yet Milne dared sit there and tell him that he, Eddington, had merely been lucky enough to assume from the start that stars were perfect gases? Whereas in fact Eddington had known perfectly well all along that white dwarfs were not perfect gases, as Fowler had shown three years earlier. Milne obviously did not have the slightest idea about building models! Once again Eddington had got the upper hand.

Nonetheless, Milne decided to present a shortened version of his ideas at the Royal Astronomical Society meeting on November 8, 1929. Eddington dismissed Milne's work out of hand. "It is difficult to discuss this paper," he wrote superciliously. "Professor Milne did not enter into detail as to why he arrives at results so widely different from my own; and my interest in the rest of the paper is dimmed because it would be absurd to pretend that I think there is the remotest chance of his being right."[21] Eddington continued to pick away at Milne in a succession of short papers published the following January.

Milne did not take Eddington's criticisms well. He wrote to his brother:

> [Eddington] uses very strong language about my work — he calls it sophistry, mysticism, suggestions without foundation. He's getting very pontifical these days, and is touchy about his old theory, which is rotten to the core. It's astounding that such a complete hoodwinking of the scientific world should have gone on for so long. His theorems about stars are mostly baseless conjectures.[22]

In his writing, Milne referred to giants and dwarfs as "centrally condensed stars" and white dwarfs as "collapsed configurations."[23] He built on the work of Fowler and of Edmund Stoner, a physicist at Leeds University. Stoner had worked out that when a white dwarf began to collapse, the pressure would be so great that the electrons would move at speeds close to that of light. On this basis he devised an equation, taking into account special relativity, that would allow a white dwarf to become rock hard provided it was below a certain mass. Milne argued that while in all stars the outer part behaves like a perfect gas, further inside there is a point at which density and pressure increase so much that the perfect gas law fails. In order to proceed further and calculate the star's internal pressure and temperature, one has to take quantum theory into account, using first Fowler's equation of state and, further into the star, Stoner's equation, which includes the effects of relativity, because here the electrons move at speeds close to that of light. It may be necessary, he argued, to find yet more equations to describe other exotic gases encountered farther inside the star, where pressures increase enormously. Milne was proposing that all stars are made up of an envelope of a perfect gas surrounding a series of rock-hard cores, each of which prevents the star from collapsing beyond a certain pressure. Because these cores were rock hard, they were called degenerate cores.

Like Eddington, Milne balked at following this line of reasoning to its logical conclusion: that white dwarfs could continue to dwindle, becoming smaller and smaller and denser and denser, to infinity. This would be a "singularity," the word used when the mathematics of a theory of physics comes up with a result of infinity. Traditionally, infinity is anathema to science; it is considered a sure sign that the theory is breaking down. In fact, singularities had been cropping up regularly in quantum physics, and physicists, less hidebound than astrophysicists at that time, were finding ways to deal with them. But the astrophysicists, including Milne, refused to consider singularities, which often made astrophysics a laughingstock for physicists. To avoid creating a singularity, Milne made the ad hoc hypothesis, with no supporting arguments, that the center of a star had to be made of some medium with incredibly high incompressibility.

In a critique published in the March 29, 1930, issue of *Nature*, Eddington poked fun at Milne's scenario. Milne, wrote Eddington, contended that "physically a star cannot 'contract indefinitely'; sooner or later it must find a (contracted) configuration of equilibrium." By the same argument, Eddington went on, "Humpty Dumpty's position on the wall was not unstable; ultimately he found a new configuration of equilibrium."[24] Even if Humpty Dumpty fell off the wall a thousand times, he would always put himself back together again and be back on the wall. Similarly, according to Eddington, Milne argued that no matter how far a star collapsed, it would never collapse completely. There would always be a core inside a core inside a core, like an infinite nest of Russian dolls.

Milne wrote to Herbert Dingle, a well-placed figure in the astrophysics community, to lodge a complaint about Eddington's caustic comments at RAS meetings: "I cannot tell you what spiritual turmoil this has caused me. I have to face being accused of talking absolute nonsense before the whole R.A.S., just because I arrived at different results."[25] Eddington, too, was reaching the end of his tether. "I almost despair of extricating my main controversy with Prof. Milne from the tangle in which it has become involved," he wrote.[26] Things were getting messy. Milne felt himself a martyr to science, a misunderstood genius. But he would go down fighting; posterity would bear him out, he was sure.

Still smarting from Eddington's criticisms, he began the most detailed exposition yet of his ideas on the constitution of stars. In his article, published in November 1930, he explained that "the present paper brings forward considerations which compel a drastic revision of our views as to the structure of the stars."[27] The standard model had to be replaced. While Eddington's basic assumption was that a star was a perfect gas, Milne's was that every star must have an incompressible core.

Nevertheless, for all the bitterness of these disputes, the rival parties concurred on one point. As Jeans put it, "Eddington, Milne and myself are of course all agreed that actually the density at the centre of a star must be finite."[28] The concept of infinite density was unthinkable.

Speculation about stars was only one part of Eddington's work. He was becoming more and more involved in his quest for one grand overriding theory that would cover everything from atoms to stars. Finding a "theory of everything" has been the Holy Grail of scientific endeavor ever since science began. Eddington called his attempt to fuse quantum theory with the special and general theories of relativity his "fundamental theory." The work absorbed him so much that it clouded his judgment on a number of matters — not least of which was the fate of white dwarfs.

Eddington saw himself as a pioneer, an adventurer in scientific research. He was prepared to push an idea to its limit, to risk failure and ridicule for the chance of taking a leap and making a great discovery. In 1926 he compared himself to Icarus, who

> soared upwards to the sun till the wax melted which bound his wings, and his flight ended in fiasco. . . . Cautious Daedalus will apply his theories where he feels confident they will safely go; but by his excess of caution their hidden weaknesses remain undiscovered. Icarus will strain his theories to the breaking-point till the weak joints gape. For the mere adventure? Perhaps partly; that is human nature. But if he is destined not yet to reach the sun and solve finally the riddle of its constitution, we may at least hope to learn from his journey some hints to build a better machine.[29]

Eddington had begun his search for the fundamental theory in 1928, inspired by Dirac's newly discovered theory of the electron. Dirac's theory was based on both the special theory of relativity and quantum theory, though it did not include general relativity. One of its weirder implications was that there had to be electrons with negative energy. But that was unthinkable, a physical impossibility, like negative time. Heisenberg complained to Pauli that the theory was the "saddest chapter in modern physics."[30]

Eddington was not particularly disturbed by the strange implications of Dirac's theory, but its mathematical structure struck him as odd. Until then, everyone had thought that the appropriate mathematical form for relativity was tensors — elaborated versions of vectors in that they can be built up from them. Central to Newtonian physics, vectors are utterly unlike ordinary numbers and do not obey the same rules. With ordinary numbers, one plus one makes two, but with vectors the sum can be zero, one, two, or a number in between. To create his new theory, Dirac had evolved an equation based on an esoteric form of numbers called spinors. Unfamiliar even to most physicists, spinors are neither vectors nor tensors. What made Dirac's theory so important was that the spin property of electrons emerged from it. Eddington, who had a deep understanding of the mathematical basis of physics, perceived that Dirac had tapped into something far more profound than simply using a new form of mathematics to provide a basis for spin.

Eddington had a strong mystical streak, derived perhaps from his strong Quaker convictions, which took the form of an awe of nature beyond science. Mysticism offered him an escape from the closed logical system of physics, but he always made it clear that physics must not be brought in to support any mystical view. Mysticism was a religious experience. Behind his imperturbable mask, Eddington agonized about the spiritual world and struggled with the dichotomy between the two worlds, equally invisible, of science and the spirit.

He was fascinated by "seen" and "unseen" worlds and the connections between them. His first book on the fundamental theory, *The Nature of the Physical World,* begins with a brilliant comparison of these two worlds.[31] In it he describes two tables that are exactly the same. We can place books and papers on both. One is an ordinary everyday table. The other is a table as perceived through the eyes of a scientist, consisting mostly of empty space crisscrossed with electric and magnetic fields, with electrons buzzing around in such a way that our books and papers do not fall through. Although both tables are real and substantial, the concept of "substance" requires redefining for the scientific table. What is the link between these seen and unseen worlds?

Eddington was convinced that mathematics was the key that opened the door between the two. Dirac's equation was based on four numbers, the four dimensions of space and time — length, width, depth, and time. Eddington used these numbers as the basis on which to create sixteen new and enlarged spinors, which he called "E-numbers."[32] From these he intended to construct a new mathematical system large enough to bring together all the parts of the universe, both on a large scale, as described in the special and general theories of relativity, and at the quantum level.

He then isolated seven "primitive constants of physics," seven numbers that remain the same everywhere in the universe. "We may look on the universe as a symphony played" on these numbers, like the notes of a musical scale, he explained — much like the music of the spheres.[33] The numbers are so fundamental that they must also underlie the scientific theories of civilizations everywhere in the cosmos, or so scientists believe. The primitive constants that Eddington isolated were the electrical charge of the electron, which is 10^{-10} in terms of size (measured in centimeters), mass (measured in grams), and time (measured in seconds); the Planck constant, a measure of scale in the atomic world and smaller still, 10^{-27}; the speed of light, a huge figure, 10^{10} centimeters per second (measured in the same units as the other constants); the masses of the proton (10^{-24}) and the electron (10,000 times smaller, 10^{-28}); Newton's universal gravitational constant (10^{-8}); and the cosmological constant of the general theory of relativity (a kind of antigravity that allows the universe to expand and that is unimaginably minuscule, 10^{-55}).

Eddington's ultimate goal was to use his new mathematics to deduce the number of electrons and protons in the universe and then to compare his result with the results obtained from experiments. As a first step, he deduced certain combinations of the seven primitive constants from his E-numbers and then compared them with actual measurements. He was particularly fascinated by something called the fine-structure constant, which can be constructed from the charge of the electron, the Planck constant, and the speed of light.

When scientists studied the spectral lines of the light emitted by atoms, they discovered that some have a fine structure; that is, they are made up of even narrower, very closely spaced lines. Sommerfeld explained this phenomenon brilliantly by incorporating relativity theory into Bohr's theory of the atom, which pictures atoms as minuscule solar systems. It turned out that the spacings between the components of a spectral line depend on the fine-structure constant. Early experiments suggested that this figure was the ratio of two integers or whole numbers: ⅟₁₃₇.

Integers lie at the heart of the quantum theory of the atom. They also played an important part in ancient Greek thought. The mathematician Pythagoras and his followers believed that whole numbers would allow them to tune in to the music of the cosmos. Even Sommerfeld, apparently a down-to-earth German physicist, rhapsodized that the "language of spectra" expressed a "true 'music of the spheres' in chords of integral relationships."[34] (The number 137 itself has mystical connotations, which were important to Eddington; in the Kaballah, certain key phrases that refer to Yahweh add up to 137.) Subsequent measurements, however, showed that the fine-structure constant is actually not ⅟₁₃₇, but ⅟₁₃₇.₀₃₆. But Eddington refused to be deterred from his belief that it was the ratio of two whole numbers. He also idiosyncratically insisted on referring to the fine-structure constant as 137, not ⅟₁₃₇.

A key element of Eddington's argument was that particles can never be considered in isolation and that there are no systems in nature in which the interactions between particles can be ignored. According to the general theory of relativity, it is gravity — the gravitational tug between particles — that gives each particle its mass and sculpts space and time. Besides being a measure of the fine structure of spectral lines, the fine-structure constant is also a measure of how strongly two electrons interact. For Eddington this fact provided further support for his contention that there is no such thing as an isolated particle in quantum theory either.

Because particles cannot be considered in isolation, Eddington argued, any theory of the electron has to deal with at least two electrons. Applying his new mathematics, each could be described using 16 E-numbers. Multiplying 16 by 16 gives a total of 256 different ways in which electrons can combine into a single system. He then showed that, of these 256 ways, only 136 are actually possible; 120 are not. He gave this result the mathematical signature $256 = 136 + 120$. Thus, as if pulling a rabbit out of a hat, he produced the number 136 from purely mathematical reasoning.

Now 136 was not 137, but for Eddington it was close enough. He was convinced that, as he wrote, the additional 1 "will not be long in turning up."[35] The obsessive pursuit of 137 literally took over his life. Russell recalled meeting him once, at a conference in Stockholm, in the cloakroom. As they went to hang up their coats, Eddington insisted on hanging his "hat on peg 137."[36] Dirac said, "He [Eddington] first proved for 136 and when experiment raised to 137, he gave proof of that!"[37]

Nevertheless, Eddington considered his result encouraging, for his fundamental theory was more than just physics. He saw himself as a spinner of grand philosophical systems that extended to all facets of science and the mind and to the essence of being. Every now and then he needed to confirm his intuitions by comparing them with data obtainable from experiments — such as the rate of expansion of the universe. But in general he preferred to draw his conclusions from thought and mathematical reasoning alone.

Eddington was determined to calculate the number of electrons and protons in the universe. He used quantum and relativity theories to make the calculation, to show that his fundamental theory was a union of all these, then he calculated the number again using his own new mathematics, based on the sixteen E-numbers. First he took the universe as governed by the general theory of relativity and deduced an equation containing the number of electrons and protons in it. Next he imagined this universe as a giant atom made up of electrons and protons and used quantum statistics to arrive at a second equation. From these two equations he isolated the number of electrons and protons, which added up to 1.5×10^{79} (1.5×10 million trillion trillion trillion trillion trillion trillion). From this result he determined the rate at which the galaxies are receding from us, that is, the rate of expansion of the universe. To his delight, the result was fairly close to the rate actually measured by astronomers.[38] The problem was that his figure was a fraction, and a fractional number of particles is an impossibility; also, he was determined to tie in his result with the fine-structure constant, which he insisted was a whole number.

At this point he turned to his E-numbers and again worked out the number of electrons and protons in the universe. He wrote the result as 136×2^{256}, a figure that held an aesthetic and mystical aura for him. He pointed out that 2^{256} is almost exactly the same as 10^{79} and that the number he had deduced compared perfectly with the figure obtained from experiments. It seemed to prove that it was no accident that the universe contains 1.5×10^{79} electrons and protons. (The key part of the calculation was the immense figure on the right — 10^{79} or 2^{256} — which is not much affected whether it is multiplied by 1.5 or 136.)

Through mathematics Eddington had found the key to understanding the structure of the universe, or so he felt.

Unfortunately, the equation Eddington deduced from quantum theory turned out be wrong. To get the result he wanted, he had to force general relativity and quantum theory to match, which he did by using a false figure.[39] Colleagues were beginning to detect, in the books and papers Eddington was writing on his fundamental theory, a dramatic shift from the succinctness and clarity of his earlier writings on astrophysics. Chandra later quoted one acerbic critic: "His [Eddington's] papers are very clear up to a point and then at the critical moment they become obscure, to become clear again after important results have been deduced."[40]

Eddington was not alone in trying to deduce the fine-structure constant from theory. Pauli obsessed over it as well and was fascinated by the multifaceted connotations of the mystical number 137.[41] In 1958 he was in hospital in Zurich, undergoing treatment for cancer. When the physicist Victor Weisskopf, who had been his assistant in the early 1930s, visited him, Pauli suddenly blurted out, "Viki, I'm never going to get out of this room alive!" "Why?" asked Weisskopf, puzzled. "Look at the number of this room," Pauli replied grimly. It was number 137. And indeed, he died there.[42]

But Pauli's fascination was a long way from the obsessive behavior of Eddington, who was prepared to bend basic laws of physics to achieve his ends. Pauli dutifully read almost everything that Eddington produced, but he had nothing but scathing comments. "I consider Eddington's '136-work' as complete nonsense: more exactly for romantic poets and not for physicists," he wrote.[43]

But despite the growing criticism surrounding Eddington's work on a unified theory, his reputation at the beginning of 1930 was at its height. His model of the constitution of the stars was at the hub of astrophysics; his ideas on cosmology were considered intriguing; and his philosophical meditations on the universe were bestsellers, widely discussed. He had managed to tie his calculation of the number of electrons and protons in the universe to the rate of its expansion, as well as deducing it from E-numbers. And Fowler had cleared up the disturbing paradox he had uncovered in his work on the fate of stars. As far as Eddington was concerned, everything was settled. That year he received the highest possible accolade — a knighthood. It was also the year Chandra arrived at his door.

5. Into the Crucibles of Nature

HANDRA LEFT INDIA on a high note. As the city of Bombay receded into the distance, perhaps he thought back over the achievements of his nineteen years of life — a golden youth. Besides completing his university training, he had already published five research papers and favorably impressed two of the world's greatest scientists, Heisenberg and Sommerfeld. The two leading Indian physicists — Raman and Saha — were full of high hopes for him. Raman proclaimed, "This young man shows all the signs of being a genius. He will surely leave an indelible mark on physics."[1] He was guaranteed a professorship at Presidency College when he returned from England. The world of physics was at his feet. He had good reason to believe that no one could resist his charms or be anything less than deeply impressed with him. During the voyage he would ponder Fowler's white dwarf theory and remember Lalitha, the young woman he was leaving behind.

For the first few days the sea was so rough that the ship had to sail at half speed. His uncle Raman's advice had been to "keep cheerful and remain on deck if possible. You will certainly have a very pleasant voyage once you get near Aden."[2] This proved to be the case. As soon as the weather improved, Chandra separated himself from his countrymen and kept pretty much aloof, to their annoyance.

Chandra settled into a deck chair and pulled another alongside for his pile of books and papers. He had Fowler's 1926 paper "On Dense Matter" (which laid out his solution to Eddington's paradox); *X-Rays and Relativity,* by the American Nobel laureate Arthur Holly Compton; and Eddington's and Sommerfeld's books, as well as his own published papers. The sea was calm, without even a ripple. The vast sky was intensely blue and the air smelled invigoratingly salty.

Those were idyllic days, especially after the hectic pace of the previous weeks. Chandra had nothing to do but think and calculate. He knew precisely what to work on — the problem he had been forced to set aside some months earlier of "making Fowler's conclusions more precise."[3] He had already begun by tying those conclusions in with current ideas on the way to calculate a star's internal temperature and pressure from its surface temperature, mass, and radius as measured by astronomers. Eddington had given formulas for doing so in *The Internal Constitution of the Stars*. Yet no one, not even Eddington, had thought of applying these methods to Fowler's new theory of white dwarf stars. Why not? It was so straightforward. Perhaps Eddington had become complacent now that, as far as he was concerned, Fowler had straightened out the heavens for him.

Some years later, Chandra would recall what happened next as "so simple and elementary that anyone could do it."[4] Quickly he calculated the density at the center of a star as required by Fowler's result and found that for the white dwarf star Sirius B it was *1 million grams per cubic centimeter* — a million times denser than water. Chandra pondered his finding. "Interesting," he must have thought. "Very interesting." But what made this huge number so interesting? We can imagine his mind drifting away to the beauty of the sea before him.

Suddenly he remembered the story his uncle Raman often told of how he had made his most famous discovery, on a sea voyage to Europe in 1921. As Raman pondered the sparkling blue opalescence of the Mediterranean, it occurred to him that the effect must be caused by collisions between sunlight and water molecules.[5] From this insight he worked out a way to unravel the structure of molecules by examining how they alter the frequency of the light striking them — known afterward as the Raman effect.

Turning his gaze from the azure sea, perhaps Chandra glanced at the deck chair next to him, on which lay Compton's *X-Rays and Relativity*. That was where he had learned about Einstein's theory of special relativity, about what happens to objects moving at speeds comparable to that of light, an area wholly beyond the bounds of classical Newtonian physics. The speed of light plays a central role in Einstein's theory. It is always the same regardless of the motion of the source of the light, and is generally taken to be 186,000 miles per second (in empty space).

That is the cosmic speed limit; nothing can move faster. This new way of peering into nature — relativity theory — also reveals that a particle's mass is not constant; it changes depending on its speed. This can lead to surprises, as Chandra was about to discover. In a flash of inspiration, for the sake of argument, for completeness, "I asked myself what the velocities of the electrons would be at the centre of the white dwarfs."[6] That might be interesting, he thought, given the extremely high density he had just calculated. The velocity turned out to be more than half the velocity of light.

Chandra was stunned, though he was well aware that introducing relativity into an equation could have unexpected consequences. Astonishingly, Fowler had neglected to apply special relativity in his solution. He had taken for granted — mistakenly, as it turned out — that the pressure inside a white dwarf star was no greater than 100,000 grams per cubic centimeter and had therefore assumed that the electrons in a white dwarf moved slowly enough that their motion could be studied using Newtonian physics. The upshot was that Fowler's result had been only approximate, not exact, as everyone had thought.

Chandra's moment of inspiration was more than serendipity. Equations spoke to him. He could lose himself in numbers and symbols, as if in a cocoon, and he was a perfectionist — he could not stop until every detail was right. But he could also see the broader significance of his results. He had the vision to see the big picture. That was the key to his genius.

Before he went any further, Chandra had to make sure he was right by working out why these astonishingly high speeds were correct. The clue lay in the uncertainty principle, which Heisenberg had discovered three years earlier, in the spring of 1927. To Chandra, of course, Heisenberg was far more than a name on a research paper. There was a face behind the name, a handshake, encouragement.

Heisenberg's discovery was that a particle's position and its velocity are inversely related: we can measure very precisely where a particle is, but we cannot simultaneously measure with the same accuracy how fast it is moving, and vice versa. The consequence is that any attempt to force the electrons in a star to occupy a precise location results in an increase in velocity. That is why electrons inside white dwarfs move at speeds close to that of light. As an expert on quantum theory, Fowler should have incorporated special relativity into his equations right from the start.

Fowler had explored the case of nonrelativistic degeneracy, that is, degeneracy theory that does not take relativity into account. By including the theory of special relativity, Chandra could explore what he called a relativistically degenerate gas of electrons.[7]

Chandra knew that a calculation using the full paraphernalia of special relativity was just too difficult and would have to be postponed. He decided instead to work out an approximation in which he assumed that the electrons within a white dwarf moved very close to the speed of light. He calculated the relationship between the pressure and density of a relativistically degenerate gas of electrons, using as a basis an idealized quantum gas in which he ignored the electrical interaction between particles. Like Fowler, he focused on white dwarf stars that have completely cooled. Bringing in relativity, however, produced the astounding result that there was a limit to the mass of a white dwarf. Chandra calculated it as a little smaller than the mass of the Sun.[8]

But what would happen to a white dwarf which, after burning up all its fuel, ended up with a mass that exceeded the limit Chandra had just discovered? With nothing to brake its collapse, could it be that it would collapse forever? In any case, Chandra believed that he had overthrown Fowler's theory and found a new solution to Eddington's problem — one that Eddington himself had toyed with but had dismissed as absurd.

After the ship docked in Genoa, Chandra traveled overland to the English Channel, arriving in London on August 19. But the proper documents had not been forwarded from India; no one was expecting him as yet. It was an irritating bureaucratic muddle arising out of the usual morass of Anglo-Indian misunderstanding. In desperation Chandra wrote to Fowler, who was on vacation in Ireland. The kindly Fowler responded just in time to cut through the red tape. "That was all there was to it!" as Chandra recalled, remembering Fowler's concern.[9]

Chandra arrived in Cambridge in September, in a buoyant mood and with high hopes. He had visions of being received there as he had been in India, where he was feted and recognized by all "the people he came into contact with [as a] genius."[10] But he was in for a shock. The precocious youth had had almost no competition in India, but in Cambridge he found himself up against some of the most brilliant minds in science. "I had a shattering experience; to suddenly find myself in an environment where there were people like Eddington and Hardy, not to mention all the other well known names, is a very strong sobering experience," he wrote.[11]

For a while Chandra was convinced that the only reason he had been admitted to Cambridge was "purely due to the accident" that he had corresponded with Fowler for the past two years. "Why I should have written then to Fowler god alone knows. I suppose that was because Fowler is to help two years later," he wrote to his father just after arriving in Cambridge.[12] He took his connection with Fowler to be his karma.

By October 2, Chandra had written two papers, which he looked forward to presenting to his thesis adviser. One, on the density of white dwarfs, was a pure application of Fowler's solution to Eddington's problem. The other, more original and ambitious, concerned the maximum mass for white dwarfs that he had just discovered. Fowler made some suggestions about the density paper and offered to submit it for publication.

He was noticeably less enthusiastic about the second paper but suggested that Milne might take a look at it, as he was studying collapsing stars. Chandra was disappointed. He was convinced that this second, shorter paper "was more important."[13] Fowler reminded Chandra that in 1930 Edmund Stoner had also arrived at a limiting mass for white dwarfs by including relativity. Chandra dutifully cited Stoner's work in his 1931 paper, pointing out once again that his own results were based on a more realistic model of a star and so had far broader implications.[14]

Stoner's result applied only to a star that was the same density all the way through, whereas Chandra assumed the more likely scenario that a star became denser toward the center. Moreover, Stoner was concerned only with the question whether the temperature and pressure inside a white dwarf became unexpectedly high as the star's mass reached a certain critical point — Chandra's upper limit. Chandra, however, was prepared to follow the idea of a maximum mass to its logical conclusion — that a white dwarf of a greater mass than this might eventually collapse to nothing. Chandra could not have missed seeing Stoner's footnote thanking Eddington for conversations about his research.[15] He must have realized from this that Eddington had more than a passing interest in relativistic degeneracy.

In the end, Chandra recalled, "Fowler and others were not doing anything about the paper I gave them in September." He was tired of hearing them simply comment "Very interesting."[16] Sir James Jeans, however, took notice. Chandra wrote excitedly to his father: "I have proved that the *MAX* Mass a Dwarf star can have is just that of our Sun. This result I communicated to Jeans. He has written quite a nice letter and thinks the result 'quite important.'"[17] Thus encouraged, in mid-November Chandra sent his paper on the maximum mass to the American *Astrophysical Journal.* It was initially rejected but finally published there in March 1931.[18]

And yet, despite the staggering implications of his discovery, Chandra had the distinct feeling that "neither Fowler nor Milne thought of [it] as very important . . . Very few people were interested in what I was doing." The young man's confidence was ebbing away. "Somehow or other, I felt I didn't belong there. It seemed to me that there were far too many big people, far too many people doing important things, and what I was doing was insignificant in comparison. I suppose I was afraid."[19]

Like all Indian students newly arrived in England, Chandra had to contend with loneliness, homesickness, bland food, and the difficulty of keeping to a vegetarian diet, not to mention worries about what was happening back home and the racial prejudice that was as prevalent in Cambridge as anywhere in England. Fiercely proud of his Indian heritage, Chandra strongly objected to any hint of discrimination. Babuji, no matter how uncaring a face he presented to his children, was concerned for him. "I am feeling very lonely here . . . suffocatingly lonely," Chandra wrote to him.[20] Once term began, however, he became immersed in his studies and felt a little better. But he remained a loner.

Fowler, his thesis adviser, was frustratingly difficult to pin down. He seemed always to be absent from Cambridge or busy in his study at Trinity or in the library. Chandra would spot him going into the library and wait patiently outside, hoping that he would emerge and invite Chandra along to his office to talk. But he never did. Chandra soon gave up.

The young man's feelings toward his new home oscillated violently. Rubbing shoulders with people not much older than he who had made revolutionary discoveries was a thrill. "I saw Dirac yesterday," he wrote to his father:

A lean, meek shy young "fellow" (F.R.S.) who goes slyly along the streets. He walks quite close to the walls (as if like a thief!), and is not at all healthy. A contrast to Mr Fowler . . . Dr. Dirac is pale, thin, looks terribly overworked. Tomorrow his lectures begin. I may get introduced to him sometimes. The place is so big that it takes a long time to get to know these people.[21]

A month later, Chandra's excitement had palled. "The formality of introduction is so great and even then it is not worth the trouble of getting introduced," he wrote sourly to his father.[22] Compared with the warmth and openness he was used to from southern India, the smug reserve of most English people he encountered was such a severe culture shock that he withdrew into himself. It was grindingly difficult to get used to living alone in this dull, gray country, far from the comfort and support of his family. Here there was nothing to do but work.

Isolation was fine for him, he wrote to his brother Balakrishnan. "There is nobody to talk to, and mathematics has become my morphia. Dull and stupid, it helps me to forget everything else. I take some walk in the evening, and Cambridge is so small that you come back to the same place in ½ an hour."[23] All day long he would work alone in his rooms. "As it is, life is so lonely. Confined to one's study all day long and even during walks not to get rid of oneself! It's all so sad, so oppressive!" he complained bitterly to his father.[24] By November he was writing that he felt like a "single electron in deadly free space" — as if Babuji would have the faintest idea what his son was talking about. He waxed enthusiastic about the papers and ideas he was working on; he had become a "physics animal."[25] Three months earlier he had been immersed in Shakespeare, "now the diversion is quantum mechanics or stars to 'chemi-luminescence' or 'theory of valence.'"[26]

"It makes me sad to think you are feeling a bit lonely," Babuji wrote to his son on New Year's Day, 1931. Chandra had been away for five months now. "The feeling will come on, as the human ties press on the mind. Hereafter I will keep a diary." A stickler for grammar and spelling, Babuji added, "Please also keep for reference an Oxford dictionary as your letters are full of *spelling* mistakes in English." In reply, Chandra apologized for being an "unlessoned boy." He was writing, he added, about Dirac's recent relativistic electron theory.[27] Rather than astrophysics, he was trying to work on pure physics like Dirac, whom he idolized. But when he handed in the paper he had written, Fowler told him in no uncertain terms that it was unsatisfactory. "One can laugh only bitterly!!" Chandra wrote to his father. He decided that "feeling lonely" was a weakness; after this, he would no longer display his feelings outwardly.[28]

Chandra was also attending Eddington's lectures on relativity. "He is in contrast to Dirac," he told Balakrishnan. "Dirac is always serious — 'a martyr to meditation.' . . . But Eddington is always funny and humorous. On the whole the lectures this term are very useful and good."[29] Chandra was in the minority in his opinion of Eddington's lectures: most students complained of their dullness. But Chandra's brilliance led him to see many things differently.

One astrophysicist, at least, did take an interest in Chandra. From the beginning, Milne took Chandra under his wing, sometimes too much so. He made his first surprise visit to Chandra's rooms in October 1930, not long after the young Indian had arrived. But then, as Chandra recalled, Milne talked almost exclusively "about his own ideas as to how all stars must have degenerate cores."[30] The surprise visit made a permanent impression on the young man. Shortly afterward, Milne wrote him a letter, suggesting that Chandra's results on the limiting mass of white dwarfs were "particular cases of my analysis of 'collapsed stars'."[31] Chandra was piqued, for he thought the exact opposite. "Prof. Milne's results are in a sense a generalisation of mine own," he wrote to his father on November 14, 1930. In point of fact, Milne's analysis was no more than a pastiche of Fowler's and Chandra's results, but Chandra was wrong in claiming that Milne's theory was simply an expanded version of his own. Chandra's letter to his father was written from Burlington House, where he was attending the Royal Astronomical Society meeting at which Milne presented yet another paper critical of Eddington. "Others (including me) believe that he has completely *destroyed* Eddington's view of the interior of stars," Chandra wrote. He added that he still felt far too nervous to join in the discussion with Eddington, Jeans, and the others.

By this time Chandra had begun to suspect that his own work, on the limiting mass for white dwarfs, was being ignored for some reason. The only way to proceed seemed to be by melding his result into Milne's idea that all stars had an incompressible core. But that ran directly counter to Eddington's standard model, which assumed that most stars behaved like a perfect gas. The young man was about to enter a bruising battle with giants.

Over at the RAS the run-ins between Eddington, James Jeans, and

Milne were becoming vicious. Eddington and his colleague and archrival Jeans disagreed on just about everything, most notably on what stars are made of and what their energy source is. The disputes at the RAS had become famous for their vociferousness. Like Eddington, Jeans had been fascinated by numbers from a very young age; at seven he was memorizing logarithms. But unlike Eddington, he came from a privileged middle-class background. In 1905, when he was just twenty-eight, he was already a star lecturer at Trinity College. Woodrow Wilson, then the president of Princeton, lured him to take a professorship at what was then little more than a posh small-town college. There Jeans enjoyed a salary much larger than the pittance that teaching staff at Cambridge received. By 1910 he was back in Cambridge. The holder of the Plumian Professorship was on his deathbed, and Jeans knew that he was next in line. But as a result of some Machiavellian internal politics, the professorship went instead to Eddington, a mere Junior Fellow at the time.

Disgusted with academia, Jeans resigned. While at Princeton he had married Charlotte Tiffany Mitchell, of the Tiffany jewelry empire. The couple lived in regal splendor in their villa, Cleveland Lodge, in Dorking, just outside London. There Jeans speculated about the universe and turned to popular science writing. His book *The Mysterious Universe* was a bestseller. A superb musician, Jeans played Bach brilliantly on a magnificent organ. Chandra once visited him and was very impressed. Compared with the regular denizens of high table, Jeans was an "aristocrat," Chandra felt, though most thought otherwise and were "sarcastic" about him. "Anyone who can play all of Bach's organ compositions cannot be a trivial person" was Chandra's opinion.[32]

On January 9, 1931, the dispute at the RAS came to a head; the gloves were off. Tucked away at the back of the room, Chandra must have been aghast, seeing these mighty minds slug it out. Launching the proceedings, Milne took a swipe at Jeans's hypothesis that at high pressures a star's interior became an exotic liquid in order to avoid complete collapse. He accused Jeans of trying to account for observed facts by "introducing hypotheses which are extrapolations of the laws of Physics. . . . To be perfectly frank," he added, "I don't suppose that anyone really believes his theory."[33] Milne then turned his fire on Eddington, reiterating his criticisms of the mass-luminosity relation and the fact that he ignored the issue of the source of a star's energy. They all, of course, agreed that the density of the core of a star was finite; infinite density was unthinkable and impossible. They were simply arguing about how and why that was true.

Then Eddington rose with a magisterial air. He brushed Jeans aside with the comment that any difference between the liquid and gas theories was "purely a matter of plausibility of hypotheses" and therefore not serious. Milne's case was different: it concerned the "correctness of a mathematical or logical deduction." Eddington went for Milne's Achilles' heel, his feelings of inferiority about his mathematical ability. Milne had argued that Fowler's theory of white dwarf stars implied very high temperatures and pressures inside stars. In Fowler's theory, a dense gas of electrons moving at speeds much less than that of light, and thus nonrelativistic, reach a point at which they cannot be further squeezed by gravity. The electrons retaliate by generating an outward nonrelativistic degeneracy pressure, and the core becomes as hard as rock. Milne had pointed out that the temperatures and pressures inside stars would be even higher if the electrons were moving at speeds comparable to that of light (as Chandra's theory of white dwarfs predicted). In other words, a star with an incompressible core must be extraordinarily hot and dense inside, especially if its electrons are moving at such high speeds that they are governed by the theory of relativity. Eddington disagreed with Milne: such high temperatures and densities were simply impossible, he claimed. He reminded his audience that the opacity discrepancy — the nagging difference between the opacity measured in the laboratory and the opacity expected for Eddington's standard model of the stars — that arose in his model was even more severe in Milne's. The only way for Milne to avoid the discrepancy would be to apply his model to other types of stars. But if he did so, he might find that the internal pressure of the star was a negative quantity, and that could not possibly be true for any real star. "So the star must be wrong as well! . . . Professor Milne is between the devil and the deep sea — or rather between me and the deep sea," Eddington concluded triumphantly.[34]

Jeans waded in. "As Prof. Milne has been so frank in the expression of his views as to my theory, I feel that I can be equally frank about his work," he snapped. "In brief, I do not think it gets anywhere." He then took a sideswipe at Eddington, saying, "I don't think that Eddington's gaseous model is tenable any longer." Then, as the coup de grâce, he concluded, "It seems to me that [Milne] is not studying stars at all, but solutions of a differential equation. In saying this I am being as frank as Prof. Milne has been about my work."[35]

W. G. M. Smart, a renowned celestial mechanician at Cambridge, reminded everyone that stellar models have to be simplified to be "amenable to mathematical treatment . . . The stars themselves must be the final referees in a matter of this kind, and personally I think that the stars behave very well according to Sir Arthur Eddington."[36] The great Cambridge mathematician and dandy G. H. Hardy had joined the RAS specifically for this sort of row. He came up with the showstopper:

> As a mathematician I don't care two straws what the stars are really like. I have a mild preference for Milne's theory, since it takes account of all solutions of the differential equations concerned. [But] I am particularly interested in Mr. Fowler's paper [which] is probably the only one of the collection which is of lasting value, for he is certainly right, whereas it is extremely likely that everyone else will be shown to be wrong.[37]

Milne sank deeper into gloom. In the end Hardy proved to be right, at least as far as this particular argument went. Thomas Cowling, a former student of Milne's, recalled that the atmosphere was "diamond cut diamond. I think everyone was absorbed, sitting at ringside seats."[38]

As for Chandra, his recollection was that "the controversies between Eddington, Jeans, and Milne led most onlookers to say 'a plague on both of your houses!' Essentially this sentiment was expressed by P. W. Merrill in his [book] 'Variable Stars' where he writes that the papers on stellar constitution are no more than 'exercises in differential equations.'"[39] Merrill's exact words were these:

> Technical journals are filled with elaborate papers on conditions in the interiors of model gaseous spheres, but these discussions have, for the most part, the character of exercises in mathematical physics rather than astronomical investigations, and it is difficult to judge the degree of resemblance between the models and actual stars. Differential equations are like servants in livery: it is honourable to be able to command them, but they are "yes" men, loyally giving support and amplification to the ideas entrusted to them by their master.[40]

It was beginning to seem very likely that Eddington, Jeans, and Milne were busy solving complicated equations to create models of stars that bore virtually no resemblance to the real thing.

February was a black month for Chandra. "I take walks every day along the 'Trumpington' — the long road which leads to London —

alone every evening," he wrote to his father. "The road is usually very lonely. I go 'down' 2 miles along the footpath and back, 4 miles walk which usually takes an hour or more." He took a break from time to time, though he felt very guilty about it. "Occasionally on a Sunday — once in a month or sparser — I take a holiday. After breakfast I go on a long long walk across the fields. As for instance last Sunday — after a month of more or less uninterrupted steady work — I took a holiday. I am ashamed to confess that I needed it, but one wants a recreation once in a while."[41] That day he strolled alone through frozen fields along the winding Madingley Road.

A main cause of Chandra's gloom was that Milne was putting pressure on him to formulate his results so as to support Milne's own. Milne welcomed Chandra's theory because it revealed the unexpected properties of gases at the extremely high densities in a star's interior. Milne expected to find more surprises as astrophysicists delved further into the interior. Yet he did not want to accept Chandra's concept of a maximum mass, above which a white dwarf, after burning out its fuel and becoming dark, would collapse to nothingness. Milne wrote to Chandra in January 1931: "Your conclusion in its present form arises from the curious properties of [relativistic degeneracy] but I think you have fallen into the Eddingtonian error of inferring physical consequences from what can be only an incomplete algebraic treatment."[42] Among those "curious properties" was the upper limit to the mass of a white dwarf.

Milne then raised an intriguing issue. Take the case of a white dwarf with a mass just under Chandra's upper limit and therefore stable. It will not collapse: its electrons, moving at speeds close to that of light, can no longer be compressed by gravity; they generate an outward pressure — relativistic degeneracy pressure — to form a rock-hard core. But what would happen if just enough inert matter were sprinkled onto this white dwarf to push it over the maximum mass? Would it begin an infinite collapse? "Until you have solved this question your publication of your paper would raise more questions than it would answer," he told Chandra.[43] In Milne's view, it was simply impossible for a star to shrivel to nothingness. Chandra now had to ponder whether stars could be stable at higher densities — a problem that would take a new turn the following year with the discovery of the neutron.

Four weeks later, Milne was urging Chandra to explore how the different possible conditions for the interiors of stars (perfect, nonrelativistic, and relativistic degenerate gases) might "transform one into the other."[44] Chandra was beginning to feel like a pawn in the bruising battle of egos Milne was engaged in. He must have wondered when his own turn would come. When it did, he was determined not to suffer Milne's fate, although he was not sure how he would manage this.

At the RAS meeting on March 13, 1931, Eddington, Jeans, and Milne returned to the attack. Eddington kicked off the proceedings with a paper in which he calculated the maximum pressures and densities both of stars that obey the perfect gas law and of nonrelativistically degenerate stars (low-density stars with incompressible cores, such as Fowler had dealt with in 1926).[45] The figures he came up with were 1 billion degrees kelvin for the maximum internal temperature of a star and 10 million grams per cubic centimeter for its maximum density. These figures were clearly much too high, he said, a result of considering ideal rather than real stars and "not likely to be attained" in reality. Nevertheless, he wanted to present these results because he thought that they "may be of use in curbing riotous speculations."[46] He was referring, of course, to Milne's and Chandra's results, which included relativistically degenerate stars and which predicted even higher pressures and temperatures.[47]

Eddington even expressed doubts about Fowler's result, which also implied unfeasibly high internal temperatures, declaring that it was far from certain that the "formula of Milne's [originally Fowler's] for the pressure of matter in a completely degenerate state is correct."[48] Chandra was shocked. Not only was he sure Fowler's argument — that a white dwarf can never collapse — was right as far as it went, but his own work was an extension of Fowler's, and based on it. Eddington was indirectly attacking *him*. In a letter to Stoner, Eddington went so far as to dismiss Fowler's results as "conjectural and fishy."[49]

Fortunately for Chandra, he had one close friend, Sarvadaman Chowla. Short and jovial, he was just the sort of person Chandra needed to help preserve his sanity. A graduate of Government College in Lahore, Chowla had gone to Cambridge to study number theory with the famous mathematician J. E. Littlewood. "Life is so lonely here, that it is some consolation to have somebody," Chandra wrote to his father.[50]

In May came the devastating news that Chandra's beloved mother had died. "Oh! My Heavens, little did I dream that when I left her at the station that it was my last farewell to my mother. I am alone in my grief and it paralyses me to stay in my rooms and shed tears," he wrote.[51] For two weeks he could not work at all. Then, with a mighty effort to make work his consolation, he returned to his calculations. He worked six to seven hours a day — which he considered "taking it easy." In June he began actively collaborating with Milne and ratcheted up the hours to eight or nine a day — though he claimed that Milne worked faster.

Chandra worried about his father's state of mind and also tried to encourage his brother Balakrishnan, whom Babuji had just put into medical college. A dreamy sort, thin and ascetic, Balakrishnan aspired to become a writer. Instead of applying himself to his medical studies, he spent his days in cafés, talking and reading. He had begun to feel he had dropped behind so badly that it was too late ever to catch up; life was passing him by. Chandra exhorted him to do his best, in the process revealing a great deal about his own state of mind. "Age does *not* matter. It is *never too late* to begin," he admonished, with an older brother's sternness. Then his own self-doubts began to color his remarks:

> Do you think that just because one read a little calculus and conics when his equals did not, read a little statistical mechanics which others of his age did not, happened to publish a few papers, Do you think that he is better than his friends, Do you imagine that he has in any case proved *his intrinsic merit?* You are woefully mistaken if you think so, and I will cry shame on him . . . I wish I could divulge to you the sorrows of my heart, and tell you how I feel at times that my heart will break by the oppression of my ignorance . . . I cannot repeat too often that nobody can be so dissatisfied with oneself, *as* I am of myself.[52]

Chandra sympathized with his brother, forced to do medicine when he longed to be a writer. "*That* surely was not my feeling when I took physics in preference to mathematics," he wrote to him.[53] In subsequent letters, he patiently counseled his younger brother to follow Babuji's advice and reminded him that "Chekhov — or was it one of the other great Russian writers? once said that every artist must undergo the training of a medical doctor to fully understand human suffering in the large."[54] Chandra, certain that Balakrishnan was destined to be one of the best of India's new wave of writers, tried very hard to persuade him of this. Yet from Balakrishnan's letters one senses that he was still jealous and in awe of his brother.

Milne, under pressure from Eddington and with little support from his colleagues, badly needed an ally. As a powerless graduate student, Chandra was the perfect candidate. As far as Milne was concerned, Chandra's calculation of an upper limit for a white dwarf was simply part of his own research program: when a star contracted so much that its internal pressure exceeded the pressure allowed for by Fowler's theory, Chandra's theory came to the rescue. Milne now wanted Chandra to prove that some of Eddington's basic assumptions were wrong — that in the case of white dwarfs, for example, Fowler's solution was valid right to the core. He wrote to Chandra:

> From the point of view of general science it is important that where a mistake is made it should be pointed out, as courteously as possible, and as there has been more than enough of controversy between Eddington and myself, it is desirable that it should appear that the pointing out of the error is not merely the consequence of pre-existent antipathies between A.S.E. and myself.[55]

Milne seemed to be lining Chandra up as his front man to draw some of Eddington's anger away from himself.

Milne now saw errors everywhere in Eddington's work. He wrote to inform Dingle, who was on the editorial board of the *Monthly Notices of the Royal Astronomical Society*, that he had found a crucial flaw in Eddington's paper on the limits of temperature in a star which made the whole argument "absolute nonsense."[56] He even wrote up the finding for publication, then two days later had to withdraw it because he realized that Eddington's calculation was correct after all. He complained to Dingle that whatever he wrote on the blackboard during his lectures at the RAS was "categorically ignored" by his audience, particularly by Eddington, who never left him in peace.[57] He intended to write lengthy replies to Eddington and insisted that Dingle publish them. Milne submitted another paper to the *Monthly Notices of the Royal Astronomical Society*, but then had to withdraw it and pay the printer's costs. Possibly he had tried to publish the same paper twice. He wrote to Dingle: "You may have felt afterwards that I had not played the game . . . Since members of the Council think the direct controversy has gone on long enough, I will cease writing controversial papers."[58]

Milne was also pushing his impressionable young colleague to "set about tackling some of the harder problems of this subject . . . You should tackle something more ambitious" than the fate of white dwarfs, he insisted. "For example suppose you take my rotating star paper of

1922 or so and re-do it."[59] Chandra, in fact, did just that for his Ph.D. thesis. But he resisted taking Milne's advice any further than that.

By July 1931, his friend Chowla had completed his Ph.D. and left for India. "So my only friend* [*rather acquaintance] here is gone," Chandra wrote gloomily to his father, demoting Chowla. He added stoically that he "had grown accustomed to being lonely."[60] Chandra felt that he needed to get away from Cambridge, with all its pressures, at least for a while. The same month that Chowla left, he set off for Göttingen, Germany, one of the centers of the new quantum mechanics. There he planned to study under the renowned German physicist Max Born, who had expressed interest in Chandra's work and had extended an invitation to him.

Born, then forty-eight, was a very distinguished-looking man with fine, sensitive features. In the 1920s he had been a pioneer of quantum mechanics. He had done important work in electromagnetic theory, applied mathematics, astronomy, acoustics (while serving in the German army during the war, when acoustics was used to spot enemy artillery batteries by their sound), relativity, crystallography, and atomic physics. He was particularly renowned as the mentor of such important physicists as Heisenberg and Pauli. Heisenberg had developed quantum mechanics while working at Born's institute in June 1925. Born's major discovery was the deep meaning of the wave function, a key and much misunderstood aspect of the Austrian physicist Erwin Schrödinger's wave mechanics. In 1926 Born realized that the wave function describes the probability for finding an electron in a certain region of space, whether it is in an atom or moving freely. Einstein, a friend and former colleague of Schrödinger's at the University of Berlin, utterly repudiated this nonclassical view of science, according to which an electron's whereabouts cannot be predicted with absolute certainty but is merely probable, a theory linked to the electron's wave-particle duality. "God does not play dice," Einstein insisted.[61]

For Chandra, Göttingen offered a thrilling opportunity to hobnob with members of the international physics community. Besides Born, he met the twenty-three-year-old postdoctoral student Edward Teller, a handsome young man with floppy black hair. The two young men enjoyed discussing astrophysics and later renewed their friendship when Teller visited Cambridge in 1933–34. Teller was born in Budapest in 1908, and his scientific brilliance was noted even when he was a boy. At the end of the Great War, the Austro-Hungarian Empire was broken up and Hungary became independent for the first time in centuries. In the resulting political chaos, Teller decided to leave to study chemical engineering at the University of Karlsruhe in Germany. But he soon discovered that his true love was physics and moved to Sommerfeld's institute in Munich. In 1928 he was hit by a tram and lost his right foot. He learned to walk at amazing speed with a prosthetic foot and went on to become an excellent skier. Perceiving that the real discoveries in atomic physics were being made in Leipzig, where Heisenberg had just opened an institute, Teller transferred there and completed his Ph.D. in 1930. He later became internationally known as the "father of the H-bomb."

A week after he arrived, Chandra wrote excitedly to his father that he was hoping to see Heisenberg shortly. "I will see if he remembers me. Possibly not!"[62] (He did.) Heisenberg frequently dropped in at the institute at Göttingen, though mainly to see the visitors, not Born, who was a complicated, rather tormented man and had made many enemies. Chandra confided that he found Born "very unpleasant."[63] While on the continent, Chandra took the opportunity to visit the art museum in Kassel, with its collection of Dutch masters; his favorite was Rubens. He went on to Berlin to see the National Gallery and the zoo and then to Potsdam to visit Erwin Finlay-Freundlich, an astronomer of mixed German and Scottish descent and a close friend of Einstein. Freundlich invited Chandra to give a lecture on his recent work, which in his opinion dealt the "death blow to Eddington's theory," as Chandra reported to his father.[64]

Chandra spent the summer in Europe. But when he returned to the dank fens of Cambridge, nothing had changed. He found himself plunged into Milne's increasingly desperate struggle, and once again the gloom descended. "On Monday next, I am 21!" Chandra wrote to his father. "I am almost ashamed to confess it. Years run apace, but nothing done! I wish I had been more concentrated, directed and disciplined in my work."[65] Babuji replied immediately. How deeply he felt for his son, he wrote, so far away, and so lonely. Chandra, he said, should go to London more often. "You need *not* save money from your scholarship . . . You have become an *adult* and should start to have your own experiences (emotional and otherwise) . . . I daydream of the near future when you will return to India, [to] take up your professorship (like your grandfather)."[66]

As well as handing out sagacious advice, Babuji was beginning to remind Chandra that it would soon be time to return home. As Amar

Singh wrote in a study of Indian students in Britain, "To the traditional three don'ts — wine, women and meat — one more seems to have been added, i.e. don't stay in Britain after completing your studies."[67] In India an English education brought enormous cachet; to be an "English returned" ensured instant access to the highest levels of society.

Things had fallen so flat for Chandra that he was reduced to asking Fowler for something to do. At the time, he was devouring a series of papers by Heisenberg and Pauli on the interactions between electrons and photons, a subject that laid the foundations for the brand-new theory of quantum electrodynamics. Chandra was very excited by these papers and, desperate to "attempt something 'substantial,'" as he wrote to his father, began to think he should shift to theoretical physics.[68]

Then he received a letter from Milne, who had again found an error in Eddington's work — but again it turned out to be a false alarm. Milne decided to give up corresponding with Eddington, who, he complained, "always grossly misrepresents my arguments."[69] Shortly afterward, Milne paid another unannounced visit to Chandra at 10:30 in the morning. "We discussed for a long time together and he left! . . . my general quality of work is still low! (too low in fact!)," wrote Chandra gloomily of Milne's blunt assessment.[70]

Chandra pressed on with his calculations and, late in 1931, discovered another astounding implication of relativistic degeneracy: a star whose radiation pressure is more than 10 percent of its total pressure cannot develop a degenerate core to prevent it from collapsing completely. No matter how densely its electrons are crowded, the star's internal temperature will always be so high that it will remain a perfect gas. Without degeneracy pressure, a rock-hard, incompressible core cannot form, so there is nothing to prevent the burned-out star from being crushed by gravity into an infinitely dense, infinitely tiny point. This finding was another way of demonstrating that such stars could shrink and shrink, becoming denser and denser, forever. But Milne forced Chandra to put this work aside. "Milne refused to accept the result and the reason was simply that it violated his thesis" that every star must develop a degenerate core or some sort of incompressible core and will never disappear into nothingness.[71]

Superficially, at least, 1932 started off on a high note. On January 8 Chandra presented a paper at the RAS on a topic guaranteed to be free of controversy. It extended Milne's research on the Sun's atmosphere and had nothing to do with the internal structure of stars. "After I spoke, Sir A. S. Eddington made some 'fine' remarks about my work and Professor Milne was in his usual raptures!" Chandra wrote to his father. "The whole thing was quite a success! But somehow the whole thing fell flat on me." He felt that he was losing his "original cheerful and proud view of things. As one ages so one gets 'sour' goes the saying!"[72] Only twenty-one, Chandra already felt he was getting old. In the year since his arrival in Cambridge, he had published eleven long and complex papers containing important conclusions and had spent over a month in Göttingen. Yet he was far from satisfied.

The following week, classes began. Chandra was in almost daily contact with Dirac. "He is *just* wonderful!" he wrote excitedly to Babuji. "His philosophical insight into the general formalism of theoretical physics, his mathematical profundity to penetrate with ease any region of unexplained physical or mathematical thought and with all this what humility!"[73] Dirac did the sort of pure physics that Chandra aspired to do. Fowler, however, encouraged him to continue working on astrophysics with Milne. Perhaps Fowler felt that Chandra was aiming too high. But Chandra struggled to avoid immersing himself completely in astrophysics; he wanted to consult Dirac. While in Germany he had decided that what he really wanted to do was study with Bohr in Copenhagen, the world center for quantum mechanics.

During Eddington's course on the structure of stars, Chandra had asked several incisive questions. Eddington "was quite 'floored!'" Chandra wrote to his father. "Intend to continue the discussion this evening." He later continued, "I met Sir A.S.E. but nothing much came out of it."[74] Eddington asked Chandra to do some calculations for him, but Chandra said he was too busy with the work he was doing for Milne. He was taking a risk: it was poor strategy for an aspiring graduate student to refuse to help out with a calculation when an eminent scientist asked him to.

Chandra was beginning to worry that he was "making a mess of the whole thing!" He was becoming more and more convinced that the answer was to change fields: "My Astrophysical work has become largely a matter of numerical work and by such work one never learns anything."[75] In his spare time he had begun feverishly reading up on quantum electrodynamics and on group theory. It was Heisenberg who had first told Chandra about the latter when he visited India in 1929. Group theory is a beautiful and rather esoteric branch of mathematics that is used in tandem with the Pauli exclusion principle to find the correct mathematical equation to represent the behavior of collections of electrons.[76]

Then Dirac and Fowler returned "from a short visit to Copenhagen where, in residence, were Heisenberg, Pauli, Darwin, Kramers, Meitner, Kronig, Coster, an intellectual aristocracy of a breadth and magnitude hardly ever 'written' in History," Chandra breathlessly told Babuji.[77] Fired with excitement, he was determined that he too would breathe that rarefied air and mix with these gods of science. Such contacts were bound to advance his work and his career. Perhaps he also thought that in Copenhagen he would be appreciated as he had been in India. He was finally able to consult Dirac, who fully supported his plan.

By the end of June, Chandra was ready for his trip. In preparation he had been trying to solve problems using group theory, but with no success. Then Dirac proposed a problem, and he dropped everything to work on it. Still in his twenties, Dirac had just been elected to the coveted Lucasian Chair, the most prestigious professorship in Cambridge, once held by Newton. Nevertheless, Chandra could not wait to leave. "Cambridge now — in spite of Dirac! — gets on my nerves and the same little ten feet square — my garret! — chokes me," he wrote to Babuji.[78]

In August he was finally on his way. Bengt Strömgren, of the University of Copenhagen, met him at the station when he arrived. Two years older than Chandra, Strömgren was already a well-known astrophysicist. The son of Elis Strömgren, the director of the Copenhagen Observatory, Bengt had been a student of the stars since boyhood and had been trained in observational astronomy by his father. His work on the abundance of hydrogen in stars had changed Eddington's mind on the subject, forcing him to rethink his standard model. But he had also become an expert in quantum mechanics through spending time at Niels Bohr's institute. He was an astrophysicist who could speak the language of physicists.

For Chandra the atmosphere of high intellectuality in Copenhagen was inspiring. Bohr, a titan of physics, had won the 1922 Nobel Prize for his discovery of the first modern theory of the atom. In prestige he was second only to Einstein, who by then was no longer working on cutting-edge issues concerning the quantum physics of atoms and nuclei. Bohr was an imposing figure. In his youth he had been an exceptional athlete and a world-class soccer player, and he still skied, cycled, sailed, and played formidable Ping-Pong. By the time Chandra met him he was in his late forties, a heavy-set man with unruly black hair smoothly combed back, a high, domed forehead, heavy features, and large hands, one of which usually held his trademark pipe. He spoke in low tones in heavily accented English. Listeners had to strain to hear him. His hesitant mode of speech reflected his unrelenting desire to look ever more deeply into the essence of quantum physics, which for him meant probing beyond science into thought itself. He liked to quote the words of the eighteenth-century German philosopher and poet Friedrich Schiller: "Only fullness leads to clarity / And truth lies in the abyss." Einstein once described Bohr as being "like an extremely sensitive child who moves about the world in a sort of trance."[79]

"In Munich and Göttingen you learned to calculate. . . . In Copenhagen you learned to think," the young physicists who visited Copenhagen often said.[80] Ideas were thrown out, critically examined, and debated in group sessions daily, in a common quest to understand the increasingly complex mysteries of the atom. The more bizarre an idea seemed, the more enthusiastically it was received. Students, postgraduates, visitors, and established scientists all mixed. Everyone interacted directly, without nuances or stage antics, and with no annoying English understatement. The Germans in particular argued fiercely. After debates there was none of the animosity that drove so many of the Oxbridge circle to despair. Bohr's famous response to the wildest ideas that came up was "That's a crazy idea. The trouble with it is that it's not crazy enough!" To Chandra, Cambridge seemed dull by comparison.

But there was a problem: Bohr and his colleagues were not interested in astrophysics. Physicists and astrophysicists were poles apart. It was not just that physicists focused on what made stars shine while astrophysicists built models of their structure. Their philosophies were diametrically opposite. In his astrophysics research Eddington insisted on playing down mathematics, using just enough to analyze data (though, ironically, his research on general relativity and the fundamental theory was strewn with abstruse equations). Physicists, however, used mathematics — a very strange hybrid variety — as their launching pad to explore unknown territory full of ambiguities, counterintuitive concepts, and singularities. Chandra recalled Hans Bethe, who finally solved the problem of how stars shine, saying that the "attitude of physicists towards the astronomers was one bordering on contempt." Eddington was the "sole exception," due to his "enormous prestige."[81]

"I didn't belong to the scientific community any more in Copenhagen than in Cambridge," Chandra later recalled. "For example, I remember that every Friday in Bohr's house there used to be a tea [after which Bohr] and others would go with Bohr to his study to work. I used to stay behind and play with [Bohr's little boys]. But the atmosphere was better [than Cambridge] and I made so many lasting friends."[82] He even relaxed enough to join the boisterous young scientists in a "harmless bit of girl-watching [and] soon developed a system of mentally classifying them, according to their looks, under three heads — Alpha, Beta and Gamma!"[83]

Inspired by the liberating atmosphere of Copenhagen, in September Chandra boldly submitted a paper to *Zeitschrift für Astrophysik* that included the result Milne had rejected so vehemently when he had proposed it the previous year: that stars whose radiation pressure is greater than 10 percent of their total pressure cannot develop a rock-hard core and will eventually shrink to nothingness. The idea was to bypass the Cambridge circle, for the *Zeitschrift*'s editorial offices were in Potsdam. But as luck would have it, Milne happened to be passing through Potsdam when Chandra's paper arrived. The editor, Erwin Finlay-Freundlich, whom Chandra had met the previous year, asked Milne to look it over. The response was predictable. "Milne wrote a long letter of some twelve pages to me arguing that my conclusion was fallacious and that I would do harm to my 'growing' scientific reputation by publishing the paper," Chandra recalled. "But I stood my ground with the encouragement particularly of L. Rosenfeld. And the paper was eventually published. This experience left me sceptical, already in 1932, of Milne's objectivity."[84]

Chandra recalled somewhat bitterly that he "modified the paper a bit to mollify Milne,"[85] reducing the impact of his conclusion with the caveat that "we are unable to avoid the central singularity by appeal to Fermi-Dirac statistics [the quantum statistics used for electrons] alone."[86] The key word was "alone," suggesting that there might be some way, as yet undiscovered, other than Fermi–Dirac statistics, of avoiding the seemingly inevitable conclusion that the star would eventually collapse to a singularity. Forced to try to save Milne's theory, Chandra waffled about the possibility of "*incompressibility setting in* later."[87] Maybe somehow, at the last minute, the star would develop an incompressible core. Nevertheless, even Milne had to concede that his theory was doomed. Chandra's latest discovery proved incontrovertibly that in certain cases stars did not need to have degenerate cores and thus could remain a perfect gas — exactly as Milne's archenemy Eddington had postulated in his standard model. The paper was totally devastating to Milne: as Chandra wrote, "My work did not alter [Milne's model] in a trivial way: it eliminated all of it!"[88]

Meanwhile in Copenhagen, the magic was beginning to wear off. "Bohr is difficult to pin down for a conversation. I have had opportunities to speak to him on various occassions," Chandra wrote to his father — once again misspelling "occasions" — "rather hearing him speak to me!"[89] This souring of relations began with the quantum mechanics problem that Dirac had given Chandra, concerning the statistical mechanics of electrons and photons. Unthinkingly, Chandra wrote up his results and gave the paper to Bohr to submit to the *Proceedings of the Royal Society*, completely bypassing Dirac. He mentioned Dirac's name in the acknowledgments, so the editor sent it to Dirac to look over. Dirac wrote directly to Bohr, "It seems to me that the result of the paper must be wrong." Chandra had overstepped the bounds of mathematics as it is applied to physics. It was an old habit of his; Heisenberg had caught him out on a similar point two years earlier. In essence, he had tried to deduce "purely from mathematical considerations" — namely, group theory — that there were other possible kinds of statistics besides the known ones that apply to electrons and photons. Dirac added that if Chandra's paper were published, he would have to append a note stating his disagreement.[90]

Chandra insisted he was right and even claimed that Dirac had never read his paper. "This means another long drawn out controversy," he wrote dejectedly to his father.[91] In fact, what happened was rather embarrassing for all concerned. Chandra had omitted the impact of spin on his calculations, an error Bohr and his colleagues had failed to notice. "Our only excuse for overlooking so trivial a point may perhaps be that we were under the impression that you, at any rate for a moment, had considered the question worth investigating. But even that may be a mistake," Bohr wrote to Dirac.[92] He discussed the matter with Chandra, who gloomily agreed to withdraw the paper. Downcast over what had happened, Chandra told Babuji that, "Quantum mechanics paper is WRONG . . . 6 months of hard work is all SMOKE."[93]

The failure put an end to his thoughts of becoming a theoretical physicist. "I have become a confounded pessimist — for instance I hold the view that 'things never get better, they always get worse' — believing in nothing," he wrote to Balakrishnan from Copenhagen. The only avenue open now seemed to be to concentrate on astrophysics and deal with his backlog of work, though it would mean more squabbles with Milne. On the bright side, he was working hard on his Ph.D. thesis. He reassured Balakrishnan that although in the West he had acquired "new habits of thought," this applied only to his intellectual life. If Balakrishnan were to ask whether he had changed his ways regarding "smoking, drinking, dancing, going out with girls, eating meat (or even putting on a hat!) then the answer is negative," he told him firmly.[94]

Chandra left Copenhagen on March 17, 1933. On his way back to Cambridge, he passed through Hamburg to see the mathematician Emil Artin, an important contributor to the highly abstruse branch of mathematics known as number theory. Chandra's friend Chowla knew Artin and wrote that he was "astonished that [Chandra had] gone so far into the algebraic theory of numbers."[95] The visit with Artin went well. Chandra still dreamed of becoming a mathematician and wrote excitedly to his father, "Might study with him in July instead of Milne!"[96] But by now everyone was aware that Germany was on the path to war, so Chandra had to abandon his plan.

Back in Cambridge, he looked into sources of funds to extend his stay in England. His father was putting relentless pressure on him to return to India after he had completed his Ph.D. "Your *mathematical* work *can be done in India*," he wrote. "There are more things in this world to experience and realise, than *merely 'intellection'*. There is the experience of love of a woman, the love of a child. These two in due course enrich a man's life [because] she is a complement to your mental and moral character."[97] He urged Chandra to think about it. After all, he declared, a father knows best. But Chandra dreaded the thought of going back to his homeland.

Springtime brought a round of garden parties, including one in May at the Eddingtons', where Chandra met Henry Norris Russell. He had finally finished his thesis, for which he had chosen to elaborate and extend Milne's results on the shape of stars, taking the conventional route of focusing on a traditional problem area. He described his work to Russell, "who was frightfully enthusiastic about it all."[98] He asked Fowler if he wanted to see it. Perhaps wary of the amount of detail involved, Fowler's response was "No, definitely no! Just hand it to the registrar."[99]

Chandra's Ph.D. orals, with Fowler and Eddington as examiners, took place on June 20 in Eddington's study at the observatory. As Chandra later recalled, it was almost comic. Fowler roared up in his car fifteen minutes late. While Chandra and Fowler wore formal academic gowns, Eddington was in everyday clothes and slippers. He looked at them and muttered laconically, "I cannot rise to the occasion." The two professors then quizzed Chandra on his thesis and on astrophysics in general, interrupting his answers to argue ferociously with each other over obscure details. Forty minutes into the examination, Fowler suddenly looked at his watch, exclaimed, "Goodness, I'm already late!" and abruptly left. As Chandra stood there, at a loss as to what to do, "Eddington simply said, 'That is all.'"[100]

Shortly afterward, Babuji wrote that he was lonely and wanted to come to London. A visit from his father was the last thing Chandra needed. He wrote back quickly, listing all the drawbacks: the poor quality of English vegetarian food, the rice half-cooked and the curries frightfully hot, and the dismal lack of sunshine in winter, when his father wanted to come.

With the situation in Germany growing steadily worse, many of that country's leading scientists, most notably those who were Jewish, were fleeing. Chandra's uncle Raman had high hopes of expanding India's scientific community, declaring that "Germany's loss is India's opportunity."[101] Chandra disagreed: "For one thing all the 'great men' have (or can) found (or find) positions in Europe or America (Max Born for instance is coming to Cambridge)," he wrote to his father. "Only the younger men — the Jews! — will at all think of going to the tropics and C.V.R. [Raman] by over looking them is shooting off his mark."[102]

Determined to stay on, if at all possible, after completing his Ph.D., Chandra had decided to go through the arduous process of applying for a fellowship at Trinity. Fowler was quite discouraging about his chances of getting one, but he persevered nonetheless. After all, he had published more than a dozen papers as a graduate student, on almost every facet of astrophysics. He had received rave reviews when he lectured on his thesis research at the RAS. Milne declared his work "beautiful," while the formidable Russell congratulated him on the "lucid presentation he has given of an intricate problem," adding that "his work is going to be of great practical value." Although Eddington had some doubts about Chandra's models of certain stars, he too found it highly interesting.[103]

But despite the depth and breadth of his publications and accolades from the most important astrophysicists of the day, weighing heavily in the balance against him was that he was Indian. Only one Indian had ever received a fellowship: the great mathematician Ramanujan. Chandra was so convinced that he would not succeed that on the day the fellowships were announced, he had packed up and was preparing to leave his lodgings to go to Oxford and work with Milne for a term before returning to India. Purely out of curiosity, he stopped at Trinity to see who had won. "I was astonished to find my name among the people who were elected," he later recalled. "Well, this has changed my life!" he told himself. "I do not know what my future would have been, except to say that it would have been very different."[104]

Chandra wrote rather movingly to his father about his many conflicting feelings on accepting the honor, which would mean that he would not return to India for several years:

> I may as well mention (to be fair to myself) that this final decision has not been made without great pain, and indeed during the whole of Sir J.J's [Thomson] speech last evening, and during all the time when so many were congratulating me there was only one picture before me — my mother in her silk saree tying with trembling hands a thread round my head praying God to look after me, and blessing me with all the force of love to '*go forward*'. Indeed I always have this vision which has been a great source of inspiration — intensely saddening yet stimulating.[105]

Milne wrote to congratulate Chandra, predictably taking the lion's share of the credit for himself. Later that year, he wrote warmly to his young colleague, "Perhaps you would do me the honour of dropping the title 'Professor' in our future correspondence. It used to be a good rule at Trinity that once a man became a Fellow he dropped all titles with other members of the high table."[106] At last Chandra had arrived.

Babuji's reply to Chandra's letter was full of joy, pride, and happiness mixed with sadness. "Your decision is not news to me," he confessed. "What I feared, you have stated." Nevertheless, he could not help but be happy for his son: "What you have achieved is only my dreams or hopes come true, of my son — obtaining intellectual 'laurels.'" He was, he added, definitely going to travel to Europe and visit his son.[107]

To everyone's surprise, Babuji's trip was very successful. He set sail on November 9, 1933, and had such a wonderful time that he extended his leave of absence from the accounts office and did not return until July 9 the following year. A violin player of some renown in India, he had arranged to give several private recitals, in London, Paris, Vienna, Florence, and Geneva, among other places; one performance was broadcast on BBC radio. Father and son spent the Easter holidays together in Munich. Chandra was far happier to see Babuji than he had expected. They talked far into the night about many things. Chandra had to apologize for seeming perhaps too argumentative: "It's just the excitement."[108] In March Babuji performed in Aberdeen, Scotland, which he particularly enjoyed, and then in northern England. In May he arrived in London, and he and Chandra spent more time together.

Shortly after Babuji left, at the beginning of June, Chandra wrote him a moving letter. His father's trip, after all, had been sparked by his own decision to stay on in England. Chandra assured him that he had maintained the ethical and moral standards instilled by his parents. "I have learned to know you better during your stay in England and I am sincerely glad that you did finally come to Europe," he wrote. "Finally the kind and sympathetic way in which you understood me on the last night of your stay in 7 Worsley Road moved me very much and when I went to my room I prayed: 'Oh! God, Let me be the Son of my Father.'"[109]

That February, Chandra had published a short paper in *The Observatory* summarizing the results he had originally dared publish only in the *Zeitschrift für Astrophysik*, which had effectively demolished Milne's work. Now that he was a Trinity Fellow, he felt more sure of himself. "The general evidence then is in favor of Eddington's perfect gas hypothesis for ordinary stars," he concluded, perhaps to draw Eddington's attention to the fact that he was switching camps and joining the winning side.[110] White dwarf stars, however, were an entirely different matter, as Eddington agreed.

In July of that year, B. P. Gerasimovich, director of the Pulkovo Observatory in Leningrad, invited Chandra to visit the Soviet Union. The two had corresponded since Chandra's days at Presidency College, and Gerasimovich had expressed great interest in his progress.[111] Like other Western scientists who visited the Soviet Union in the early 1930s, Chandra was taken on a special tour that ensured he would take away an impression of a free and dynamic country; the darker side was concealed. Chandra was very impressed. As he told Balakrishnan, "Russia gives the impression of a young man, full of ideals, who has such indomitable moral courage and indefatigable physical strength to go forward in spite of setbacks and who with his hopes derives consolations from his ideals during times of adversity."[112] He did not realize it at the time, but this idealistic personification of Russia would come to fit Chandra himself all too well.

In Leningrad he met the Russian physicist Lev Landau as well as the Armenian astrophysicist Viktor Ambartsumian, who invited Chandra to give a lecture there about his results on white dwarfs. Much impressed with Chandra's talk, he urged him to write up the results in as broad and rigorous a manner as possible. Until that point Chandra had lacked a firm mathematical basis for the astounding implication of his theory of the upper limit for the mass of a white dwarf. Ambartsumian suggested that he deduce this result from the mathematical formalism that astronomers used to examine the structure of spheres of gas no matter which gas law they obeyed — perfect, imperfect, classical, or quantum.[113] If this exact version of his theory was right, it would be expected to yield Fowler's result for white dwarfs, which holds for electrons traveling at speeds much lower than that of light. An important step in validating and clarifying a new result in theoretical physics is to demonstrate that it reverts back to previously accepted cases, which, it would turn out, were valid only within certain limits. That is the case with Einstein's theory of special relativity: for speeds much lower than that of light, it becomes Newton's theory of motion. Chandra thought Ambartsumian's idea an excellent one and resolved to turn to it when he returned to Cambridge.

Among those in the audience for Chandra's lecture at the Sternberg Astronomical Institute at Moscow University was the Russian explosives expert Yakov Zel'dovich, who was then developing an interest in astrophysics. Lev Landau may also have been present. Chandra's reception in the Soviet Union was a welcome change from Cambridge, where everyone either ignored his work or claimed it as a special case of his own.

On his return to Cambridge, Chandra turned his attention to a personal matter that had been on his mind. He wrote to Babuji, "I wonder if you would excuse me if I made bold to be impatirent [impertinent] enough to ask what you thought of Lalitha — she wrote to me that you had met her."[114] Babuji wrote back, saying he had met "Miss D. L . . . on *three* occasions." Lalitha had helped him with the sound quality of his performances of Indian music. His sense of propriety prevented him from saying anything very specific, but he was prepared to concede that his "general impression is that she is a modest and quite reserved young lady."[115]

Chandra's response stunned Babuji. Throwing aside Indian tradition, he did not ask his father's permission but came straight to the point. "It was very kind of you to have understood me and written to me of your conversations with Lalitha," he replied. "Actually now Lalitha and I have come to *definite* mutual understanding that we should get married on my return. L knows too that you are aware of the fact that L and I have been interested in each other for these years."[116]

Babuji wrote back rather touchingly: "On receipt of your letter, I could not exactly express the feelings which arose in me. I was incredibly more sad than happy — the reason which caused the sadness to arise in me was that your mother was not alive to welcome home your bride-elect." He confessed that he had had some reservations. "My impression of the alliance was not favourable" because of Lalitha's aunt, who had "married a certain gentleman — who had already a wife and some children." But of course, he went on, "present-day morals are different from the past generation. All this flashed in my mind in a space of 2 minutes or less . . . I only pray to God that the union you propose should be blessed and happy." He ended on a sentimental note, quoting from Tennyson's *Idylls of the King* (Tennyson had been Babuji's favorite poet when he was a romantic young man of Chandra's age). "I do hope and pray that 'this maiden passion for a maid' will bring all happiness that a father can wish for his son."[117]

At the end of October 1934, Chandra submitted a paper on some work in progress, summarizing three major papers he was preparing to publish the following year.[118] It was notably more self-assured than the earlier articles he had written on the subject and promised broader results than any before.[119] He wrote that he had deduced the precise equation governing the structure of spheres of degenerate gas made up of slow and fast-moving electrons, as Ambartsumian had suggested. It was so straightforward, he added, that "it is surprising that it has not been isolated by anyone previously."[120]

In the months that followed, Eddington visited Chandra's rooms regularly, two and sometimes three times a week. He even arranged for Chandra to be given the fastest mechanical hand calculator available.

Something strange was going on — perhaps Eddington smelled blood. Chandra had convinced him that the seemingly endless controversy with Milne was finally to be settled, and in Eddington's favor.

As Chandra was to state in the summary of his paper at the RAS meeting of January 11, 1935, his thesis was that "for a star of small mass the natural white dwarf stage is an initial step towards complete extinction. A star of large mass [greater than the upper limit for white dwarfs] cannot pass into the white dwarf stage, and one is left speculating on other possibilities."[121] He put together his results in a graph that he exhibited at the RAS, which showed very clearly that as the mass of a white dwarf (M) reaches the maximum (M_3), the star's radius (R) becomes zero and the star shrinks to nothing.[122]

Chandra had already notified Milne of his results. Milne congratulated him: "The concept of a 'complete' equation of state [for relativistic degeneracy] is valuable and new to the subject formally," he wrote. But he was still unclear about Chandra's concept of an upper limit to the mass of a white dwarf. He repeated his query: "If we add a piece of dust to the star" whose mass is just under the maximum, will it become unstable? And he added, "You may think me very conservative and stick-in-the-mud."[123] Lack of stability was anathema to a classical physicist.

The battle lines were drawn. Chandra had shown Eddington and Milne the manuscripts of his first two 1935 papers, which were highly detailed, dense with mathematical formulas and numerical results. In November 1934, Chandra invited Milne to be his guest at the Trinity Commemoration Dinner. "During this visit, I almost persuaded him to concede that his work . . . was essentially left without any basis," he recalled. "I was hopeful that my work would dissipate the Eddington-Milne controversy."[124]

The energetic director of the Harvard Observatory, Harlow Shapley, a former student of Russell's, had also begun to take an interest in Chandra, suggesting that he spend some time at Harvard. Babuji became even more concerned. An invitation by someone of Shapley's stature had to be taken seriously. Chandra's return to India seemed to be receding ever further into the future.

The day of the all-important meeting at which Chandra was to present his result was fast approaching. "I am overpowered with the work I have to do before January on my stellar structure," Chandra wrote to his father, adding, "As the meeting day approaches I have to work more and more and faster and faster." He was running out of time. Along with his calculations and new results, he was preoccupied with his impending marriage to Lalitha and the need to schedule a trip to India. "I feel I am pretty deep in a mess," he wrote:

> I should feel much worse if it were not that I am now passing through an essentially successful period in my scientific life. When I am deeply absorbed in my work I often recall a statement which Bohr made when he heard of the engagement of Rosenfeld — "It is a great pity that human beings cannot find all their satisfaction in scientific contemplativeness."[125]

Chandra was sure that the originality and truth of his discovery were about to be recognized at long last, even by the stern old men who populated astrophysics at Cambridge. He should have felt euphoric, yet he could never put aside the darkness in his mind. As he wrote to Balakrishnan in his last letter before that fateful Friday in January: "Involved in the puzzles of the interior of stars, battered by differential equations, boxed by numerical calculations, impeded by ignorance, rushed on by the dawning of the New Year, I have at last emerged not indeed with the anticipated joy with which I began [diving] into the Crucibles of Nature, but burnt and smoking, dissatisfied, tired."[126]

6. Eddington's Discontents

O N THE MORNING of January 12, 1935, Chandra awoke into a different world. The previous day he had been full of hopes and fears, apprehension and excitement, daring to imagine a glorious future for himself after he revealed his momentous discovery. Perhaps he hoped that even though he was young and Indian, he would finally be accepted into the inner circles of Cambridge academia. Instead, Eddington had utterly humiliated him.

Eddington's rebuttal had been delivered with far more than his usual venom. For Chandra it had been a complete shock. The essence of the great man's argument, delivered with superlative arrogance, was "There is no such thing as relativistic degeneracy!"[1] Relativistic degeneracy was, of course, the bedrock of Chandra's theory. Without it the situation reverted to Fowler's theory, which applied only to a dense gas of electrons moving much more slowly than light. In Fowler's theory there was no limit to the mass of a white dwarf. No matter how massive, such a star would always die a peaceful death. Eddington was claiming that Chandra was totally wrong and his work worthless.

At the time, astrophysicists assumed that most stars turned into white dwarfs at the end of their life, except for a few that blew themselves to pieces. All stars, they thought (mistakenly, as it turned out), went through the same life cycle, zigzagging across the H-R diagram: they began as brightly burning giant stars, then joined the main sequence as dwarfs, and continued to contract until they finally burned off all their fuel and expired as white dwarfs, hot but dim, tiny but astoundingly dense. But here was Chandra, daring to suggest there might be another end to the life of the stars.

Eddington continued his rebuttal with the convoluted argument that although he agreed with Chandra's mathematics, he considered the physics incorrect because it was based on a combination of two diametrically opposed theories — relativity and quantum theory — "and I do not regard the offspring of such a union as born in lawful wedlock."[2] Chandra was aghast. The man must have his head in the sand. Had not their colleague at Cambridge, Dirac, been awarded a Nobel Prize in 1933 for formulating a relativistic version of quantum mechanics? What was going on? But worse was to come. Next Eddington claimed that the Pauli exclusion principle, the very basis of Chandra's theory of white dwarfs and of quantum theory itself, was not a general law of nature. As Pauli himself would have said, "Why, that's not even wrong!"

Eddington argued that Chandra's mathematics might be correct, but the result — relativistic degeneracy — had nothing at all to do with the stars; it was just playing with numbers. That is what Eddington had meant by saying that "there is no such thing as relativistic degeneracy." He added that Fowler's theory of white dwarf stars, which did not agree with relativity theory, as Chandra had discovered, actually did take relativity into account. In order to prove this Eddington resorted to arguments so convoluted and obscure that they would eventually wreck his reputation as a scientist.

Eddington, of course, was fully grounded in relativity theory — more so, perhaps, than anyone in the world other than Einstein. He was also deeply involved in the effort to construct a theory that would combine general and special relativity with quantum theory — namely, his fundamental theory. So to refer to such a union as not "lawful wedlock" was inconsistent, to say the least.

In *The Internal Constitution of the Stars*, Eddington had happily considered the implications of Einstein's general theory of relativity for the study of stars. According to the general theory — which, unlike special relativity, encompasses the effects of gravity — space (or, more properly, spacetime) is distorted by objects to a degree determined by the object's gravity. Spacetime has a structure, a fabric, and the extent to which that fabric is distorted depends on how massive the object is. Just as a stone placed on a rubber sheet creates a hollow, an object will create a well in spacetime; the heavier the object, the deeper it sinks. Conversely, the curvature or warping of spacetime around a massive body is a measure of the object's gravity — the stronger its gravity, the deeper the well. At a great distance from massive objects such as stars, space-time is more or less flat.

Eddington had used the general theory of relativity to try to understand what happens when stars collapse under their own gravity. According to general relativity, he pointed out, a small, dense star would create a far deeper well than a star like the Sun. It would distort space around it so much that no light could escape and it would disappear; it would be, in his phrase, "nowhere."

But to Eddington this theorizing was mere fanciful speculation. So he pooh-poohed Chandra's argument that if a star burned up all its fuel and still had a mass exceeding the upper limit Chandra had discovered, it would inevitably collapse to a singularity, infinitely small and infinitely dense. If that was the case, Eddington could only conclude that "the star has to go on radiating and radiating and contracting and contracting until, I suppose, it gets down to a few kilometres in radius, when gravity becomes strong enough to hold in the radiation, and the star can at last find peace." But for Eddington this apparently absurd scenario was unthinkable. "Dr. Chandrasekhar has got this result before, but has rubbed it in in his last paper," he concluded dismissively.[3]

As Chandra recalled in later years, Eddington was the only one to understand, with his "enormous physical insight," that the "existence of a limiting mass implies black holes" and that this must occur "once one accepts the physics. If [he] had accepted that, he would have been 40 years ahead of anybody else."[4]

Yet despite the flimsiness of Eddington's arguments, the established scientists chose to support him against Chandra, the outsider. What must have been bitterest of all for Chandra was that he knew in his bones that he was right. But who could he turn to for advice and support? Certainly not Babuji; he wouldn't understand. McCrea had disappeared after the session and in any case seemed to be a blind follower of Eddington. Chandra's close friend Léon Rosenfeld came to mind.

At thirty-one, only six years older than Chandra, Rosenfeld had a sterling pedigree. A warm, easygoing Belgian physicist, always immaculately dressed, with glasses and a high, domed forehead, he had completed his Ph.D. at the University of Liège in 1926. He then spent two years in Paris with the maverick physicist Louis de Broglie — an outsider in the science community, perpetually at odds with the interpretation of quantum theory hammered out by Bohr and Heisenberg. Rosenfeld moved on to become an assistant to Pauli, and after 1930 began a lifelong collaboration with Bohr. Chandra had first met him in Copenhagen in 1932. The two became firm friends, particularly after Chandra went to stay with him and his family in Liège. Rosenfeld knew how to handle himself in the intellectual rough-and-tumble of Copenhagen, and Chandra thought he might have some helpful advice.

He wrote to Rosenfeld immediately, telling him what had happened and asking him to pass the news on to Bohr. "Eddington sprang a surprise on everyone," he declared, explaining that he had dismissed Chandra's method for deriving his new equation as completely wrong and, worse still, had claimed that Chandra had used two fundamentals of physics — Pauli's exclusion principle and the theory of relativity — incorrectly. Instead of feeling satisfied with the fruits of working twelve hours a day for the last four months, Chandra said, he was "terribly worried."[5] Could all his work have been for nothing? Over the next few months a flurry of letters went back and forth between Chandra and Rosenfeld. Rosenfeld fully supported Chandra; Eddington's remarks, he declared, were "utterly obscure."[6]

One assertion of Eddington's particularly astounded Rosenfeld and Bohr: that Pauli's principle depended on considering electrons only as standing waves as opposed to progressive or traveling waves. If you attach one end of a rope to a doorknob, then shake the other end, you generate a wave that appears to move but in fact simply goes up and down — a standing wave. Progressive waves move forward, like those in the sea. To analyze electrons in an atom, you can imagine them to be confined and therefore at rest, like standing waves. Eddington declared that this was the correct way to apply the Pauli exclusion principle. But Chandra, claimed Eddington, treated electrons as progressive waves because they are in motion inside a star, then transformed them into standing waves by applying a mathematical technique from relativity theory. According to Eddington, such a transformation was illicit; Chandra should have treated electrons like standing waves from the start in order to apply the Pauli principle correctly. Instead he had combined special relativity with a version of quantum theory that did not agree with it — the "unholy union" Eddington had referred to at the RAS. However, in any quantum calculation, the type of wave is totally irrelevant to the end result, just as doing calculations in the metric or American system makes no difference to the result. Eddington's line of argument was entirely specious.

"Nobody ever dreamt of questioning Pauli's principle," Rosenfeld wrote to Chandra, adding that his use of relativity theory was also beyond question. "These quite obvious arguments would seem to settle the question beyond any doubt. So I think you'd better cheer up and get on with your own work instead of losing one's time with fruitless arguments."[7] Bohr and Dirac agreed. Chandra, however, was not prepared to listen to advice, even from such eminent men.

Soon after their confrontation, Eddington sent Chandra a manuscript with the same title as Chandra's lecture, "On Relativistic Degeneracy." He went into elaborate detail to argue his claim that Fowler's result for white dwarf stars, which did not take into account electrons moving at speeds comparable to that of light, and so did not include relativity theory, actually did. He accomplished this remarkable feat by playing fast and loose with relativity in ways that, although mathematically correct, have no meaning in physics. In other words, he brashly cooked up his own version of relativity theory and claimed it was the correct one. He even went so far as to insist, with his customary aura of authority, that Fowler himself should have realized this in 1926. Eddington's reasoning was so confused that at first even Fowler was unsure of what he himself had done in 1926.[8]

From Eddington's remarks on January 11, it was clear that he had already been preparing this paper behind Chandra's back. When he visited Chandra's rooms, ostensibly to offer encouragement, his real motive had been to check on what Chandra was doing. His sheer mean-spirited duplicity caught Chandra totally unawares.

In early 1935, Eddington published a series of papers attacking Chandra. They were a hodgepodge of dubious claims and poorly argued assertions. He belittled Chandra's theory as a *"reductio ad absurdum,"* meaning that although relativistic degeneracy might be an implication of the mathematics of Chandra's theory, it did not actually exist.[9] The only reason Chandra had deduced it, he argued, was because he was using the "wrong" relativity — that is, not Eddington's version, which included his own mathematical formulation of Dirac's theory. Had he used the "right" one, he would have ended up with Fowler's result — that white dwarfs remain stable regardless of their mass when they have exhausted all their fuel. What Chandra had done, in Eddington's view, was to defeat the "original purpose of Fowler's investigation."[10]

But what was Eddington's real objective? Chandra agonized over this question. Perhaps everyone was missing something in Eddington's argument. Even Fowler had wavered for a moment, though he soon came around. He told Chandra, "Don't be worried. I'm sure Eddington is wrong," then, noticing how miserable the young man looked, added in his usual genial way, "Be cheerful!"[11] But he never confronted Eddington on Chandra's behalf.

Chandra continued to pester Rosenfeld: "Honestly! Is it possible that there is anything" in Eddington's papers?[12] Rosenfeld replied emphatically, "After having courageously read Eddington's paper twice I have nothing to change my previous statement: it is the wildest nonsense!"[13]

In fact, the real reason that Eddington was so opposed to Chandra's theory was that it entirely undermined his fundamental theory, which he had been honing with obsessive care for seven years. Eddington's aim was to unify the general and special theories of relativity with quantum theory using a mathematics that could also produce the fine structure constant. The crux of the fundamental theory was that there are no systems in nature in which the interactions between particles can be ignored and that therefore electrons never operate in isolation. But now Chandra had produced a far-reaching, experimentally verifiable theory by studying an idealized quantum gas of *noninteracting* electrons.[14] If Chandra's theory was valid, then Eddington's fundamental theory and his whole mathematical system of E-numbers would come tumbling down. Instead of being, as he thought, a unique way to deduce the physical world from its mathematical structure, the theory would be relegated to mathematical game playing, as would his calculation of the number of electrons and protons in the universe, which connected so perfectly with its rate of expansion.

Determined to save his theory, Eddington suddenly and with great ferocity began including white dwarfs in his writings on it. He also began to make more and more fantastic claims about relativity theory, trying to rescue Fowler's more acceptable theory on the fate of the stars while attacking Chandra's as mere mathematics having nothing to do with the true nature of the heavens. Like everyone else in the world of science, Chandra was aware of Eddington's determination to develop his fundamental theory and of the seriousness and importance of the project. He was beginning to see that defending the theory was at the heart of Eddington's objections.

Rosenfeld declared Eddington's two papers of 1935 to be "nonsense,"[15] while Chandra wrote more temperately, "We are quite unable to follow his arguments."[16] Yet Eddington's authority and his uncanny skill at arriving at correct results from dubious premises meant that few astrophysicists were prepared to reject his arguments out of hand. Besides, the result of Chandra's theory was beyond anyone's imagining.

Edmund Stoner, of course, had also come up with the idea of a limiting mass for white dwarfs, though he had never pushed it to its logical conclusion. Strangely, while Eddington chose to attack Chandra, he encouraged Stoner's research. However, he kept tight control over the work, advising Stoner to do no more than explore what happened at the region where Fowler's theory broke down, thus making sure it did not threaten his own work.

Milne dealt with Chandra's result by suggesting that stars contain an absolutely incompressible core of some unknown material, which would ultimately prevent their complete collapse. Eddington's route was more direct. He simply tossed out Chandra's theory by declaring that it bore no relation to physical reality, then went on to create an unstoppable mathematical juggernaut to force everyone to accept the universal validity of Fowler's result. Fowler himself was not privy to Eddington's approach; in fact, as Chandra recollected, Fowler had assured the young man that he supported his conclusions. But he was not prepared to stand up in public and say that. To cap it all, Eddington hauled in Dirac's theory, rewritten in his own E-numbers. By 1935, in response to Chandra's work, Eddington had tied up his fundamental theory so tightly that Chandra's result absolutely had to be incorrect — otherwise the whole structure would come tumbling down. Eddington was playing for the highest of stakes. His fundamental theory was the culmination of his life's work, which he had always conceived of as the search for truth.

Rosenfeld was sympathetic to Chandra's predicament. Both he and Bohr, he wrote, "realise your troubles and feel very sorry for you."[17] He suggested forwarding Eddington's paper and Chandra's letters to Pauli, by then one of the leading figures in physics. Chandra did so, and Pauli replied that Eddington was completely wrong and appeared to be driven by the desire to mold physics to fit his theory. But he would not go so far as to make a public statement supporting Chandra. Chandra had also been corresponding with his friend McCrea. Writing five days after the fateful meeting, McCrea was still undecided. He offered to write a "(polite) discussion of E."s paper sentence by sentence" and send it to Eddington, but he never did. He added, "No matter how [Ed-

EDDINGTON'S DISCONTENTS ✦ 111

dington] may antagonise me on some points he never fails to entertain me, too."[18] This was probably no consolation to Chandra.

A pattern was beginning to emerge. Senior figures like Bohr, Fowler, Dirac, Rosenfeld, Pauli, and McCrea were sympathetic, but they were not prepared to get involved. Chandra kept running into colleagues who said, "I'm sorry. I'm sorry."[19] Chandra had helped Fowler prepare the second edition of his book on statistical mechanics, which he had originally read as a teenager in India. He could not help noticing that Fowler had added a note saying that Eddington considered the relativistic degeneracy formula to be incorrect, without any further comment.[20]

As for Milne, on January 13 he sent off a letter to the editors of *The Observatory*. In view of Eddington's comments, he had recalculated some of his results but this time had ignored relativistic degeneracy, he wrote, adding wryly, "whether it exists or not." He concluded, "Sir Arthur Eddington's investigations may now confer on our work a justification to which it is only accidentally entitled."[21] To Chandra it must have seemed as if the world had turned upside down, with Milne now agreeing with Eddington.

For Milne the situation had turned out surprisingly well. Ever since Chandra first proposed his theory, Milne had done his best to bully him into not making it public. He tried several times to persuade him not to publish his early papers, driving Chandra to declare that he had "lost all respect for Milne's objectivity." Chandra knew very well that his theory demolished Milne's. In fact, he had imagined that Eddington would be overjoyed by his theory because it supported his standard model, demonstrating that in certain cases stars remain perfect gases and do not develop incompressible cores. Eddington's rebuttal of Chandra's theory thus actually provided backing for Milne's research, putting Milne unexpectedly on the same side as Eddington. It was too good an opportunity to miss. Despite his and Eddington's ancient antagonism and his friendship with Chandra, Milne seemed almost gleeful to find himself on the side of the enemy.

Chandra poured out his troubles to Babuji:

> My last paper on stellar structure involved me desperately in the rival jealousies of Eddington-Milne-Jeans. I am taking care to be scrupulously polite to all of them. Fortunately Fowler and Bohr are on my side. I cannot properly go into these matters in a letter. It is the continuation of the his-

tory of my differences in attitude in results with Milne which has been brewing for the last three years. The explosion hasn't yet occurred. The whole thing would have been smoothed over had it not been for an awful howler Eddington has started. He (ASE) thinks that Pauli's principle is wrong! I do not know what he is up to. But the scientific politics in stellar structure in the triangular contest Eddington, Jeans, Milne strains me as I am in the middle and refuse to take anybody's side and E, M and J are all quarrelling with my work!!* [*I won't let them have any.] But Fowler supports me from the top![22]

Chandra was no longer a bystander caught between warring giants; the controversy had made him a giant too. Babuji replied immediately. "Your letter of 9th Feb has interested me very much," he wrote. "You are apparently standing at the median point, according to your diagram. You may let me know further developments."[23] It was not all bad news, however. Chandra was now giving three lectures a week on "special problems in astrophysics." He told Babuji proudly, "By the way I am the first Indian to have given University lectures at Cambridge — I mean no other had had the *opportunity*."[24] As far as he was concerned, it was a small step against the racism at Cambridge.

Toward the end of February, Milne wrote to Chandra reasserting his opposition: "Your marshalling of authorities such as Bohr, Pauli, Fowler, Wilson, etc., very impressive as it is, leaves me cold."[25] His comment was no doubt colored by his dislike for these scientists, who dismissed his own research as mere mathematical game-playing. For Chandra, Milne's antagonism was very hurtful. He often quoted those words of Milne's and, in the 1970s, even had the letter transcribed and sent to those who were interested in the controversy with Eddington.[26]

In India, Chandra's Uncle Raman was also proving to be a foe. Chandra told his father, "I have been told that CVR [Raman] has said recently that I 'am wasting my time in astrophysics and that there will be no astrophysicists miles near Bangalore.' I wonder if anything could be done to stop him talking about me. He can be assured anyway that I will not get so much as scores of [miles] near Bangalore."[27] By then Chandra was in Copenhagen, where he spent March and April conferring with Rosenfeld on the Eddington situation. He wrote again from there that Cambridge had just granted him fifty pounds to buy a Brunsviga 20 hand calculator, one of the best available. "I am good enough for Cambridge and they are sympathetic to my work and do not regard me as 'wasting my time,'" he complained to his father.[28]

Babuji was upset that Chandra was traveling so much, especially as India never seemed to be on his itinerary. "DON'T postpone your marriage too long," he wrote. "I do not think you ought to become an 'astrophysics' animal. Come over in x-mas week of 1935 and get married and 'let all intellectual ambition or work go a hang.'"[29]

A month later, Chandra dropped a bombshell:

> I had not written during the past weeks, partly as I was travelling and partly as I was thinking over my "personal matters" . . . I realised there that my relations with Lalitha was purely illusory and that I really had not known her at all. I had seen her in college but that was five years ago. I was simply deluding myself. So I have written to her breaking our understanding and this is final so far as this matter goes.[30]

Babuji replied that he was sorry to hear about Chandra's decision, but perhaps it was for the best. "There are ever so many girls, who are reading in the BA class, much younger — whose inclination for family life are better." Babuji, concerned about his son's state of mind, begged him to come home the following April for an extended visit. "I do not like your way of letters since the incident," he wrote. By "incident" he meant the run-in with Eddington, of course, and Chandra's increasingly cynical tone in the letters that followed. "The fact of it perhaps is that Miss D. L. has not sufficiently *known* and loved her father to be able to transfer that love to a husband in due course." (Lalitha had only hazy memories of her father, Captain Doraiswamy, who had been away fighting throughout the Great War and died a year after returning.) He concluded with some fatherly advice:

> If you stay in Madras for about 4 months you would be able to get to know some girls, then choose! The Indian way of love is far superior than the romantic attachments of the European: The wife you need is one who will be a comrade to you as well as see that your home interest will be attended to. That woman should be partly *potential* in her talent and not entirely *kinetic* is the best for married happiness.[31]

The last sentence was no doubt intended as a witty piece of physics punning.

The true story turned out to be quite different from what Chandra had written to his father, and rather mysterious. Two months after the dramatic exchange with Babuji, he mentioned to Rosenfeld, in the middle of a technical discussion on relativistic degeneracy, that "there is another 'interesting' fact. Since I came back [to Cambridge from Copenhagen] I have got engaged with an Indian girl who was here. Lalitha by name . . . And I shall try to make her happy at least."[32] What was going on? While Chandra continued to write to Babuji right into the summer of 1936, bewailing his broken romance, he was simultaneously telling his friend and confidant Rosenfeld that he had become engaged to a woman and — yet more mysterious — that she had been "here," in Cambridge. Perhaps Lalitha, sensing that Chandra had developed cold feet, had taken the first boat to England and turned up on his doorstep. Given the rigidity of Indian social mores, such a daring step would have required a great deal of courage and commitment. Lalitha was a freethinker, quite different from most other Indian women. She did what she pleased. In later years she often said that Chandra was the love of her life. Perhaps she felt it was worth making a dramatic move to ensure that everything worked out between them.

Meanwhile the dispute with Eddington simmered on. At a meeting of the RAS that May, Milne repeated the comment he had made in his letter to the editors of *The Observatory:* "Sir Arthur Eddington is justified in his argument that the 'relativistic' degeneracy formula leaves us with the old difficulty concerning white dwarfs, whereas the 'non-relativistic' degeneracy formula does not lead to such a difficulty."[33] Milne had clearly made a complete about-face and was now with Eddington. Each man, for his own reasons, needed to discredit Chandra's theory. Eddington rose for a minute and quipped, "I have only to add that my degeneracy formula seems to have raised a hornet's nest about me. In my opinion, however, I have not yet been stung!"[34] Chandra decided to keep quiet. "Eddington is behaving in the most obscurantist fashion," he wrote to his father. "Though Fowler, Dirac, Bohr and many others agree with me, it is still very annoying to have such strained scientific relations with E. Indeed my differences with him (and also Milne) has to a considerable extent made me unhappy during the past months."[35]

At the time, Chandra was talking with John von Neumann about how to extend his results on white dwarfs.[36] The thirty-two-year-old Hungarian-American mathematician, on leave from the Institute for Advanced Study in Princeton, was spending 1934 and 1935 at Cambridge. He often visited Chandra's rooms, and the two became close friends and confidants. They made a formidable pair in professional terms but a strange duo socially. Both were brilliant at numerical calculations and obsessed with detail. Von Neumann had the broader scientific base. Along with Edward Teller, he was a leading light in the galaxy of brilliant Hungarian émigrés. Fritz Houtermans, a German nuclear physicist who made important contributions to the problem of what makes stars shine, joked that the Hungarian scientists "were really visitors from Mars."[37] Von Neumann had been one of the first professors at the Institute for Advanced Study, along with Einstein, for whom it had been founded. He had virtually invented game theory — the field that investigates how to arrive at the best possible decision from the information available — and also did original work in computer science, economics, mathematics, and physics. Despite the strong anti-Semitic climate in Germany in the late 1920s, his decision to leave in 1929 was purely for professional advancement, as he often said. But he loved nightlife and was a regular fixture at parties, with an endless repertoire of scatological jokes, which he told hilariously. Chandra recalled with some bitterness that "Von Neumann was one of the people who privately supported me against Eddington. Of course all these people who supported me never came out publicly. It was all private."[38]

Chandra was becoming profoundly depressed. Milne wrote to him:

> I was much distressed to hear from you that you have discontinued dining regularly in Hall at Trinity owing to the damping influence of Eddington. I think we must all struggle against this feeling of discouragement. We have a responsibility to our scientific posterity not to give up our efforts merely owing to the dominating influence of one man; Posterity will put us all in our proper, humble, places, and will judge us chiefly for our courage in stating the truth as we see it . . . I do beg you not to discontinue dining in Hall. The older fellows don't like it and it destroys the college spirit.[39]

In June 1935, Chandra ventured to refer to the dispute publicly. In the third installment of his papers on the structure of stars, he mentioned in a discreet footnote that Eddington had questioned "the validity of the relativistic equation of state for degenerate matter which is still generally accepted."[40] He was afraid that the *Monthly Notices of the Royal Astronomical Society* would refuse to publish the paper because he had dared to be critical of "High Priest Eddington." So he took care to mention Jeans's work on radiation theory right at the beginning. Gleefully he explained to Rosenfeld why he had included Jeans:

> I have been nice to Jeans. It was all politico! . . . The R.A.S., I knew, would have been reluctant to publish it (my paper), but I knew also that to save

their faces they would send it to Jeans hoping to get a bad report! — then they can turn down the paper with no qualms! I was aware of this, and so I referred to Jeans very nicely. The trick worked! Jeans was emphatic about the publication! — It is all really sickening — these underhand methods, but what can one do?[41]

The dispute with Eddington had convinced Chandra that success in the world of science was not simply a matter of the brilliance of one's ideas; even more important, it seemed, was playing the game and scheming along with everyone else. Or perhaps he was becoming paranoid, fearing enemies everywhere. Maybe the *Monthly Notices of the Royal Astronomical Society* would have published his paper with or without this subterfuge.

This "politico," as he called it, forced Chandra to reconsider the scientific process. He mused in a letter to Babuji, "By the way I do not agree with your statement that original thinkers *must* disagree. The whole quantum mechanics has been built by the beautiful conformity of opinion of some of the greatest thinkers of our time — Dirac, Heisenberg, Bohr, Pauli. In astrophysics, the differences are of a 'political' nature." It was a very important point, and not fully clear to anyone why that should be the case. "Prejudices! Prejudices!" he went on:

> Eddington is simply stuck up! Take this piece of insolence. "If the worst comes to the worst we can believe your theory. You see I am looking [at] it from the point of view *not* of the stars but of Nature." As if the two are different. "Nature" simply means Eddington personified into a Lady!! Milne is more reasonable. He is a sport, and he is frank.[42]

Chandra's position was complicated by the fact that his allies and supporters were all abroad, in Germany and Copenhagen. The astrophysics community in England steadfastly refused to speak out on his behalf. But he was determined to continue the fight.

That same month, Raman offered him a job as a nuclear physicist. Chandra was furious. "He [Raman] first says that he will not have 'an astrophysicist miles near Bangalore' and now he wants to give me a clue that if I worked on nuclear physics, he might *tolerate* me — I suppose — some furlongs near his Institute," he wrote to his father.[43]

Chandra had not yet responded to Eddington in print. The many colleagues he consulted had advised against it, though they had acknowledged in private that they were on his side. Now, several months on, his confidence had returned. On June 7, he and the Danish physicist Christian Møller submitted a paper to the *Monthly Notices of the Royal Astronomical Society*. Cleverly, they used the same mathematical representation of relativity theory that Eddington had used. But they used it in the accepted manner, putting in the solutions from Dirac's equation in his theory of the electron based on relativity. Using this equation, they were able to deduce Chandra's theory, including the upper limit for the mass of a white dwarf. In fact, they did the calculation twice, using the solutions from Dirac's equation for electrons as standing waves and then for electrons as traveling or progressive waves.[44] They presented their result without comment, except to add pointedly that "we wish to state that we do not intend this note as a reply in any sense to Eddington's papers."[45]

It was like waving a red rag at a bull. Eddington fired off a note to Chandra, again stressing what he saw as Chandra's confusion between electrons as standing and as traveling waves, still refusing to concede that in the end it was a totally irrelevant issue. That November Eddington published a new paper in which he wrote that he had been led to "amplify [his] attack on the relativistic degeneracy formula by showing why I am unable to accept Møller and Chandrasekhar's proof."[46] They were, he wrote, violating Heisenberg's uncertainty principle — unthinkable in the world of physics. The two chose not to reply, at least in print.[47] "He is completely crazy," Chandra wrote to Rosenfeld. "I recently met H. N. Russell. He was frightfully enthusiastic. He finally whispered to me 'Out there, we don't believe in E.'!! ... So finally I do feel a bit relieved. Met Milne recently. He really hates Eddington."[48] So even Russell, that grand old man of astrophysics, was on Chandra's side, though only in private. Rosenfeld replied with a jocular pun: "The story of Eddington's degeneracy (if I may use such an ambiguous expression) takes the shape of the *Iliad*, with the various gods and heroes coming in."[49]

As for Milne, he had set off on a quixotic quest for a new theory of gravity to supersede Einstein's general theory of relativity, even though Eddington had twice confirmed Einstein's theory, in his eclipse expedition in 1919 and in his work on the effect of gravity on the light from white dwarfs in 1924, and no one else questioned the theory at all.[50] Milne's aim was to deduce a new law of gravity that would explain the shapes of galaxies. His book on the subject was published that year. As with his theory of stellar structure, he was full of bravado, declaring that his new cosmological theory "destroys at one swoop much of the recent work of Einstein, Jeans and Eddington. How far it will ultimately reach I cannot tell, but I believe it to be the only really new idea in relativity since Einstein. And the joke is that Einstein (who has written to me about it) doesn't really see the point!"[51] Einstein dismissed the theory out of hand: "Concerning Milne's ingenious reflections I can only say that I find their theoretical basis too narrow. From my point of view one cannot arrive, by way of theory, at any at least somewhat reliable results in the field of cosmology, if one makes no use of the principle of relativity."[52] Some of Milne's ideas were certainly original. But as Chandra put it, they would have been more fully appreciated had they been presented "in a more modest framework."[53]

Chandra wrote in exasperation to Rosenfeld:

> Oh! Milne's book — it is a bitter disappointment. Can you imagine the following appearing in a scientific book? "One can say if one pleases that we have found God in the universe — The physicist and the cosmogonist then need God only once to ensure creation. For the biologist, the world provides further opportunity for Divine planning . . . for a man, as more than a cosmologist, as more than a biologist, as possessing mind . . . an immortal soul . . . God is perhaps needed always."!!! Hardy asked me the question "Is Milne deteriorating?" . . . "It looks as though the question is at least pertinent" that was my answer! — You see in Cambridge one learns to use bad language *but* politely![54]

In mid-July Chandra went to Paris for a meeting of the International Astronomical Union. His reputation was soaring. "At the moment I am thoroughly enjoying myself in the company of some 'great' astronomers," he wrote to his father. "I have already had hearty discussions with H. N. Russell, H. Shapley, [Adriaan] van Maanen, [Bertil] Lindblad."[55] It was the first time that Chandra had ever flown, and he was thrilled. He defended his choice of transport against Babuji's dire warnings. Even Raman agreed that he should not take such risks.

The meeting was a splendid affair, attended by three hundred astronomers from thirty countries. Eddington too was present, and in fine form. The French president opened the proceedings, and there were sumptuous banquets and sightseeing tours. The gala dinner to celebrate July 14, Bastille Day, was held on the bottom platform of the Eiffel Tower. Afterward, guests climbed to the second platform to view the city of lights magically illuminated for France's national holiday.

McCrea was one of those who attended. The Commission for Stellar

Constitution, he wrote, "met under conditions described by its president Sir Arthur Stanley Eddington (who knows about these things) as similar to those obtaining in the inside of a star!"[56] Perhaps the blazing heat Eddington alluded to was a reference not just to the summer weather but to the fiery intensity of his remarks about Chandra, who was in the audience. McCrea reported that "Sir Arthur also reiterated his conviction of the validity of his recent conclusions concerning degenerate matter and their astrophysical significance."[57] He did not mention Chandra's contribution.

Chandra remembered the meeting with some pain. "Eddington gave an hour's talk, criticising my work extensively and making it into a joke. I sent a note to Russell [who was presiding], telling him I would wish to reply. Russell sent back a note saying, 'I prefer that you don't.' And so I had no chance even to reply; and accept the pitiful glances of the audience."[58] Once again, Eddington had humiliated him and he was not permitted to respond. Chandra soldiered on, perpetually frustrated in his yearning to engage in a verbal jousting match with Eddington. As always, the great man had had the last word.

7. American Adventure

CHANDRA'S STAR was rising; his brilliance was finally being recognized. Even before the momentous confrontation of January 11, Harlow Shapley had invited him to spend the following summer at Harvard. Chandra decided to accept, though he would continue to work at Cambridge for the remainder of his fellowship at Trinity, which was supposed to end in December 1936. He spent the end of the summer vacation of 1935 in Scotland, at a farmhouse in Kirkmichael, Perthshire, where he enjoyed "long walks and the fresh air, the cool breeze and the open space — all violently(!) invigorating."[1] He was engrossed in reading Dostoyevsky, Hardy, and Balzac. "Because of lack of 'paper and pencil,' I have been able to *think* more about stars, than I would have — as paper and pencil necessarily lands one in the intricacies of the calculation!" Full of new ideas, he was suddenly "homesick for Cambridge."[2] The world looked brighter.

Then in October, Shapley offered him a lectureship in cosmic physics at Harvard for the next three months. Chandra was rather apprehensive about taking up the appointment, but his friends assured him that it would be exciting. On November 29, 1935, he took the train from Cambridge to Liverpool and the following day boarded the *White Star Britannica* and set off on the eight-day journey to the New World. He spent the time absorbed in a volume of Ibsen's plays, a pleasant diversion from his usual preoccupation with the life of the stars.

The ship steamed into Boston harbor at midday on Sunday, December 8. To Chandra's surprise and delight, Shapley himself had come to the dock to meet him. That evening the director invited Chandra to tea at his home, where he was introduced to several distinguished astronomers. One was Gerard P. Kuiper, who had been observing white dwarfs. To Chandra's excitement, Kuiper said he had "obtained some remarkable confirmation of [Chandra's] last year's work on Highly Condensed Configuration."[3] Then thirty-three, Kuiper had won an impressive reputation for his work on binary star systems, particularly ones that included a white dwarf star. He went on to make a series of remarkable discoveries. In the late 1940s he discovered one of Uranus's moons, which was named Miranda, and the second moon of Neptune, Nereid. In 1951 he predicted the existence of a belt of chunks of ice, debris from the formation of the solar system, swinging through space just outside the orbit of Neptune. The first of these "dirty snowballs," in astronomer Fred Whipple's colorful phrase, was actually discovered in 1992 and is about a hundred miles across. Since then hundreds have been found, and astronomers suspect that there may be tens of thousands more. This icy swarm is called the Kuiper Belt.

For the first time, to his great pleasure, Chandra had an office of his own. He enjoyed the social life of the astrophysics community, including what he referred to in a letter to Babuji as the "bachelor-club(!)."[4] He complained that the scientists spent too much time talking shop, in contrast to high table at Cambridge, where this was frowned upon. "On the whole I prefer Cambridge — England . . . Here in spite of the extraordinary friendliness of the observatory people — I feel a stranger," he wrote.[5] Once again an outsider, he often quoted Nehru's words, "At home nowhere, a stranger everywhere."

He was still haunted by the shadow of January 11. In December he wrote to Balakrishnan, "As for my work, it is going on but unfortunately, some controversies with Eddington and Milne during the past year has upset my enthusiasm a great deal. . . . My controversy with Eddington has poisoned me and my peace quite considerably."[6] But there were also opportunities for fun and some mild revenge. At Christmas Shapley arranged a revue entitled *The Follies of 1935*. Chandra was asked to make a "scientific" contribution. "I took my 'revenge' on Eddington by telling some funny stories about him which were quite well appreciated," he told Babuji. No doubt Eddington's story about the technological importance of women's zippers was one. Chandra closed his letter in downbeat style: "There is nothing particularly new except that 1936 begins tomorrow!"[7]

Cambridge remained on his mind, though he was always ambivalent about the place where he had experienced both triumph and despair.

"My visit to Harvard has upset my plans and quite a considerable work must await me at Cambridge on my return. . . . One always misses something, though at Cambridge everything is so beautiful that one misses only — I mean one misses the least number of things — one does not in any case miss Cambridge!"[8] Thus he expressed the confusion in his mind.

The news from India was irritating. Once again Raman was making Chandra's life difficult. Despite sterling references from the great Cambridge mathematicians Hardy and Littlewood, Raman had rejected Chandra's old friend Chowla for a post. Chandra expressed his outrage to Babuji:

> Oh! I can't stand Raman and his intrigues. He believes himself omniscient, though it would make no difference to his beliefs(!) what the physicists here and in Europe think of him and his students. Raman and his intrigues [are] 'dirty double crossing' to use an Americanism . . . My desire to settle finally in India and be of some service to Indian Science seems to dwindle day by day.[9]

In the New World, Chandra was going from success to success. In February 1936, Otto Struve, the director of the Yerkes Observatory, part of the University of Chicago, invited him to give a lecture there. At the time, Yerkes was one of the most important observatories in the world, thanks entirely to Struve's dynamic leadership.

At thirty-eight, Struve was the most successful director Yerkes had ever had. Russian-born of German descent, he was the fourth generation of a distinguished family of astronomers. In 1916, after two years at Kharkov University, he had joined the Russian army and fought the Turks in the Caucasus. After Russia and Germany made peace, he returned to the university, only to be swept up in the revolution; he fought for the White Russians, the losing side. Wounded and seriously ill, he finally managed, after many trials and tribulations, to obtain a position at Yerkes through family connections. He completed his Ph.D. there in 1923. Along the way he established a reputation as an indefatigable researcher and went on to produce important papers in the field of stellar spectroscopy. In 1932 he was appointed director. The observatory's reputation had declined severely under his predecessor. Thanks to a no-nonsense managerial style that demanded a level of research excellence and productivity which he himself adhered to, he succeeded in reestablishing Yerkes' preeminence. Struve, with his

Prussian bearing and purposeful demeanor, his crystal-clear directives, and his vision of Yerkes as a great observatory, was a formidable character.

He was also greatly impressed by Chandra, and immediately after his lecture offered him a position as a research associate. Harvard too had offered Chandra a post, a three-year appointment as a member of its prestigious Society of Fellows, to commence in February 1937. That position would have enabled him to pursue his research full-time, but the Yerkes offer was more attractive. Struve had written:

> It seems to me that your brilliant theoretical work could be made even more valuable to astronomers if it could be combined with practical investigations by observers. This would require close cooperation between yourself and the astronomers at the great observatories of America and might lead to new types of investigations of a non-routine character that would be especially valuable in the present state of astrophysics.[10]

Kuiper, who had already agreed to go to Yerkes, played a part in convincing Chandra to join him. Strömgren too had been hired. It was all most exciting. Chandra and Strömgren would be instrumental in leading the development of theoretical astrophysics in the United States, which had been lagging behind Europe.

The problem was how to break the news to Babuji. Considering all the intrigues in India and the "underhanded dealings going on in Indian scientific circles, it may after all be best that I spend some time in America," Chandra wrote to him.[11] He also promised to return home for a visit before settling in America. As Chandra had feared, Babuji was deeply disappointed. He warned his son of the shallowness of American culture and reminded him that settling there might ruin his chances of becoming a Fellow of the Royal Society (FRS) and thus of winning a top position in India. He concluded with stern optimism: "I believe you will after all, return to India."[12]

Chandra went back to Cambridge in April. He had accepted the position at Yerkes, which he would take up on January 1 of the following year. "I had carefully considered the situation and Eddington fairly strongly advised me to accept [it]," he explained to Babuji. Perhaps Eddington relished the prospect of Chandra leaving England. The time in America had renewed Chandra's confidence. "FRS etc., are completely trivial," he continued exuberantly. "I may get it sometime within the next 5 or 10 years, but the chances are not renounced by my going to

America. In any case I am quite indifferent to it." As for Lalitha, "I do not think that marriage for me is an important consideration."[13]

Chandra was fully aware of the stark contrast between his welcome in America and his treatment in England. "On the whole I am convinced that America has been the first in recognising me sufficiently to consider me worthy of an annual salary with a definitively senior position in a university," he wrote to his father. "In India they are blissfully unaware of my existence, and in England . . . there is some prejudice in giving Indians a definite appointment, tho' at Oxford, they have now appointed Radu Krishnan."[14] It was indeed striking that the Cambridge heavyweights, Eddington and Fowler, had never considered offering Chandra a position, while in the United States universities as prestigious as Harvard and Chicago had done so without hesitation. Small wonder that when given the opportunity, he chose to go to America.

Chandra was also well aware that he was doing very little to raise the profile of science in India. Max Born spoke approvingly of Raman's efforts to promote science there. Chandra began to wonder if he had been wrong in his criticisms of his uncle. A deeply divisive rivalry had arisen between the two stars of Indian science, Raman and Saha. Having talked to Saha, Chandra wrote to Babuji of "his scheming, sneering attitude — an attitude which desires to destroy all things productive — an attitude which blurts out when in inviting me to come back to India he asks me to 'join in the politics.'"[15]

And why was Chandra stalling over his marriage? In India it was the custom for the eldest son to marry before his brothers. Chandra's hesitation made it difficult for Balakrishnan, for whom Babuji wanted to find a wife. He also wanted to find bridegrooms for his younger daughters. "I think I shall have to postpone the girls' marriages," he wrote in despair to Chandra. "I do not have any suitable offers — in fact, you know I have to go begging in the present state of social affairs in India, and I do not feel much up to it."[16]

Chandra kept at least part of his promise to Babuji: that July he sailed for India, arriving in Bombay in the full heat of August. It had been six years since he left. As soon as he arrived, he wrote to Lalitha. According to Chandra's official biographer, she had not heard from him for a year, though she had seen his picture in the paper and knew he was in India.[17] But in reality she had secretly visited him a year earlier in Cambridge. So perhaps the two had kept up a surreptitious correspondence — or perhaps, in his old age, telling the story to his biographer,

Chandra wanted to make things appear as respectable and aboveboard as possible and therefore did not mention the secret visit. In any case, the two met in Madras. When Chandra saw her, whatever fears he might have had — that she might be a burden in some way or perhaps hinder his work — melted away completely. They agreed that they had to marry, and as soon as possible. Together they walked along the Marina, discussing their future in America.

Chandra informed Babuji that there was no need for any of the traditional preliminaries, such as consulting an astrologer on the compatibility of their horoscopes. More important was to change his reservation on the ship back to England from a single to a double cabin. A marriage of love was deeply unconventional for India. The wedding, on September 11, was very simple, quite different from the traditional elaborate ceremony, which would have entailed vast expense for Babuji. Lalitha's unconventional family was very pleased. After a few happy days in Bangalore, the newlyweds sailed from Bombay on October 13. Chandra had been in India for just two months.

On their way to the United States the couple spent a month in Cambridge. Eddington and his sister, Winifred, invited them to tea. It was Lalitha's only encounter with the man who had caused her new husband such pain. He was all charm; her memories of the meeting are warm.

Eddington was just back from Harvard, where he had been invited as an honored and popular guest on the occasion of the university's tercentenary; the college had been established in 1636 by John Harvard, a graduate of Cambridge. The celebrations were splendid, beginning with a spectacular fireworks display on the Charles River on September 17. Half a million spectators lined the banks. The next morning, a distinguished audience of 17,000 took their seats in Harvard Yard, where the fledgling Continental Army had drilled in 1776 and George Washington had been headquartered. In a grand ceremony, sixty-two of the world's leading scholars were awarded honorary degrees, among them the psychologist Jean Piaget, the psychoanalyst Carl Jung, and Hardy and Eddington.[18] Harvard's dynamic young president, James Conant, read out Eddington's citation in ringing tones: "Sir Arthur Stanley Eddington: Doctor of Science. A student of the cosmos who peers within the atom and surveys the expanding universe, an expounder to the multitude of the poetry of modern science."[19] It was a fitting accolade for a lifetime's work.

Eddington lived up to his star billing. His subject for the Tercentenary Conference was white dwarf stars. He began his address with a witty barb directed at Ralph Fowler: "My colleague Fowler was in his youth a pure mathematician, and I am afraid he has never really recovered from this upbringing."[20] His criticism of Fowler, he said, was that he had not pursued his result further using relativity — by which he meant, of course, *his* brand of relativity. Had he done so, Eddington argued, Fowler would have understood that his nonrelativistic result was actually relativistic.

Next Eddington turned his guns on Chandra, who, he said,

> put the stars back in precisely the same difficulty from which Fowler had rescued them. The small stars could cool down all right and end their days as dark stars in a reasonable way. But above a certain mass . . . the star could never cool down, but must go on radiating and contracting until heaven knows what becomes of it. That did not worry Chandrasekhar; he seemed to like the stars to behave that way, and believes that is what really happens. But I felt the same objections as earlier to this stellar buffoonery; at least it was sufficient to rouse my suspicion that there must be something wrong with the physical formula used.[21]

He continued, using his favorite image: "I was not surprised to find that in announcing these conclusions I had put my foot in a hornet's nest; and I have had the physicists buzzing about my ears, but I don't think I have been stung yet. Anyhow, for the purposes of this lecture, I will assume that I haven't dropped a brick."[22] He went on to deliver a spellbinding lecture, giving his audience what they took to be the last word on the fate of the stars, conveyed with enormous authority, humor, and panache. He had proved beyond all doubt that he deserved his reputation as the greatest astrophysicist of the day, recognized as such by all his peers.

As an honored guest, Eddington was invited to stay with Harvard's recently retired president, A. V. Lowell. He also went to see the Red Sox play at Fenway Park with his Cambridge colleague Hardy, who shared his love of baseball.[23]

Back in Cambridge that November, Eddington published a new book, *Relativity Theory of Electrons and Protons,* the first detailed exposition of his fundamental theory. Chandra's work, he wrote dismissively, was a mere study in paradox. The ordinary degeneracy formula, he wrote, was first applied by R. H. Fowler [in 1926]. But for many years it has been discarded by astronomers in favour of a supposed relativistic degeneracy formula. The "relativistic" degeneracy formula appears to be without foundation. The difference is important in the theory of the evolution of white dwarf stars; and it was the paradoxical results of the "relativistic" formula, disclosed in an investigation by S. Chandrasekhar, which led me to examine its validity.[24]

Having thus disposed of Chandra and his theory, Eddington had put the heavens back in order, as far as he was concerned. Rosenfeld told Chandra, "I had a look through Eddington's new book. I find he is not only stupid, and irritatingly conceited, but most unfair especially toward you. But I suppose it is not worthwhile to come back on that subject!"[25]

By the end of 1936, Chandra and Lalitha had settled into their daily routine at Yerkes. The observatory is beautifully situated on the shores of Lake Geneva, with lawns in front and rambling woodland at the back, in the small township of Williams Bay, Wisconsin, eighty miles from Chicago. It was the perfect site for an observatory, well away from the smoke, haze, fog, and bright lights of the city. The couple were to live in this idyllic setting for twenty-seven years.

They quickly bought a big, old-fashioned wooden house only a few minutes' walk from the observatory. Built at the end of the nineteenth century by Edward Emerson Barnard, a famous observational astronomer at Yerkes, the two-story house provided ample room for a couple without children. The Chandrasekhars kept it intimidatingly clean, "very inhibiting to American graduate students whom the Chandrasekhars occasionally invited in for afternoon tea," with white rugs and plain white walls.[26]

Their days followed a regular pattern. Every morning they rose at seven. Chandra was in his office by nine. He usually walked home at lunchtime, stayed for half an hour before going back to his office, then returned home for dinner at six. In the afternoon Lalitha would read, join him in his office, or attend lectures. In the evening they went back to the observatory, where Lalitha read while Chandra worked for a couple more hours. Bedtime was usually at eleven or eleven-thirty. On Sundays they relaxed. "There is not a great deal of social life here," Chandra wrote to his father.[27] They spent time with the Strömgrens, the

Kuipers, and sometimes the Struves, drove around the countryside in Chandra's recently acquired secondhand Dodge, or went for picnics.

They were eager to start a family. Early in 1944 Lalitha discovered that she was pregnant at long last. Delighted, she and Chandra began making preparations. But to their grief, she miscarried. They continued to be utterly devoted to each other, but remained childless.

Raman by now had decided that the time had come to repair his relationship with his nephew, now a major player on the astronomical scene. He wrote to Chandra in humorously conciliatory terms: "You are right in deciding not to return to India except as a full professor. Your instinct for choosing Astrophysics for your special field was fundamentally right. I see papers on white dwarfs being solid neutrons, or double stars as cyclotrons, etc. etc. Apparently Astrophysics $\approx$ [approximately equals] nuclear physics."[28] The times had indeed changed. As Raman well knew, nuclear physics had begun to figure prominently in astrophysics, helping to solve the puzzle of where the Sun's energy came from.

In 1939 Chandra published a monograph entitled *An Introduction to the Study of Stellar Structure.* It was his farewell to white dwarfs. He had had enough of fighting with Eddington and of having his ideas misconstrued by the astronomical community. According to Eddington's intimate friend Charles Trimble, the great man's only comment on the book was a sardonic "How nice to have all the wrong things in one place."[29]

Chandra's second book, *The Principles of Stellar Dynamics,* came out in 1943. Eddington was asked to review it, and, predictably, he jumped at the opportunity to demolish Chandra once again. One of his criticisms was Chandra's excessive use of mathematics. Eddington's review was cutting:

> As a subject progresses the attractive simplicity of the early researches gives place to laborious elaboration. In the last three years, Dr. Chandrasekhar has been very active in the mathematical development of stellar dynamics. The trend of this work may be judged from the fact that one contribution alone contains more than 1800 numbered formulae. There is no denying that this heavy method of attack can be justified; but it leaves us with the depressing feeling that the subject which began thirty years ago as a joyous adventure has reached a stage of uninspiring ugliness.[30]

He had made a similar point in 1939, referring to both Fowler's and Chandra's work: "Almost as important as the elimination of actual error is the restoration of simplicity to a subject which had become highly and, as it now appears, unnecessarily complicated."[31] Chandra did use a lot of mathematical detail to buttress his argument. His goal was always to investigate the structure and dynamics of stars as systematically and rigorously as possible. It is also true that some of Chandra's pages are overburdened with mathematics. Nevertheless, his books marked significant advances. Every astrophysicist to this day has a copy of *An Introduction to the Study of Stellar Structure* on his desk, and for all its ponderous mathematics, *The Principles of Stellar Dynamics* was an invaluable contribution to the subject. Chandra shrugged off the review as "typically Eddington."[32]

Eddington and Chandra met once again — for the last time, as it turned out — at the Colloque International d'Astrophysique in Paris in July 1939. The theme was novae and white dwarfs. By then Chandra had been at Yerkes for more than two years and had gained confidence. In group photographs he stands out, handsome and aristocratic, in an immaculate suit, utterly unlike the rumpled Cambridge dons. Despite his time in the United States, he still spoke with a refined Cambridge accent, tinged with an Indian lilt.

There had already been signs of trouble to come between the two men. Initially Chandra was scheduled to speak on novae, white dwarfs, and planetary nebulae, while Eddington was supposed to address theories of white dwarfs.[33] Eddington informed Chandra in January that he assumed Chandra would stick to his assigned topic — Wolf-Rayet stars and novae — "because I have no ideas at all" on these subjects.[34] Wolf-Rayet stars, which are in an advanced stage of evolution, are extremely dense, hot, and massive (more than twenty-five times the mass of the Sun), and they eject very hot gases at high speeds. Chandra had conjectured that these stars could not pass directly to the white dwarf stage.[35] He himself, Eddington continued, would focus on planetary nebulae, since the "subject has been obscured by the prevalence of the 'relativistic degeneracy' heresy."[36] A planetary nebula is a colorful, glowing shell of gas ejected in the explosion of a star of relatively low mass, some four times the mass of the Sun. Astronomers already suspected that the burnt-out core left behind after the explosion was a white dwarf. In using the term "heresy," Eddington was throwing down the gauntlet. He added a veiled threat: "If we stick to our original subjects, I do not think we shall overlap, as if you refer to details of white dwarf theory at all, I expect we shall completely contradict one another!"[37]

The colloquium was sponsored by the Paris-based Fondation Singer-Polignac. Amos J. Shaler, a scientist from the Massachusetts Institute of Technology, was the secretary-general in astrophysics. The planning committee included Henri Mineur, an eminent astrophysicist at the Paris Observatory; Shapley, director of the Harvard Observatory; and Donald Menzel, also from Harvard Observatory and a former student of Russell's.

It transpired that Mineur, expecting to spark a debate, had deliberately put Eddington and Chandra together in the white dwarf session without clearing it with anyone. Shaler then tried to shift Eddington out of the white dwarf session to deal instead with cosmological issues, but he encountered fierce opposition from Eddington. Shaler wrote to Chandra, "Please do not think for a moment that you should abandon the white dwarfs, as a large number of the participants have voiced their desire to have you treat them rather than Sir Arthur. If you have any trouble with Sir Arthur, do not fail to 'pass the buck' back to us, as it is not fair to have you shoulder any such burden."[38] Here was ample evidence that Chandra's contemporaries now regarded him, not Eddington, as the authority on white dwarfs, from which Chandra could conclude that they also now accepted his discovery of the upper limit to a white dwarf's mass. No one, however, was yet prepared to accept the logical conclusion: that a burned-out star far more massive than the Sun would collapse completely to an unimaginably tiny and dense point.

A few days later, another letter came from Shaler: "I was very much afraid that Sir Arthur Eddington would put up a fight . . . I shall ask Dr. Mineur to do the dirty work right away. He started it anyway, by misinforming us in the first place, and he can damn well finish it."[39] Having brought about the unpleasant confrontation in the first place, Mineur had to take on the awkward task of insisting that Eddington switch sessions. It was an inauspicious beginning.

Despite his apprehension over the forthcoming confrontation, Chandra was exuberant about the opportunity to travel in considerable luxury to Paris. "My real object, however, is to get away from work — to be on vacation!" he wrote to Balakrishnan en route, adding, "I am leaving my wife behind in Yerkes."[40] Their finances did not allow Lalitha to accompany him.

On his way to Paris, Chandra stopped off at Cambridge and took dinner at high table. He was seated with Dirac, Eddington, and Maurice Pryce, a young Cambridge physicist who had just married Max Born's daughter. Years later, Chandra could still remember that dinner in vivid detail:

> Pryce expressed surprise at seeing me and asked me whether I would join them in discussion with Eddington after Hall on the matter of relativistic degeneracy. After Hall, we adjourned to Pryce's room in Neville's Court. The discussion began with Pryce trying to tell Eddington his version of Eddington's arguments against relativistic degeneracy, so that Eddington could be satisfied that he, Pryce, understood Eddington's arguments. After Pryce had completed his narration, Eddington remarked that Pryce's account was entirely fair and accurate, and asked, "What was the argument about?" [In other words, what was all the fuss about?] Pryce turned to Dirac and asked him, "Did you agree with any of the things I have said?" Dirac said, "No." And Pryce added, "I do not either." Eddington became very angry — in fact, it was the only occasion when I saw him really angry. He got up from his chair, walked back and forth, and said, "The matter is not for joking!" And he went on finding fault with Pryce's argument even though he had agreed with it a moment earlier. And for the next hour or so, it was Eddington's monologue. Next day, after Hall, Eddington came up to me and said that he was very disappointed that Dirac did not seem to understand the implications of his own relativity theory of the electron [that is, the importance of Dirac's spinors, which played a central role in Eddington's fundamental theory]. I did not assent or dissent with Eddington's remark but asked instead, "How much of your fundamental theory depends on your ideas on relativistic degeneracy?" He replied, "Why, all of it!" And since I did not react to that remark, he asked me why I had asked that question. My response was, "I am only sorry"; — not a polite remark to have made; but by that time I was really enraged with Eddington's supreme confidence in himself and in his own ideas.[41]

Thus Chandra forced Eddington to acknowledge that his fundamental theory depended entirely on his repudiation of relativistic degeneracy. If Chandra's theory was right, Eddington's whole intricately constructed theory — his life's work — would crumble.

Russell, the grand old man of American astronomy, presided over the Paris colloquium. At the conference luncheon, held at l'Hôtel de Ville, "all of the Paris great were there . . . at high table," Chandra remembered, adding glumly, "and I was way off in the corner somewhere."[42] In his opening remarks at the colloquium, Russell referred to Eddington's discovery that inside stars high-energy x-rays almost completely strip the atoms of their electrons, thereby increasing the volume available to squeeze in more atoms. The discovery had, he said, gone some way toward solving the mystery of why white dwarfs were so dense and had laid the groundwork for Fowler's results. Russell did not mention Chandra. He had led Chandra to believe that he was on his side, but he was not prepared to stand up for him in the presence of his friend Eddington. It was another slight for Chandra to add to his list.

Kuiper, the next speaker, introduced the subject of white dwarfs. Besides the three "classical" white dwarfs, o^2 Eridani B, Sirius B, and Van Maanen's Star,[43] he had personally identified sixteen more. As he reminded his audience, it was straightforward to calculate the mass of white dwarfs within a binary system and to deduce their radii from their luminosity and temperature.[44] Chandra's theory, he pointed out, was the only one that matched the observational evidence. Most white dwarfs were astoundingly small, little larger than Earth, one hundredth the size of the Sun. Furthermore, every white dwarf that had ever been found had a mass less than the Sun's, exactly as would be expected from Chandra's theory of relativistic degeneracy — that a white dwarf having more than 1.4 times the mass of the Sun would collapse. Fowler's theory of nonrelativistic degeneracy, on the other hand, predicted that the smaller the radius of a white dwarf, the greater its mass, totally contradicting what was observed.[45] If Fowler was right, we would expect to see white dwarfs of a thousand or more times the mass of the Sun.

Then Chandra rose. Despite all the wrangling about who would speak on what, he ignored his assigned topic and devoted his entire talk to white dwarfs. He began with an audacious statement: "It is now generally agreed that the equation of state appropriate for the discussion of the structure of white dwarfs is the degenerate form of the equation of state."[46] In other words, his own theory of white dwarfs was the correct one. The perfect gas law from the era before quantum theory was definitely not appropriate for white dwarfs. And by not including the theory of relativity, Fowler had missed out on the appropriate equation of state, even though he had used quantum theory. In the discussion that followed, Eddington restricted himself to disarmingly simple questions.

Chandra's lecture had been on a Friday; Eddington's was the next morning. "Our beliefs on Saturday must be different from our beliefs on Friday," he said, as Chandra later recalled.[47] He began by summarizing results he had published since 1935, without mentioning his fundamental theory. He stated firmly that the contraction of stars "to a diameter of a few kilometres, until, according to the theory of relativity, gravitation becomes too great for the radiation to escape, is not a fatal difficulty, but it is nevertheless surprising."[48] Once again he refused to accept that a star might keep collapsing until its gravity became so great that not even light could escape. Fowler, he said, had resolved the question, while Stoner and Anderson had reopened it.[49] He himself, after reexamining everything, had rejected the concept of an upper limit. Instead of crediting Chandra with the new equation of state, Eddington attributed it to Stoner and Anderson, who had simultaneously and independently discovered the equation but had never examined its ramifications or placed it on a rigorous mathematical basis, as Chandra had. Eddington added that it was impossible to devise an experimental test to verify whether Fowler's theory or the Stoner-Anderson theory was correct. In any case, it was unnecessary. Relativistic degeneracy was strictly a mathematical result with no astrophysical meaning — a *reductio ad absurdum*. However, further observational data would show whether modifications should be made to Fowler's theory, which predicted a mass for white dwarfs far greater than had been measured.

Though everyone was keen to bring the hostilities to an end, Kuiper was not going to let Eddington off the hook. Were there "any observational tests that would permit a choice between rival theories," he asked.[50] Kuiper himself, of course, had just provided one. Eddington replied that it was incorrect to say that there were rival theories and reiterated his claim that the Stoner-Anderson formula (Chandra's) existed only mathematically. Then Chandra leaped to the attack, arguing that he and Eddington differed "in their fundamental premises."[51]

Eddington suggested one area of compromise. Perhaps they could investigate some of the very small white dwarfs that Kuiper had found. If Fowler's theory was correct, they should be extremely massive, with extremely high densities and pressures. These "extreme cases," as Eddington called them, might lead to the formation of neutrons and a rock-hard neutron core, which would prevent such massive stars from collapsing.[52] According to the conference minutes, published in *The Observatory*, "The meeting closed with a moderately outspoken duel between Sir Arthur Eddington and Dr. Chandrasekhar concerning the correct formula for degenerate matter under stellar conditions. Opinion was divided as to the moral victory."[53] Chandra remembered the debate quite differently. As far as he was concerned, he was absolutely right. As he recalled, "Eddington and I really talked to each other in strong language."[54]

Just before leaving the conference, the two had a brief moment alone. Eddington attempted a reconciliation. "I am sorry if I hurt you this morning. I hope you are not angry with what I said." Chandra looked at him and replied, "You haven't changed your mind, have you!" It was a statement, not a question. "No," Eddington retorted. Seething with frustration, Chandra looked him straight in the eye and demanded, "What are you sorry about then?" and strode off. Eddington stood for a few moments, then shrugged his shoulders and went on his way.*[55]

Chandra sailed back to the United States on the *Normandy*. It was the ship's last voyage before being converted into a troop transport vessel. That September hostilities broke out. What had been known as the Great War was renamed World War I as people began to see that a second world war was looming.

Chandra wrote to Babuji to report on the Paris conference. It had been, he said "a relatively dull affair. I had a nerve racking controversy with Eddington who showed himself thoroughly opaque to reason. I had a trying time avoiding him, but he was equally after me — we were playing blind man's bluff!"[56] Babuji struggled to work out what had happened and what was going on in his son's mind. He had noticed a reference to Chandra in *Relativity Theory of Protons and Electrons*, in which Eddington had written, "The relativistic degeneracy formula appears to be without any foundation."[57] Babuji wrote back to his son:

> Dr. Krishnan was telling me that in [the conference summary in *The Observatory*], your words were construed as an *insult* to Sir Arthur Stanley Eddington. I wonder why you have acted so. There must be always some genteel geniality in our speech . . . (I am told that this was at the International Conference at Paris — where you said you played "hide and seek" with Sir Arthur Stanley Eddington.)[58]

* The story of Chandra and Eddington is to a great extent also the story of Sirius B. Appendix A brings that story up to date.

Dr. Kariamanikkam Srinivasa Krishnan was one of India's most distinguished physicists, whose work with Raman had been critical to the discovery of the Raman effect. He had also made important contributions to the study of the magnetic properties of crystals. Chandra had no choice but to reply to his father's criticism. "You have referred to my 'insulting' Eddington," he wrote:

> Actually the President of the Conference in Paris (H. N. Russell) referred to my restraint under extreme provocation . . . I have never been accused of using insulting language. Indeed, just today, I read an account of the Paris meeting, where my controversial discussion with Eddington is referred to as "stimulating." And in my "dynamics" where I find and prove conclusively Eddington's work to be totally wrong, I refer to the circumstance in footnotes and obscure corners. If I had followed the traditional methods (set by Eddington among others) I should have made a real capital out of his errors — I should have obtained greater publicity in that way. Actually competent mathematicians [such as] von Neumann [and] Heisenberg have congratulated me on my dynamics [and] have actually accused me of just this restraint in exposing the Eddington fraud![59]

In 1942 Dirac, Peierls, and Pryce at last published a paper that defended Chandra and was highly critical of Eddington's views on relativity and quantum mechanics.[60] They pointed to Eddington's adroit use of mathematics to achieve his own ends. "The issue is a little confused because Eddington's system of mechanics is in many respects completely different from quantum mechanics, and although Eddington's objection is to an alleged illogical practice in quantum mechanics he occasionally makes use of concepts that have no place there," they wrote.[61] The "alleged illogical practice" was Chandra's derivation of the relativistically degenerate equation of state, which Eddington claimed was incorrect because Chandra had represented electrons as standing waves.

One of Eddington's key criticisms of Chandra's theory was that Chandra assumed that electrons in a gas behave like free particles, not interacting with one another or with nuclei. He made that assumption to simplify a difficult mathematical calculation. Dirac, Peierls, and Pryce showed that allowing for electrons to interact with nuclei made no substantial difference to Chandra's results for white dwarfs. Their arguments turned on certain basic concepts of relativity theory, which Eddington did not agree with. In his determination to disprove Chandra's result on white dwarfs, Eddington had tried to force radical changes in the relativity and quantum theories.

Eddington replied to the paper by fudging the issue, maintaining that what was needed was a theory that incorporated relativity and took account of interactions among particles, instead of ignoring them as Chandra had done.[62] In the 1960s, scientists were finally able to calculate exactly the effect of interactions among the electrons in a relativistic degenerate gas. Far from disproving Chandra's theory, as Eddington had hoped, this calculation led to a fine-tuning of his equation of state.[63] Dirac, Peierls, and Pryce's paper was important and widely read, but in later years Chandra always chose to forget it. He continued to complain bitterly that the leading physicists had never come to his defense in his battle with Eddington.

Though Chandra was now well settled at Yerkes, Raman continued to pursue him. From time to time Chandra was tempted to return to India, but he was wary of the limited research positions available there and of the incessant internecine rivalries — he had had enough of that in Cambridge. When Raman offered Chandra the professorship that had previously been held by Saha at Allahabad University, Chandra turned it down. Struve wrote to Raman in Chandra's support: "I thought that you might perhaps be interested in learning what we, his astronomical associates in the United States, think of him and of his place in our science." He had, he said, invited Chandra to come to Yerkes because he was "one of the most brilliant astrophysicists in the world. Chandrasekhar is in my opinion the only person who can successfully undertake the task of developing the theoretical side of American astrophysics. While he is here, no one could be a more effective 'ambassador of good will' from India to the United States than Chandrasekhar."[64] This was high praise indeed, especially from as demanding a taskmaster as Struve.

Foiled in his plan, Raman proposed instead to nominate Chandra for a fellowship of the Royal Society. That was something he could do to benefit both Chandra and Indian science. He asked Chandra to suggest five names from whom he could choose someone to second him.[65] Chandra proposed Milne, Fowler, Eddington, Jeans, and Sir Edmund Whittaker, an applied mathematician at Edinburgh University. Raman chose Milne and requested supporting letters from Jeans, Fowler, Harry S. Plaskett (a Canadian astronomer who had known Chandra at Cambridge), and Eddington. Eddington wrote warmly of the fine work that Chandra had done in "dynamical astronomy" (but making no reference to stellar structure) and said they had been "such good friends ever since [Chandra's] first coming to Cambridge from India."[66] Chandra was elected a Fellow in March 1944.

Even though Raman had set the wheels in motion, in later years Chandra always downplayed his role. The reason seems to be a misunderstanding that took place in 1961, when Chandra was in India. Chandra never forgot a slight, even an imagined one. He walked into his uncle's office one day just as Raman was unwrapping Chandra's opus, *Hydrodynamic and Hydromagnetic Stability*. Raman remarked carelessly that "the only book of this size I have seen before is a novel by Anthony Trollope — absolute trash."[67] How had Chandra found the time to write such a tome? he asked, adding that he himself much preferred research. In fact, in the late 1920s, he said, his archrival in research had taken time off to write a book, and during that time Raman discovered the Raman effect and won a Nobel Prize. Infuriated, Chandra snapped that by that standard he had already lost four Nobels. Raman was furious. Of course he had not intended to insult Chandra — it was Trollope's book he had called trash, not Chandra's. Ramaseshan, Chandra's cousin and a distinguished scientist, tried to convince Chandra of Raman's good intentions. Chandra responded icily, "You have the privilege to hold on to your views."[68] Raman went so far as to write Chandra a note reiterating his high opinion of the book. He sent it to a forwarding address in India, but according to Ramaseshan, Chandra never received it. Others believe that he did but did not care. He had made up his mind about Raman.

Raman made several more attempts to right matters between them. Having already nominated Chandra to the Royal Society, in 1948 he proposed putting him forward for a Nobel Prize. Chandra replied courteously that he was most flattered. But astronomers were not eligible, he said, since at the time there was no Nobel Prize for discoveries in astronomy, in which he included astrophysics. In any case, his interests in physics leaned toward the "less fashionable."[69]

Upon Chandra's election to the Royal Society, Milne sent him a letter of congratulation. The contest had been close, he wrote; Chandra's name had not been as high as Milne had hoped. The aspect of Chandra's work that Milne had emphasized was his achievement in establishing theoretical astrophysics in America. Eddington, too, said Milne, was impressed by this and felt that it merited Chandra's election. "I thought you would be interested in this tribute to you from your and my ancient enemy," he concluded.[70]

For Chandra, the move to America was a new start in many ways. He had a new career at Yerkes, where he was universally respected and admired, and a new marriage. His election to the Royal Society — a signal honor for any British scientist and considerably more so for an Indian — gave him newfound confidence, as did the long-awaited public support of Dirac and others for his theory. He had every reason to put the feud with Eddington behind him.

8. An Era Ends

EDDINGTON, JEANS, MILNE, and Fowler were never again to be major players in Chandra's life, but the reverberations of his early humiliation never died away. By 1950, all four were dead, but even in death the dispute smoldered on.

Jeans died in 1946 at the age of sixty-nine. In an obituary written for the *Obituary Notices of the Royal Society*, Milne could not resist the opportunity to return to the burning controversies that had so marked their lives. Eddington, he noted, had never really come up with a convincing reply to Jeans's criticisms of his standard model, which resembled Milne's own misgivings. Eddington, arrogant to the end, simply "made debating replies, and stood by his results . . . Like many of us at the time [Jeans] partly fell under the spell of Eddington's investigations."[1] Perhaps the secret of Eddington's power was his sheer charisma. Milne described Jeans as someone who "thought of the world as a mathematician thinks of the world; he tidied it up in his imagination."[2] It was an apt description of Milne too. Chandra made a similar remark in his own obituary of Jeans, published in the journal *Science*. He also recalled that G. H. Hardy once told him that he had asked Eddington if he had ever bet on a horse. Eddington gave him one of his sardonic looks and replied, "Just one. I bet on a horse named Jeans." Hardy laughed and asked whether it had won. "Eddington, with his characteristic smile, replied with the one word, 'No.'"[3]

The dispute consumed all of them. Chandra was convinced that Eddington "effectively destroyed Milne."[4] "Milne's enormous originality," he recalled, was squandered on an obsessive desire to come up with a result "which would contradict Eddington."[5] Milne truly hated him.

Their wrangling sent him into the blackest gloom. Tragically, his persistent unhappiness had a devastating effect on his wife, Margaret. Already in a delicate mental state, she fell into a deep depression and suffered bouts of mental illness, which finally led to her suicide in 1938. Whenever Milne recalled those years, he "could hardly control his tears," Chandra recalled.[6]

After Margaret's death, Milne was overcome by great despair. The following year, hoping to cheer him up, Chandra invited his old friend to join him and Lalitha in Chicago. Together they traveled to Austin, Texas, and up into the Davis Mountains to attend the opening ceremonies for the McDonald Observatory, spectacularly located on a mountaintop plateau. There Chandra took a photograph of Milne, with his round boyish face and thick glasses, smiling broadly. The visit, it seemed, had had the intended effect.

Buoyed by his travels, Milne returned to England by steamer. On the voyage he met a woman named Beatrice Brevoort Renwick. Fascinating, beautiful, and strong-willed, she came from one of New York's grandest families. Her paternal grandfather was the celebrated architect James Renwick, who designed St Patrick's Cathedral on Fifth Avenue and the Smithsonian Institution in Washington, D.C. Milne and Beatrice fell head over heels in love and decided to marry. He was deliriously happy. His children were astounded that their usually reserved father was remarrying so quickly, and someone he had just met. Beatrice went back to the United States to sort out her affairs.

But by the end of May 1940, when she left New York to rejoin her fiancé, war had broken out in Europe. She sailed on an Italian ship because Italy had not yet entered the war. After landing in Genoa, Beatrice had to make her way overland to Saint-Malo on the coast of northern Brittany. The timing could not have been worse. Traveling through France, she found herself just one step ahead of the advancing German army. It was a terrifying journey. After many harrowing experiences, she and Milne were finally reunited in Oxford, and they married immediately. But Beatrice too suffered from depression: she never fully recovered from the trauma of her flight through France.

In 1939 Milne went back to work on munitions research, as he had during World War I, once again under the aegis of Archibald Hill. "To think of doing the same job for the same purpose 25 years later over again!" he wrote grimly to Chandra.[7] He updated his previous work, including theories of shell stability, calculations of the optimum distribution of antiaircraft guns, and analysis of the results of experiments on armor penetration. His achievements remain valuable to this day.

Chandra sent food parcels to the Milnes, who were eking out a meager existence in wartime London. In May 1944 Chandra was promoted to professor. Milne wrote to congratulate him and also asked after Lalitha, who was then expecting a baby. But that was the year of her miscarriage. Milne wrote, "My wife and I are deeply grieved to hear of the trouble and disappointment that befell your wife," when he heard the sad news.[8]

Both Milne and Beatrice suffered greatly during the war. He was exhausted by the long hours of war work and the Parkinsonian aftereffects of the encephalitis he had contracted in 1924. For Beatrice, who had moved from high-society New York to wartorn London in the midst of the Blitz, the strain was even worse, and she now had a one-year-old daughter to take care of. In the summer of 1944 a flying bomb destroyed their home. It was more than Beatrice could bear. She became deeply depressed and took her own life in December 1945.

Colleagues were dying, too. Milne was called upon to write obituaries first of Hardy and Jeans, then of his old friend Fowler. "Fowler's death was a great shock and grief to me," he told Chandra. "He had looked ashen and very old of late." Milne could not even attend the funeral because "travelling is so appalling nowadays."[9] He had great difficulty writing the obituary because his files had been lost when his house was destroyed. He had to send letters to colleagues asking for details of Fowler's personal life.

Typically, perhaps, of men of that time, Milne and Chandra's closeness seems to have been largely professional. Many years earlier, Milne had written to Chandra that he "learned later that you had brought a bride from India."[10] Though they were friends, Chandra had apparently never told Milne that he had married or even that he was considering it.

Sometimes Chandra wrote about Milne as if with a checklist in hand, alternating compliments and criticisms. Milne's backing of "Eddington's conclusion (now considered false)"[11] after January 11, 1935, and the letter he had written saying that Chandra's "marshalling of authorities such as Bohr, Pauli, Fowler, Wilson, etc., very impressive as it is, leaves me cold"[12] were etched in Chandra's memory. He also could not forget that Milne had tried to prevent him from publishing his 1932 paper showing that stars with a radiation pressure of more than 10 percent of their total pressure will collapse to an infinitely dense point. That was the paper that threatened to undermine Milne's theory, which was based on the premise that stars will invariably develop an incompressible core to keep from collapsing into nothingness. It was this clash that first gave Chandra doubts about Milne.

Milne's obsessive anger after his showdown with Eddington in 1929, when Eddington had publicly demolished his arguments at the Royal Astronomical Society, had, in Chandra's opinion, been "bound to have a detrimental affect on [his] scientific work."[13] Some of Milne's ex-students and colleagues disagreed. Thomas Cowling and Harry Plaskett, a contemporary of Milne's, insisted that Milne and Eddington remained good friends and that Milne's lack of success in later years was caused not so much by resentment as by exhaustion and problems in his personal life.

Chandra's answer was that the fact that they remained "personal friends [was] not contrary evidence. I can substantiate the role of 'evil genius' which Eddington played in Milne's life by innumerable quotations from his letters to me and what he has said to me."[14] Much the same applied to his own relationship with Eddington.

Students recalled Milne's passionate and involved lecture style. He would cover the blackboard with equations, a piece of chalk in each hand. "When he lectured at Cambridge he finished with himself and all his students out of breath (I can believe it!)," wrote Cowling, recalling a story that McCrae had told him.[15] Once, when Chandra was visiting Milne at Oxford, he had a strange premonition. When he appeared as usual at Milne's study at the Radcliffe Library, "I found that he had listed on the blackboard the names of the great mathematicians starting from Archimedes, through Newton, Hamilton, Lagrange, Laplace, Poincaré, and Einstein, and in the end his own name. I was shocked on seeing this and vaguely felt that Milne's scientific career must end in tragedy, as in fact it did."[16]

In September 1950 Milne was on his way to an astronomical meeting in Dublin. He had told Chandra how much he was looking forward to it and "how, after years of depression and many personal tragedies, he was at last gaining serenity and some poise."[17] He could not, however, pass up the opportunity to make a critical comment on Chandra's recent work on how light and heat travel through matter — radiative transfer. In a letter, he made the "now oft-quoted statement that I have the habit of 'choking the cat with the cream,'" Chandra remembered, in reference to his habit of swamping his books and papers with mathematical detail.[18]

But before he reached Dublin, Milne had a heart attack and died, at the age of fifty-four. Plaskett wrote in an obituary that Milne "died as he had lived, undefeated."[19] There were always tears in Chandra's eyes when he reminisced about the man who had been his first friend at Cambridge and who had suffered so much throughout his life. The photograph he took of Milne in the carefree summer of 1939 was his favorite picture of his "very great personal friend . . . the first and the best."[20] The jutting veins on Milne's temples and neck show his passionate intensity. But the laughter lines around his eyes show that he retained his love of life, despite his great personal suffering.

And what of Sir Arthur Eddington, the "evil genius" who drove Jeans out of academia and had such an impact on the lives of Milne and Chandra? For many years Eddington continued to work on his fundamental theory. "In fourteen years I have never had the smallest doubt that the direction which I took in 1928 was the one which leads to the unified relativity and quantum theory," he declared with supreme confidence.[21] His lectures at the Royal Astronomical Society on his theory were utterly spellbinding. The president of the RAS, Harold Knox-Shaw, described them as a "truly amazing performance," and even Jeans conceded that they were "remarkable."[22] Chandra was more caustic: "Eddington became cocksure of his views on the [fundamental theory], on relativistic degeneracy, on the formation of black holes, and, indeed, on his whole approach to the 'unification of quantum theory and relativity theory.'"[23]

Eddington was consumed by the pursuit of his fundamental theory. He described it in Bottom's words from Shakespeare's *A Midsummer Night's Dream* as "a most rare vision . . . It shall be called Bottom's dream, because it hath no bottom."[24] In the summer of 1928, when he was beginning to shape his theory, he visited the monument to the seventeenth-century astronomer and mystic Johannes Kepler, at Weil der Stadt in southern Germany. There he laid a wreath, an impromptu tribute to a man who, like himself, had been a "strange erratic genius, guided by a sense of mathematical form, an aesthetic instinct for the fitness of things."[25]

The most direct way of verifying a scientific theory is to test its predictions through experiment, but Eddington was unable to produce any experimentally verifiable predictions from his fundamental theory.

Even if a particular approach is unproven or unprovable, however, it may inspire others to more fruitful lines of thought, and Eddington's fundamental theory had that effect. In 1937 Dirac, the personification of the coldly rational, austere, otherworldly physicist, surprised everyone by publishing a note on the significance of the dimensionless numbers that can be constructed from the fundamental constants of physics, such as the seven constants of Eddington's theory. The fine-structure constant, for example, has no dimensions, such as length and time, even though each of the fundamental constants that comprise it have dimensions. This had "excited much interest in recent times," he wrote, adding that "Eddington's arguments [were] not always rigorous" but were nevertheless suggestive.[26] Even more out of character, Dirac had recently announced his wedding, prompting Bohr to remark, "Look what happens to people when they get married!"[27]

Chandra, who at that time had just settled at Yerkes, was much taken with Dirac's paper. He decided to write up his own thoughts on fundamental constants and send them to Dirac for his opinion. Later he paid homage to Eddington, "the originator of this group of ideas." Dirac and Eddington, he noted with respect, "were playing for high stakes."[28]

Dirac's paper had jogged Chandra's memory regarding "certain 'coincidences' which I had noticed some years ago, but which I have been hesitating to publish from the conviction that purely 'dimensional arguments' will not lead one very far."[29] Dirac's extension of Eddington's concepts inspired Chandra to take a fresh look at his own results, rewriting some in a way that pointed out their dependence on fundamental constants of nature. Developing the analogy with the fine-structure constant, he investigated how he could end up with the maximum mass for a stable white dwarf star by using a combination of the fundamental constants. Quantum considerations for the proper equation of state meant that he would have to use Planck's constant as well as the speed of light and the mass of the proton. The universal gravitational constant also had to be included.[30] Dirac changed one or two sentences in Chandra's letter and submitted it for publication.[31]

Chandra did not discuss fundamental constants again in any detail until his Nobel Prize lecture in 1983, when he argued that just as the conditions that make atoms stable can be expressed in terms of fundamental constants, the stars too should be described in this way.[32] In his foreword to a new edition of Eddington's *Internal Constitution of the Stars* that appeared in 1988, Chandra took Eddington to task on this point. According to Chandra, the main point of Eddington's famous discussion on how stars form — from the compression of interstellar gas until eventually "what 'happens' is the stars" — was that if a star is indeed a gas that obeys the perfect gas law, then it should be possible to express the ratio of its gas pressure and radiation pressure in terms of fundamental constants, namely the Planck constant, the speed of light, the universal gravitational constant, and the mass of the proton. That was surely the sort of argument Eddington would have loved, he said. But strangely, it was a point "on which Eddington is silent, Eddington does not isolate it — a surprising omission in view of his later preoccupations with material constants."[33] For once, in his foreword, Chandra made no mention of the events of January 1935.

By 1937 Eddington had become increasingly isolated. He made eccentric comments, such as that "one almost wishes that the Lorentz Transformation had never been invented; it is so continually applied without attention to its significance"[34] — which is akin to saying "one almost wishes" that Einstein's relativity theory "had never been invented." Throughout the early 1940s, the war years, Chandra maintained a cordial correspondence with Eddington and kept him up to date on his latest work in stellar dynamics. Eddington reported back that the atmosphere in Cambridge had changed: there were hardly any graduate students, and the Observatory Club had stopped meeting. He tried to make light of the terrible effects of food rationing, joking that it was so bad that "the regular dinners [at the RAS] have been dropped this session."[35] Chandra sent Eddington parcels of rice — evidence, according to Lalitha, that the two were close.

Life at Cambridge became worse and worse. In 1941 the town was bombed, causing some damage. Eddington wrote that the "Christmas mail from America [to] here was lost" and that everyone at Cambridge "missed a great many of the greetings from American astronomers." "My most severe deprivation is absence of the usual relaxations; I long to see a baseball match again!" he joked.[36] But he was still hard at work. He mentioned that his research on interstellar gas was going well. He remained cocksure: "In the case of Einstein and Dirac people have thought it worth while to penetrate the obscurity. I believe they will understand me all right when they realise they have got to do so — and when it becomes the fashion 'to explain Eddington.' "[37]

During the summer and autumn of 1944, Eddington put everything aside to complete his book on the fundamental theory. Perhaps he had a premonition. He was now sixty-one and had begun to suffer bad stomach pains. He persevered stoically, not even informing "his most intimate friend," Trimble, that he was unwell.[38]

Everyone thought he was overworking again. Doctors were busy coping with all the casualties from the war, and Eddington had to wait several weeks for an x-ray, which revealed a large tumor in his stomach. He had an emergency operation, but it was too late: the tumor was too far advanced. Eddington's last days were pitiful. He was put into a nursing home, where he received very little care. Winifred wrote to Shapley at Harvard, telling him of her brother's illness and adding, "My hands are so cold so do excuse the writing — we have to save fuel!"[39] Eddington died on November 22, 1944, leaving his book on the fundamental theory unfinished. Russell wrote in his obituary in the *Astrophysical Journal,* "The death of Sir Arthur Eddington deprives astrophysics of its most distinguished representative."[40]

Chandra and others turned out obituaries that naturally were full of praise, dealing only with Eddington's achievements, which were indeed formidable. Nevertheless, some questions of a more personal nature remained. What was at the root of his bristly, antagonistic nature and willingness to cause pain? Was it anything more than the insensitivity of a certain class of Englishman at that time? What of his personal life? Was there any significance in the fact that Winifred destroyed all his personal papers after his death? Such questions might seem irrelevant were it not for his treatment of Chandra, which threw everything into sharp focus.

On January 11, 1935, Eddington went beyond the lambasting he routinely meted out to Jeans and Milne, turning the full force of his venom on a young man from a culture in which the verbal pyrotechnics practiced in English academic and intellectual circles were completely alien. Yet Chandra's offense was that he had solved a critical problem that Eddington himself had posed. Chandra was taken completely by surprise. Until then Eddington had been willing to consider anything that would advance the study of astrophysics, but after that he went on to pervert the theory of relativity to support his view. Chandra's theory threatened the very basis of the fundamental theory, which had become an obsession. Scientists have been known to back down when shown to be indubitably mistaken. Why did Eddington refuse to let go?

As Chandra put it, "Eddington worked on the universe."[41] His dream was to put together, in one beautiful, harmonious package, a theory that would encompass all aspects of the universe, from atoms to stars. It was crucial that his estimate of the number of electrons and protons in the universe agree with the rate of expansion of the universe. But the numbers matched only if Chandra's theory of white dwarf stars was wrong and Fowler's correct. Eddington had fine-tuned his own version of Einstein's relativity theory to "prove" that Fowler's theory agreed with the standard interpretation of relativity, while Chandra's theory, he claimed, was merely an exercise in mathematics. Chandra's results and his refusal to admit defeat threatened to scupper everything Eddington believed in and strove for.

In Eddington's life personal relationships took a distant second to his research, to which he devoted himself with single-minded passion. Over the years his work provided him with friends, fame, travel, and considerable wealth. His books were bestsellers; when he died he left £47,000, a fortune in those days. By the time he was knighted he had become a household name. Yet what the physicist W. H. Williams called his "incommunicativeness," his stiff British diffidence, condemned him to a life of isolation. After his mother died, his only family was his sister, Winifred. His friend Trimble was the one person with whom he could relax. We have no idea how far their friendship went simply because in those days intimate male relationships were not just forbidden but illegal. If he was homosexual, as many of his colleagues were, he would have had to lead a life of concealment, a life on the edge. Perhaps when Chandra inadvertently undermined his fundamental theory it also threatened to topple his fragile psychological balance.

This was one tile in the mosaic of Eddington's brutal treatment of Chandra. But a healthy dose of colonialism, an essential element in the British attitude toward Indians, was involved as well. It was expressed with panache and subtlety, but it was there all the same. Herbert Compton, an old India hand, wrote in 1904:

There is no assimilation between black and white. They are, and always will remain, races foreign to one another in sentiment, sympathies, feelings, and habits. Between you and a native friend there is a great gulf which no intimacy can bridge — the gulf of race and custom. Amalgamation is utterly impossible in any but the most superficial sense, and affinity out of the question.[42]

Gilbert Slater spent the years from 1915 to 1921 at the University of Madras as its first professor of Indian economics. In his book about his experiences there, he describes the Madras Club:

> Some time about the end of last century a move was made to establish a club in which Europeans and Indians could mix on equal terms, and which, it was hoped, they would join in equal numbers. The club-house was built, and the club started, with every influential backing, as "The Cosmopolitan." But in the evening, after the day's work is over, men need to relax in the company of their own kind . . . While I was in Madras the Cosmopolitan was flourishing, but its membership was exclusively Indian. On one occasion I received and accepted an invitation from a member to dine with him there, but he gave me an ordinary European dinner, and I met none of the other members.[43]

Chandra recalled a story that Raman had told about a visit he made to Cambridge in the twenties. Rutherford was his host there. As they walked around, Raman noticed that all the young men seemed to be playing tennis. "Do they ever do any work?" he asked. Rutherford replied with a loud guffaw, "My dear Professor Raman, we don't want bookworms. We want governors for our empire."[44] Raman, taken aback, could do no more than laugh politely. Chandra recalled that "most of the people in academic circles were very conservative in Cambridge."[45] Dirac, Fowler, and Milne were not sympathetic to the cause of India's independence and he avoided the subject with them. But Chandra was always careful to mention the interest that British scientists took in Indian science. Important figures such as Eddington and Rutherford often visited India. Chandra pointed out that "Ramanujan would have died unknown, but for Hardy."[46] Left unsaid was that Eddington was far from a Hardy for Chandra. Moreover, Ramanujan never wanted to stay in England, whereas Chandra did. Chandra dreamed of becoming Lucasian Professor of Mathematics at Cambridge, following in the footsteps of Newton and Dirac.[47]

John Littlewood described an unpleasant episode in 1919, when the Trinity dons were discussing the issue of electing Ramanujan to a fellowship. He had already been elected a Fellow of the Royal Society, so election to Trinity should have been a foregone conclusion. Yet there was a great deal of opposition. One voter said bluntly that "he wasn't going to have a black man as Fellow."[48] Nevertheless, Ramanujan's application was successful simply because "you can't reject an F.R.S."[49] Fourteen years later, in 1933, such crude and overt racism had vanished. Chandra was elected a Fellow despite Fowler's pessimism. Yet no British university offered him a position, even though there were openings available. Nor did any of Chandra's colleagues at Trinity think of recommending him as a Fellow of the Royal Society. Chandra wrote to his father that he was the first Indian ever to teach a course at Cambridge, adding that there was "some prejudice giving Indians a definite appointment."[50]

Chandra also experienced racism in Chicago and in the Deep South, when he worked at the Aberdeen Proving Grounds during World War II. In New York the Barbizon Plaza, a first-class hotel, once refused the Chandrasekhars accommodation.[51] Unpleasant as this was, in some ways it was easier to deal with overt racism. At least you knew where you stood.

When Chandra wrote of Eddington as working on the universe, he meant the universe of the mind as well as the physical universe. Eddington had a dream that goes back to the very beginning of Western science in ancient Greece: of finding one theory that would explain all of nature. He thought that he had discovered this through pure mathematics.

Eddington described his books about the fundamental theory as "workshops," messy and cluttered with scrappy ideas waiting to be plucked out and developed — such as nature's curious choice of the number 137. His other books were "showrooms," displaying tested products.[52] Yet even the showrooms had some untidy corners, which the scientist might choose to ignore for the moment. In one of these corners was the question of white dwarfs. That was where Chandra got his start, tidying up Eddington's showroom.

Things might have been different had Eddington engaged in more rational debate. There are always byzantine machinations when an issue as fundamental as the nature of physical reality is under question. That issue was debated fiercely in 1905, when Einstein presented his special theory of relativity, and even more strenuously between 1926 and 1933, when a viable interpretation of quantum mechanics was being hammered out. It continues today. But what occurred on January 11, 1935, was singular in the annals of modern science.

Chandra's discovery might well have transformed and accelerated developments in both physics and astrophysics in the 1930s. Instead, Eddington's heavy-handed intervention lent weighty support to the conservative community astrophysicists, who steadfastly refused even to consider the idea that stars might collapse to nothing. As a result, Chandra's work was almost forgotten.

It was left to the physicists to take up the baton. They thrived on challenges and were happy to take chances, to construct apparently outrageous models of real phenomena replete with nitty-gritty detail, which most astrophysicists preferred to avoid. Coming to grips with the fundamental issue of what makes stars shine demanded an understanding of nuclear reactions, as Eddington for one had foreseen. Physicists were on the lookout for new applications for the burgeoning field of nuclear physics. The heavens furnished an untouched area in which to apply their theories. The quest to understand the universe was about to pass out of the hands of astrophysicists into the domain of the physicists.

PART II

Stars and Bombs

9. How Stars Shine

IN 1932, WHEN CHANDRA was still an eager young graduate student, physicists went neutron-crazy. This particle, electrically neutral and with a mass about the same as the proton's, was detected at the beginning of the year by James Chadwick, a young colleague of Rutherford's, at the Cavendish Laboratory in Cambridge. Physicists immediately seized upon the neutron as the panacea for all previous confusion about what the nucleus was made up of and began to formulate promising theories of how the nucleus is held together and how it decays. Nuclear physics was born. But Chandra, who was at the Cavendish at the time, and most other experienced astrophysicists never spotted that it might have any significance for their field. The Cavendish was the undisputed center for nuclear physics in England, but the schism between physicists and astrophysicists remained as wide as ever.

In fact, this remarkable advance in scientific knowledge would eventually have an enormous impact on astrophysics. The discovery of the neutron also helped open people's minds to the possibility that the demise of a star might be far more spectacular than anyone had so far dared imagine. Milne, for one, took an immediate interest in the debate on the formation of neutrons and their possible relation to the structure of stars. Perhaps, he suggested, electrons and protons form neutrons at the high temperatures and densities found inside stars; the energy liberated in this process might play an important part in how a star cools down. He did not, however, follow up on the subject.[1] Instead the physicists took the lead in applying new scientific developments to the study of the stars. It was the beginning of a long path of exploration that would lead eventually to the rediscovery of Chandra's theory and the realization that he had been right all along.

The discovery of the neutron made the study of the nucleus one of the most exciting areas of inquiry. Previously, scientists had believed that the nucleus was made up of protons and just enough electrons to give it a net positive charge, so as to balance the negative charge of the electrons revolving around it. The result was an electrically neutral atom. A conspicuous drawback to this model was that, according to Heisenberg's uncertainty principle, electrons confined in the nucleus would have more than enough energy to burst out.

But the new model of a nucleus made up of neutrons and protons also presented problems. What force could possibly bind the neutral neutrons and positive protons together? It required a considerable stretch to imagine any such force, because only particles with opposite charges were supposed to attract one another. Fortunately, one did not need to understand the nature of the nuclear force to begin speculating on how the newly discovered neutron might relate to the fate of stars. Its very existence was enough to spark the imagination of the maverick Swiss-born physicist Fritz Zwicky.

One of the photographs commemorating Albert Einstein's visit to the California Institute of Technology (Caltech) in January 1931 shows an impish man grinning joyously from the front row, two seats away from the president of Caltech, Robert A. Millikan, next to whom sits Einstein. Zwicky, one of the least popular members of the staff, had probably grabbed the chair for himself. Born in 1898, Zwicky completed his Ph.D. on crystal structure at Einstein's old university, the Federal Institute of Technology in Zurich.[2] He remained an ardent and patriotic Swiss citizen all his life, returning regularly to vote in his canton, Glarus.

Millikan, a Nobel laureate in physics, brought Zwicky to Caltech in 1925 to initiate a research program in crystallography. Then fifty-seven, Millikan was famous as a talent spotter. Between 1910 and 1921 he built up the physics department at the University of Chicago. Then in 1921, George Ellery Hale, one of the greatest fundraisers in the history of astronomy, invited Millikan to perform his magic at a small institute in Pasadena, which had recently been renamed the California Institute of Technology. In the years that followed, Millikan transformed Caltech into a mecca of science.

Zwicky — a self-proclaimed genius — had a reputation for being difficult as well as for spinning fantastic schemes. But Millikan was willing to take a chance on him. He put the word out that he was Zwicky's protector. Zwicky certainly needed one: almost immediately he got off on the wrong foot, viciously criticizing colleagues in the physics department and dismissing most of them as "spherical bastards" — that is, bastards any way you viewed them. Millikan suggested that Zwicky should switch to astronomy. He did so, and was soon antagonizing his new colleagues as much as his old. But his abilities as an astronomer could not be faulted, especially his observations of the astonishingly bright stars that astronomers had dubbed novae. Chandra attended a lecture Zwicky gave on that topic in Cambridge in the autumn of 1930.[3]

In 1931 Zwicky teamed up with Walter Baade, a new arrival from the Hamburg Observatory. Five years Zwicky's senior, Baade was an acclaimed observational astronomer. Baade and Zwicky admired each other's brilliance and were often seen in the Caltech corridors chatting away in German. Short, with a sharp, angular face and a beaklike nose, Baade described himself as having a voice like a barking dog. Talkative, stimulating, and ebullient, he had a certain quizzical look that conveyed the message "Tell me more, and with greater precision and clarity." His style was the diametric opposite of Zwicky's. Cracks in their friendship began to appear when Zwicky accused Baade of taking all the credit for their joint work.[4] Baade formally distanced himself from Zwicky in 1936 after an embarrassing incident with Cecilia Payne. She had sent them a paper she was planning to publish in which she questioned some of their results. Zwicky fired back a note heavily criticizing Payne's paper and calling her a fool, signed with Baade's name as well as his own.[5] Their final estrangement came during World War II, in a scene that showed Zwicky's hair-trigger personality all too well. Baade had been classified as an enemy alien, and one day, totally out of the blue, Zwicky accused him of being a Nazi and threatened to kill him. "They were a dangerous pair to put in the same room," recalled a colleague some years later.[6]

But in the 1930s Baade and Zwicky made a superb team. With a great leap of imagination, Zwicky saw that the newly discovered neutron might offer a way of understanding the stars. He conferred with Baade, and the two decided to present the idea at a meeting of the American Physical Society, to be held at Stanford University in December 1933. In a famous abstract for the *Physical Review,* they suggested a new way that stars might end their lives.

At the time they were trying to understand why novae suddenly flared up and became a hundred thousand times brighter, then gradually declined, sometimes over a period of months, toward their former brightness. Flare-ups had also been observed in stars so distant that they had to be enormously brighter than novae to be seen from Earth. Astronomers called them "giant novae," "exceptional novae," and "chief novae"; in their paper, Baade and Zwicky referred to them as "supernovae," a term they had coined in 1931.[7] A supernova can be 10,000 times brighter than a nova, as bright as all 200 billion stars in our galaxy put together. A supernova that appeared in A.D. 1054 was visible even during the day and could be seen for several weeks. Astronomers in China and Japan saw it, as did the Anasazi Indians of the southwestern United States. Observations made at Mount Wilson Observatory in 1921 and refined by Baade, among others, finally identified the cause. Astronomers observing the Crab Nebula, a huge cloud of luminous gas in the constellation Taurus, discovered that it was expanding, and worked out its rate of expansion. Winding back the clock, they calculated that the expansion had begun nine centuries earlier; the Crab Nebula was actually the wreckage of the massive star that "went supernova" in A.D. 1054.[8]

In a nova, the amount of mass lost from the exploding star is about .01 percent. In a supernova, however, the star can lose 90 percent of its mass in an explosion of unimaginable violence. The Crab Nebula is 40,000 trillion miles away, 60 trillion miles wide, and 7,500 times as luminous as the Sun, bright enough to be visible with a small telescope. Fragments of the exploded star are still rushing outward at more than 1,000 miles per second. From the measured distance between Earth and the Crab Nebula, we know that the star actually exploded in about 5500 B.C., some 6,500 years earlier than the Chinese observations. It took the light from the exploding star, traveling at 186,000 miles per second, 6,500 years to reach the eyes of the Chinese observers.[9]

As late as the 1960s, astronomers were content to assume that most stars turned into white dwarfs when they died, while the rest simply blew themselves to smithereens. Baade and Zwicky, however, dared to speculate otherwise.[10] Zwicky's brainstorm was to focus on what might be left behind after a supernova explosion. To find the answer he had to bring physics into the picture. With great verve, Zwicky and Baade proposed that in some cases a supernova explosion might leave a highly compact core, which they dubbed a neutron star — a star made entirely of neutrons. Perhaps "supernovae represent[ed] the transitions from ordinary stars into *neutron stars*."[11] They qualified their proposal with the phrase "with all due reserve," most likely at Baade's insistence. Zwicky was sure they were right. "Nobody else would have dared to say that at the time. I thought it was pure fantasy. How could such a thing be?" recalled Hans Bethe, a physicist at the University of Tübingen who became fascinated with nuclear physics after the discovery of the neutron.[12] What made the idea so daring was that Baade and Zwicky had thought of understanding some of the largest known objects, stars, in terms of a tiny part of the nucleus of the atom.

Between 1934 and 1939, Baade and Zwicky used data derived from observation of the stars to hone the neutron star hypothesis. Using state-of-the-art equipment at Mount Wilson Observatory, they made extensive measurements of the brightness of supernovae and produced wide-angle photographic surveys covering thousands of galaxies in a single exposure, taken at intervals of weeks. They then compared the photos, looking for any bright supernova that had suddenly flared up.[13] Zwicky was sure that the "tremendous *rate* of generation of energy in supernovae would require an explanation along new lines."[14] Indeed it did. His conclusion was that a supernova is the result of a massive star exploding with such violence that its core collapses into a highly condensed star a mere 12 miles across, the size of Manhattan, yet of nuclear density — the density at which neutrons and protons are packed into a nucleus, 100 trillion grams per cubic centimeter. This is 100 million times greater than the density inside a white dwarf. On Earth a teaspoonful of white dwarf matter would weigh more than 6 tons. The same tiny amount of neutron star matter would weigh a billion tons — probably enough to take it plunging through Earth. Over the next twenty years, Zwicky and Baade continued to seek out more observational evidence for supernovae.

Chandra was among those fired by Zwicky and Baade's discovery. As early as 1935, he mentioned the "super-nova phenomenon" in a paper published in the *Monthly Notices of the Royal Astronomical Society*.[15] He took a longer look in 1939, when he spoke on the "origin of the supernova phenomenon" at the Paris colloquium where he clashed so memorably with Eddington.[16] He tied supernovae in with his own work, suggesting that when a star that had exhausted its fuel was left with a mass more than 1.4 times that of the Sun — the Chandrasekhar limit — the outer layers of the star would collapse under the pressure of gravity, releasing an enormous amount of gravitational energy. This release would blast the star's outer layers into interstellar space, while the electrons and protons in the core would be squeezed to form an incredibly dense neutron core. Eddington, too, as we have seen, backed the scenario of a neutron star.

In the Soviet Union, Lev Landau had been speculating about highly condensed star cores since 1932, in terms that made more sense to physicists than to astrophysicists. He was fond of quipping that he had entered physics too late: "All the nice girls are already married, and all the nice problems are already solved."[17] Born in 1908, Landau was a child prodigy who could "scarcely remember not being able to differentiate and integrate."[18] He entered Baku University at the age of fourteen, then transferred to the more prestigious University of Leningrad two years later. After completing his Ph.D. at nineteen, he made a Grand Tour of the great European centers of theoretical physics. His brilliance was clear to everyone — even the hypercritical Pauli, whom he visited in 1929. Landau ended his tour at Niels Bohr's institute in 1930 and immediately began research on the foundations of quantum theory. Tall and lanky, with bushy black hair piled into an eccentric quiff, Landau was a formidable scientist whose interests spanned the whole spectrum of theoretical physics. He more than held his own wherever he went. In Copenhagen he refused to be ground down by Bohr's relentless arguing. Once, when Bohr was lambasting him after Landau had presented a lecture, Landau ambled nonchalantly over to an unoccupied front bench, where he stretched out and gazed quietly at the ceiling as Bohr stood over him arguing intensely, his face red and his veins bulging.

In 1931 Landau returned to the Soviet Union. A fervent Marxist and patriot, he was determined to bring the latest in theoretical physics to his country. He was immensely successful. Under his leadership the physics department at Kharkov Gorky State University in Ukraine became the best in the Soviet Union; in 1937 he became head of the theoretical division at the Institute for Physical Problems in Moscow. Landau also brought astrophysics to the Soviet Union. He suggested a radically new way of approaching the subject — by constructing stellar models using the "methods of theoretical physics."[19] Landau then made a brief and elegant calculation that enabled him to construct a model of a star made entirely of cold degenerate matter — highly compressed matter that no longer radiates light or heat. The model yielded the result that a star of this kind with a mass greater than 1.5 times the Sun's would be unstable and liable to collapse indefinitely to an infinitely dense point. He had rediscovered Chandra's upper limit to the mass of a white dwarf star (although his exact figure was slightly different) through the path of theoretical physics.

On reaching this conclusion, Landau immediately declared that "in reality" stars could not possibly "exhibit any such ridiculous tendencies."[20] Like Milne and the rest of the astrophysics community, Landau was convinced that at some point a star simply had to stop collapsing. He concluded that at that point it "must consist of a core of highly condensed matter, surrounded by matter in an ordinary state."[21] Along with almost everyone else, he accepted the view that as a star burned up the last of its fuel, it would somehow shed sufficient matter to bring its mass down below the upper limit and thus avoid disappearing into nothingness. As a physicist, Landau did not read astrophysical journals and so was unaware of Chandra's earlier work. For years to come, it was Landau's paper of 1932 rather than Chandra's article of 1931 that was credited with first identifying the limiting mass for white dwarfs, even though Landau himself did not initially believe his own discovery.

Six years later, Landau spun a variation on his earlier result. He proposed that somehow, deep inside a star where the pressure is extremely high, a core could develop, very much like a white dwarf. If the mass of the core exceeded the upper limit, this core would become unstable and collapse to nuclear density, 10^{14} grams per cubic centimeter. If Earth were compressed to this density, it would have a diameter of only 1,000 feet instead of 7,926 miles. The electrons and nuclei in the star would be squeezed together so tightly that they would form a neutron star. Like electrons, neutrons obey the Pauli exclusion principle and generate a degeneracy pressure that counteracts any further gravitational compression. Landau had worked out the *maximum* mass for a stable white dwarf, but he realized that the question to be asked was, What is the *minimum* mass needed to form a neutron core? Combining Newton's theory of gravity, as applied to planets, with quantum statistics, he estimated that the minimum must be around one one-thousandth the mass of the Sun.

Landau went on to develop a new theory of how stars shine, quite different from Eddington's, which claimed that subatomic processes such

as fusion were the mechanism. Landau suggested that when gas particles from the shell surrounding the neutron core fall into the core,
their energy of motion turns into heat, which we observe as starlight.

Landau was driven to this theoretical flight of fancy by more than just
a desire to push back the frontiers of knowledge. His personal survival
was at stake. Since 1931, Josef Stalin's regime had become more and
more brutal, and by 1937 life in the Soviet Union was becoming intolerable. The forced collectivization of agriculture led to the deaths of
some 7 million people, and the country's leading political, intellectual,
and military figures were being purged. As one of the country's high-
est-profile scientists, Landau found himself under scrutiny. The best
way to keep out of the dreaded political prisons, almost invariably a
one-way journey, was to be as visible as possible. Landau hoped that his
paper on neutron cores would do the trick.

Landau sent the paper to *Nature* and to a leading Russian physics pe-
riodical, and he also sent an English translation to Bohr, who was an
honorary member of the Soviet Academy of Sciences and highly respected in Soviet science circles. On hearing that Bohr was impressed
with the paper, Landau arranged for the leading Soviet newspaper,
Izvestia, to invite Bohr to publish a comment there.[22] But despite his efforts, Landau was arrested on 28 April 1938. He was taken off to
Butyrka, one of the worst prisons run by the NKVD, the forerunner of
the KGB. As always, the charges were trumped up and absurd. Although Jewish and a Marxist, Landau was accused of spying for Nazi
Germany. According to NKVD files, he had been overheard criticizing
the way the state organized scientific research as well as Stalin's reign
of terror. As far as the authorities were concerned, that was anti-Soviet
activity.

The Russian physicist Pyotr Kapitza, a discoverer of superconductivity, friend of Bohr and Rutherford, and creator of the Institute for Physical Problems in Moscow, where Landau had worked, wrote to Stalin
and to the people's commissar, Vyacheslav Molotov, on Landau's behalf. He pleaded that only Landau had the capacity to unravel the puzzle of superconductivity, which would greatly enhance the stature of
Soviet physics. Taking a chance, he even threatened to halt his own scientific research if Landau was not released. It worked. Landau was
freed a year later, but he had suffered terribly and emerged a broken
man. He would have been "unable to live for even another half year,"
he recalled in an interview in 1964 for the Soviet newspaper *Pravda*.

He reestablished himself as a scientist, but he never again put himself in a position that could be interpreted, even vaguely, as politically offensive.[23]

Across the world, two refugees from Communism, George Gamow and Edward Teller, were leading figures on the staff of George Washington University in Washington, D.C. Every spring, beginning in 1936, they organized a conference on theoretical physics at the Department of Terrestrial Magnetism at Washington's Carnegie Institution. The conference in March 1938 was on how stars shine. It marked a turning point, both in astrophysics and in nuclear physics.

Gamow, from Russia, and Teller, from Hungary, made a potent combination. Through the force of their personalities and the originality of their research, they were able to bring together key figures in astrophysics and physics who were at the cutting edge of the new field of nuclear physics. Bethe, who was to make the first breakthrough in cracking the mystery of stellar energy, declared that the meetings were the "most stimulating that I have ever attended."[24]

Chandra, naturally, was among the twenty scientists invited to the 1938 meeting. Bengt Strömgren, his old friend from Copenhagen and colleague at Yerkes, was also present. He was an astrophysicist who could speak the language of physicists, and Gamow regarded him as the "ace" of the meeting.[25] Gamow as coordinator sparked many lively exchanges in the informal discussions. There were no recorded minutes, and participants were not required to submit papers. In such a relaxed atmosphere, no one felt that they lost face in an interchange. It was different in every way from the meetings at the RAS. Einstein and Russell were invited, but neither attended. The subject matter was outside Einstein's range of interests. Russell's absence, however, was more surprising, for he had long been interested in the question of stellar energy. Perhaps speculations by fast-talking physicists such as Gamow were not to Russell's patrician tastes. One can imagine how he might have reacted to a "Gamowian" statement such as "A physicist doing stars feels happy as long as he dose not nead tuch the astronomical tables" (in Gamow's exuberant spelling).[26] It was now physicists, not astrophysicists, who dominated the field.

Gamow was one of the most colorful characters in physics. Léon Rosenfeld said of him, "How could anyone who has ever met Gamow forget his first meeting with him — a Slav giant, fair haired and speaking a very picturesque German; in fact he was picturesque in every-thing, even in his physics."[27] When he visited Bohr's institute in Copen-hagen, Gamow was a chief instigator of the high jinks that became a trademark there, with table-tennis tournaments (which Heisenberg usually won), movie showings (American Westerns were preferred), and mock theatrical portrayals of Bohr, Heisenberg, and Pauli, usually written, directed, and illustrated by Gamow.

Born in 1904 in Odessa, Gamow had always wanted to be a scientist. He spent a year at Novorossyshky University in Odessa in 1922, then moved to the better-equipped University of Leningrad. Landau was among his friends there. They formed a study group and rapidly got up to speed with modern physics. By 1926 Gamow had mastered general relativity and quantum physics. Looking for fresh areas to research, in 1928 he set off for Max Born's institute in Göttingen, where he decided to focus on nuclear physics. He set to work to investigate the nuclear processes that make stars shine.

It turned out to be a magical summer. Studying the literature, he dis-covered an intriguing article published by Rutherford the previous year, describing the extraordinary result of striking uranium atoms with alpha particles (the nuclei of helium atoms).[28] Physicists knew that uranium nuclei could emit an alpha particle in a form of radioactive decay. Rutherford found that in some cases alpha particles with twice the energy of those emitted by the uranium nucleus could not pene-trate it. The puzzle was how the low-energy alpha particles managed to escape from the uranium nucleus if particles with a higher energy could not get in. The uranium nucleus has ninety-two protons, and their combined positive charge acts as a strong barrier that the alpha particle would have to penetrate in order to escape. Gamow decided to attack this problem by applying quantum mechanics to the nucleus, something no one had previously thought of doing.

According to Newtonian physics, particles inside the nucleus are like billiard balls in a glass fruit bowl. In order to escape they have to generate enough speed to roll up and over the side of the bowl. But quantum theory states that particles can also behave like waves. An al-pha particle could therefore get through the barrier erected by the massive positive charge of the uranium nucleus much the way a light wave passes through the glass of the bowl or from water to air. Gamow dubbed this process "tunneling."[29] Alpha particles confined within a nucleus can tunnel their way out, rather like prisoners laboriously digging a tunnel to escape. Alpha particles hitting the nucleus from the outside, however, are much more likely to bounce off. Gamow was the first physicist to formulate a theory of nuclear phenomena; his astounding discovery later had an impact on verifying Chandra's theory.

The logical next step was for Gamow to apply his tunneling theory to the interior of stars to establish how stars shine. But this involved lengthy calculations, work he detested. He soon found a way around that problem.

During that incredibly creative summer, he met another maverick physicist, Fritz Houtermans. Of Dutch and Austrian parentage, Houtermans was half Jewish and a Communist, a lethal mixture in the 1930s. Increasingly uneasy about living in Nazi Germany, he emigrated to the Soviet Union in 1933, but in 1937, at the height of the Stalinist purges, he was arrested as a German spy. Houtermans spent two harrowing years in a series of NKVD prisons, including Butyrka, where Landau was interned. In 1939, when Hitler and Stalin made their pact, he was repatriated to Germany, barely alive. But he was immediately arrested by the Gestapo on suspicion of being a Russian spy and further mistreated. The Nobel laureate Max von Laue, a close friend of Einstein's and one of the very few German scientists willing to stand up to Hitler, arranged for Houtermans's release. He spent the war working in a private research laboratory.[30]

When Gamow met him in 1928, Houtermans had just received a Ph.D. in experimental physics at Göttingen. He convinced Gamow that he was Viennese at heart and that bohemian spirits such as they could do their best thinking in cafés. During their conversations, Houtermans confided that he really wanted to be a theoretical physicist. Gamow could not believe his luck. The two worked together in their favorite café, papers and coffee cups littering the table.

Houtermans introduced Gamow to Robert d'Escourt Atkinson, a Welshman who had just completed his Ph.D. at the same laboratory in Göttingen as Houtermans. He too dreamed of being a theoretical physicist. Gamow suggested that the two of them look into the problem he had isolated, of applying his tunneling theory to the interior of stars. Astrophysicists had not yet realized that stars are made up mostly of hydrogen, so Atkinson and Houtermans assumed they were made of a mixture of elements, including lithium, beryllium, boron, carbon, calcium, nitrogen, and oxygen. At the high temperatures inside a star, the nuclei of these elements have been stripped of all their electrons. The heat is so extreme that the nuclei are able to overcome the mutually repulsive force of their positive charges. They move close enough to tunnel into one another, causing nuclear reactions that release the enormous energy that powers stars. Houtermans and Atkinson concluded that nuclei with the fewest protons are the most likely to tunnel into one another; only a few at a time burn up completely (that is, convert into energy), which means that the nuclear reactions proceed slowly enough to enable a star like the Sun to shine for billions of years.

The day after they completed their paper on thermonuclear reactions in stars, Houtermans said, he "went for a walk with a pretty girl. As soon as it grew dark the stars came out, one after the another, in all their splendour. 'Don't they shine beautifully?' cried my companion. But I simply stuck my chest out and said proudly: 'I've known since yesterday why it is they shine.'"[31] Two years later, they married.

When he left Göttingen, Gamow visited Bohr's institute, then returned to Leningrad in 1931. He found the political situation there increasingly intolerable and resolved to leave as soon as possible. But international travel for Soviet citizens was severely restricted. In 1933 the prestigious Solvay Conference, in Brussels, happened to be on nuclear physics. Gamow organized an invitation for himself and his wife. They never returned to the Soviet Union. After the conference, Bohr invited him back to Copenhagen, where he met Teller.

Physically, the blond Slavic giant and the dark intense Hungarian with a prosthetic foot were polar opposites. Teller had had a bright future as a professor of physics in Germany. But that ended in 1933 with Hitler's rise to power, which was especially devastating for him, as he had witnessed the excesses of Communism and fascism in Hungary. Teller went straight to Bohr's institute at Copenhagen, where he and Gamow struck up a close friendship, both being birds of passage. After a year, Teller went to England, where he met Chandra, while Gamow moved on to George Washington University. In 1935 Gamow offered Teller a position there. They worked together closely and made several important discoveries in nuclear physics. Teller was greatly impressed by the ease with which Gamow came up with new ideas and liked to believe he could do the same.

By now physicists had homed in on the crucial problem: what was the chain of nuclear reactions that enabled stars to shine for billions of years? By 1932 it had been pretty well established that stars contained a high percentage of hydrogen, as Strömgren had reasoned. Everyone already suspected, as Eddington had proposed, that it was proton-proton fusion — burning hydrogen to make helium — that makes stars shine. Francis Aston had shown that nuclear fusion was possible in the laboratory. Atkinson returned to his original work with Houtermans and looked into several chains of nuclear reactions that might power stars like the Sun. But they were all fired by nuclei which, under further scrutiny, turned out to be unstable and would thus result in stars burning their fuel too quickly.

Another question was whether fusion could be initiated by thermonuclear reactions. That term had been coined by Gamow for the unimaginably violent reactions that occur when the intense heat and high density inside stars permit two protons to overcome their mutual electrical repulsion, smash into each other, and become glued, or fused, together.

During the summer of 1928 at Göttingen, Gamow, Atkinson, and Houtermans were the first to verify theoretically that nuclear fusion could power the stars. Ten years later Gamow's doctoral student Charles Critchfield investigated further. Before the all-important 1938 conference at the Carnegie Institution on how stars shine, Critchfield sent Hans Bethe a manuscript of his paper. Bethe made many comments and corrections, and it became known as the Bethe-Critchfield paper.[32]

Chandra recalled the impression Bethe made on everyone at the conference in Washington. He was well over six feet tall and massive in build, and he understood every aspect of nuclear physics. He grasped the essence of a problem immediately and "just bull-dozed in."[33] Born in 1906 in Strasbourg, Alsace-Lorraine, Bethe studied under Sommerfeld, Pauli, and Fermi. As the acknowledged expert on nuclear physics, Bethe was a natural for the conference. His three extremely detailed articles in the *Reviews of Modern Physics* were known as "Bethe's Bible." He did not thrive in chaotic situations where scientists have to create something "out of very little."[34] Like Chandra, he preferred to "take a subject in which the foundations have already been laid and then try to exploit them."[35]

Every important figure whose work touched on how stars shine was at the 1938 meeting in Washington. The astrophysicists were the first speakers. Chandra expounded on white dwarfs, and Strömgren on the hydrogen content of stars. Everyone agreed that fusion processes were what powered stars and caused them to shine, but "people were really at a loss as to what to do and what reactions to consider," Bethe recalled. He was struck "by the total ignorance which pervaded everybody at the meeting."[36] The astrophysicists were suddenly apprehensive that the "thermonuclear reactions in the stars were [being studied] by physicists — [Gamow] and Bethe, among them — because astronomers didn't know about nuclear physics. They were sitting on their astronomical things," as Gamow recalled in his colorful way.[37]

Bethe became hooked on the problem of how stars shine. For an entire month after the conference he obsessed over it and finally solved it, at least for stars of about the same mass as the Sun's. He worked fifteen hours a day, seven days a week, poring over his "Bible" and over papers by Gamow and Teller.[38] But one puzzle remained to be solved. What nuclear reactions powered stars that were more massive and more luminous than the Sun, such as Sirius A? After numerous attempts Bethe discovered the chain of nuclear reactions that occurred at the much higher internal temperatures of stars such as these, burning hydrogen and transforming it into helium. The key to all nuclear research concerning fusion was to find elements that would react at the exact temperature inside stars — about 10 million degrees kelvin — and would burn at a rate that agreed with their lifespan, around a million years.[39] Eddington had guessed that that temperature was necessary to initiate fusion, writing that in the early lifetime of a star there is a "small scale rehearsal of the great development which appears to set in at" about 10 million degrees kelvin. Chandra liked to point out that this was a case where astrophysicists could teach physicists something — about the necessary temperatures and conditions for nuclear reactions.[40]

Bethe published his research in an epoch-making paper in 1939.[41] But some big questions had still not been answered. What happened to a star once it had burned up its hydrogen? What sort of cataclysmic event could have generated the heaviest elements, such as uranium? In November 1938, Gamow wrote a paper summing up his thoughts on the issues that had been raised at the Washington meeting. The question of the neutron core, he concluded, was of merely academic interest. Stars would always shed enough mass to get below Chandra's upper limit.[42] But Bethe, unlike Gamow and the others, was prepared to speculate. He agreed that stars nine or ten times as massive as the Sun would be able to burn up enough fuel to get below Chandra's upper limit and expire as white dwarfs. But for more massive stars he proposed, on the basis of his new theory, that by the time such a star had completely burned up its fuel, it would be made up of heavier stable elements and would stop collapsing only "when a neutron core is formed." But, he concluded, "these questions obviously require much further investigation."[43]

During the summer of 1938, Bethe had met with J. Robert Oppenheimer and his group of brilliant young graduate and postdoctoral students at the University of California at Berkeley. The group promptly decided to turn their full attention to the fate of stars. Oppenheimer was born on New York City's Riverside Drive in 1904, into an affluent and cultured family. Two events in his early life uncannily presaged his future. In 1921, as a teenager, he made the usual Grand Tour of Europe, during which he visited the Joachimstal mines in northern Czechoslovakia. Among the many ores mined there was a dense, black, sticky material called pitchblende. In 1789 Martin Heinrich Klaproth, the first professor of chemistry at the University of Berlin, had extracted a grayish metal from this ore. The English astronomer William Herschel had recently discovered the planet Uranus, so Klaproth named his newly isolated metal "uranium." It was used mainly for coloring ceramics until Marie and Pierre Curie discovered that the ore contained the radioactive elements radium and polonium. As late as the 1940s, the Joachimstal mines were Europe's only source of uranium, whose use by then went far beyond decorative purposes.

The following year, to toughen up his six-foot-tall, gangly son, Oppenheimer's father sent him out west to a dude ranch in the Jemez Mountains northwest of Santa Fe, New Mexico, to camp and ride horses. It was there that, in 1922, the eighteen-year-old first saw the Los Alamos mesa. Both uranium and Los Alamos were to loom large in Oppenheimer's future career.

A man of great complexity, Oppenheimer hid deep insecurity behind a charismatic yet arrogant exterior. "Robert could make people feel they were fools. He made me [feel that], but I didn't mind. [Others] did," recalled Bethe.[44] After completing his Ph.D. with Max Born at Göttingen in 1927, Oppenheimer set out to learn the new quantum theory firsthand by sitting at the feet of the masters. Born, Heisenberg, and Pauli were much impressed with him, as was everyone else. Back in the United States, Oppenheimer rejected a job at Harvard and went instead to Berkeley "because it was a desert," with no school of theoretical physics. "I just thought it would be nice to start something."[45]

Berkeley was the ideal location for him. At Caltech, not far away, were superb theorists, such as Richard Chase Tolman, and the world's best cosmic ray group, led by Robert Millikan. Carl Anderson, who had discovered the positron, was a member of Millikan's group. Cosmic ray research was at the frontier of physics.[46] In the days before particle accelerators, cosmic rays provided the only way to study high-velocity particles, a subject on which Oppenheimer had become an expert. Jointly appointed to Caltech and Berkeley, Oppenheimer made use of the sagacious advice and criticism of the established physicists at Caltech while starting his own school of theoretical physics at Berkeley. Like Landau in Russia, Oppenheimer was determined to thrust his country to the forefront of physics research. He succeeded, creating the greatest school of theoretical physics ever seen in the United States.

Everyone agreed that Oppenheimer's lectures at Berkeley were difficult to follow. But they conveyed the beauty of the subject and the excitement of its development. He liked to cultivate the look of a tortured European intellectual with a cigarette or pipe hanging from his mouth, and he spoke in a velvety whisper with long dramatic pauses punctuating perfectly formed sentences. The word went out that Oppenheimer was the person to study with. A collection of brilliant young men found their way to him. Most were to leave their mark on postwar physics after earning their research spurs on the Manhattan Project.

In the summer of 1938, following the Washington meeting, Oppenheimer invited Bethe to address his students. But Bethe, looking over the sea of hungry eyes, decided not to draw their attention to the problem of what makes massive stars shine and what happens when they grow dark. Oppenheimer's students were like sharks who would pounce on whatever tidbits were offered, and Bethe wanted to hang on to his hard-won discoveries.[47] Seeking to cast a little light on Bethe's teasingly enigmatic remarks about the fate of massive stars, Oppenheimer thought of applying general relativity. He approached the general-relativity guru at Caltech, Richard Chase Tolman, whose book on statistical physics Chandra had devoured as a teenager. Tolman, a Massachusetts-born Quaker, was then fifty-seven. He had made significant contributions to Einstein's general theory of relativity and was regarded as one of the world's leading cosmologists. When Einstein visited Caltech in 1931, it was because Tolman was there. He was highly respected for the depth and breadth of his intellectual interests as well as his wisdom and urbanity. Oppenheimer frequently visited Tolman's home, but not simply to talk physics; he was said to be having an affair with Tolman's wife, Ruth.[48]

Tolman was intrigued by Landau's paper suggesting (as Chandra's earlier paper had) that a white dwarf star might implode if it exceeded a certain mass. Later, with the Manhattan Project, it was Tolman who first suggested implosion as a way to squeeze nuclear fuel so it would blow up.[49] Tolman suggested to Oppenheimer and Robert Serber, a young collaborator of Oppenheimer's, that they read Landau's paper and his more recent article, written earlier in 1938, which looked into the minimum mass needed for a neutron core to form inside a star. For Tolman, the concept of a neutron star was breathtaking, a classic conundrum that cried out for the general theory of relativity. At the density of a neutron star — 10 million times denser than a white dwarf — relativity predicts a warping of space around the star that cannot be described using Newton's theory of gravity.

Landau's 1938 paper seemed to be full of mistakes, so Oppenheimer and Serber set to work. They saw immediately that Landau had used Newton's theory of gravity instead of the general theory of relativity to estimate the lowest possible mass at which the neutron core could become as hard as rock and resist being crushed by gravity. "This figure appears to be wrong," they wrote bluntly.[50] (Had they known that Landau was languishing in an NKVD prison at that moment, they might have been more circumspect.) In their research they focused on the part played by nuclear forces. Inside the nucleus, neutrons and protons are pressed extremely close together, one ten-thousandth of a billionth of a centimeter (10^{-13} centimeters) apart. At this degree of closeness the nuclear force is much stronger than the pull of gravity, meaning that the neutron core could contain far more matter than Landau had estimated.

On the basis of the little that was known about nuclear forces at the time, Oppenheimer and Serber estimated the minimum mass for a stable neutron core at about one tenth the mass of the Sun, a hundred times bigger than Landau had thought. But if that figure was correct, Eddington's model for stars like the Sun would completely break down. If the Sun had a neutron core of 10 percent of its mass, it would give rise to effects that could be observed by astronomers, effects completely different from those we would expect to find if the Sun was, as previously assumed, a perfect gas. However, no such effects had been seen.

But what if the neutron core was a star in itself? Was there a *maximum* mass for a neutron star, as one would expect, given that neutrons function much like electrons and could generate a degeneracy pressure in the same way? And what would happen if the maximum mass was exceeded? Oppenheimer and two of his graduate students, George Volkoff and Hartland Snyder, looked into these questions. Combining general relativity with the embryonic nuclear physics, they began to lay the modern theoretical framework for our understanding of the fate of collapsed stars. Their work showed how massive stars could perish as either white dwarfs or neutron stars — or collapse completely.

Oppenheimer and Volkoff's landmark paper "On Massive Neutron Cores" was published in the *Physical Review* early in 1939.[51] With great verve and some elegantly minimalist calculations, they addressed the question that Landau's paper, and also Oppenheimer and Serber's, had failed to consider: what is the maximum mass for a stable neutron star? Working with the limited knowledge of developments in astrophysics provided by Tolman, they calculated that the maximum would be between one tenth and seven tenths the mass of the Sun. Such a star would be massively dense yet a mere 12 miles in diameter. Oppenheimer and Volkoff stopped at this point, putting aside the key question of what would happen if the maximum mass was exceeded.

There was one astrophysicist at Caltech who had been a physicist and could have filled them in more completely on developments in astrophysics: Zwicky. But the physicists generally avoided him — who needed to be called a "spherical bastard"? The sophisticated and cultured Oppenheimer kept his distance, and Zwicky returned the compliment. Four months after Oppenheimer and Volkoff's paper was published, Zwicky submitted an article to the *Physical Review* in which he tackled the problem of neutron stars head on, linking them to supernovae.[52] But he made no reference to his physicist neighbors.

Chandra recalled that during 1934 and 1935 he and von Neumann were jointly exploring topics similar to those Oppenheimer and Volkoff had considered.[53] The work on neutron stars must have been more von Neumann's than Chandra's. Chandra preferred to consider ideal systems rather than delve into the nitty-gritty of nuclear physics, and in the 1930s there was very little nuclear physics to consider. Nor was he eager to ponder the effects of general relativity, a subject he considered a graveyard for physicists, having seen what it did to Eddington and Milne. Eddington's perverse version of relativity, spun specifically to prove Chandra's theory wrong, and Milne's attempt to formulate a cosmological theory to supersede relativity did great harm to both their careers. "I did not want to go into that region; I just wanted to settle on safe grounds," Chandra recalled.[54] He refused to reformulate his conclusions along lines more familiar to physicists. Instead he continued to publish papers in the classical style of astrophysics, and most physicists ignored them.

Chandra was, of course, up to date with all the latest developments in nuclear physics. He read texts on the subject, taught a course on it at Yerkes, and was on hand for discussions at the Washington meeting. Bethe referred to conversations with Chandra, during and after the Washington meeting, in the paper he published in 1939. But Oppenheimer and his group did not take Chandra seriously; they thought he was not a "real" physicist.[55] In their paper on neutron stars, Oppenheimer and Volkoff praised Eddington for opening up the field of modern astrophysics and reviewed Landau's research, but they relegated Chandra's to a footnote.

Chandra had been in the United States since the end of 1936. Yet even after Oppenheimer's papers on astrophysics began to appear in 1938, he was not curious to find out more. He preferred to remain ensconced in the remote rural beauty of Williams Bay, in domestic peace and tranquillity. Like the great scientists he so admired — Einstein, Newton, Poincaré — Chandra preferred to deal in grand mathematical theories, leaving the details to others. The problem is that in the study of stars, the details are all-important. In his otherwise exhaustive monograph, *An Introduction to the Study of Stellar Structure*, written in 1938, the concluding chapter on why stars shine is curiously incomplete and unsatisfactory. Chandra was fully aware of Bethe's theory that it was the reactions of carbon with protons that makes stars shine, but he chose not to mention it. Instead, he discussed other theories that he already knew were wrong. He had, it seemed, absolutely no interest in collaborating with Oppenheimer's group. He simply abandoned his work on collapsed stars and let Oppenheimer steal a march on him.

No one yet had come up with an answer to the great question posed by Chandra's theory: what happens to a massive white dwarf star that exceeds the upper limit he had discovered? It was Oppenheimer who finally addressed the problem. He examined the fourth way that a star could expire: by collapsing and collapsing until it sucks in the very space around it. In considering this problem, he was attacking the very frontier of astrophysics, and he organized the onslaught as carefully as he would later plan the Manhattan Project. He broke the problem up into carefully defined units, then selected the right people to deal with each. With Serber and Volkoff, Oppenheimer had calculated the minimum mass necessary to form a stable neutron core and the maximum mass at which a neutron star could remain stable. The logical next step was to work out what would happen if this maximum mass was exceeded. Oppenheimer assigned this task to Hartland Snyder, who was adept at the difficult mathematics of general relativity.

Oppenheimer and Snyder set out to explore collapsing stars by using the general theory of relativity, with Tolman on hand to provide advice. They studied stars so massive that they are unable, even after they have burned up all their nuclear fuel, to blow off enough material to shrink below the mass limit and thus cannot form a stable neutron core. To simplify their calculations, Oppenheimer and Snyder limited themselves to a spherical cloud of gas. They traced the collapse of the cloud and explored what would happen when it shrank beyond what they called its gravitational radius, now known as the Schwarzschild radius.

Karl Schwarzschild, from Germany, was a highly theoretical physicist and a practical astrophysicist. When World War I broke out in 1914, he gave up the directorship of the Potsdam Observatory and, at the age of forty-one, volunteered for the German army. Late in 1915, while he was calculating artillery trajectories on the Russian front, he happened to read Einstein's published papers on general relativity. Einstein formulated his general theory using elegant mathematical hypotheses relating gravity to the geometry of space and time. But the equations governing the theory were extremely difficult to solve, so he assumed that broad approximations would be the best anyone would manage. Remarkably, Schwarzschild worked out the first exact solution almost immediately and used general relativity theory to investigate the effect of a spherical object on the space and time around it. Einstein was astonished that his equations could be solved so quickly and simply. He wrote to Schwarzschild that the work "appears to me splendid."[56] Schwarzschild would have made an ideal coworker for Einstein. Sadly, in March 1916 he returned to Berlin seriously ill with a rare skin disease contracted in the trenches and died soon afterward. Eddington fondly remembered meeting Schwarzschild at a scientific meeting in Hamburg in 1913. He participated in a motor race with "Schwarzschild and

Arthur Stanley Eddington at his writing table in 1931.

A portrait of Chandra as a Fellow of Trinity College, Cambridge, in 1934.
AIP Emilio Segrè Visual Archives

Marina Beach in Madras, where Chandra and his brothers often went to escape the summer heat. *Photograph by Arthur I. Miller*

Ralph H. Fowler, a "big man in every sense of that phrase," circa 1930.
Courtesy Royal Society

Chandra's favorite picture of his friend Edward Arthur Milne, taken by Chandra at the McDonald Observatory, Texas, in 1939. *Courtesy Meg Weston Smith*

James Jeans lecturing at the Royal Astronomical Society, March 1935, in the meeting room where Eddington and Chandra had clashed two months earlier.
Courtesy Royal Astronomical Society

G. H. Hardy, the "quintessential beautiful man," around 1930. *Master and Fellows of Trinity College, Cambridge*

Paul Dirac (left) and Werner Heisenberg looking dapper during Heisenberg's visit to Cambridge in 1933. *AIP Emilio Segrè Visual Archives*

Arnold Sommerfeld in 1927, a year before he met Chandra in Madras. *AIP Emilio Segrè Visual Archives*

The Solvay Conference in 1927. Front row, fifth from right, Albert Einstein. Middle row, from right, Niels Bohr, Max Born; fourth and fifth (behind Einstein), Arthur Holly Compton, Paul Dirac. Back row, second from right, Ralph Fowler, then Werner Heisenberg, Wolfgang Pauli. *AIP Emilio Segrè Visual Archives*

Left to right, Lev Landau, George Gamow, and Edward Teller with Bohr's sons Aage and Ernest in the garden of Bohr's institute in 1931. *Niels Bohr Archive*

The 1939 Paris conference on novae, supernovae, and white dwarfs. Front row, right to left: Serge Gaposchkin, Arthur Eddington, Amos Shaler, Henry Norris Russell, Cecelia Payne-Gaposchkin, Frederick Stratton. Back row, from right: Walter Baade, Chandra, Bengt Strömgren, Gerard Kuiper. *Courtesy Yerkes Observatory*

Charles Trimble with students at Christ's
Hospital School in 1923. *Courtesy Christ's
Hospital*

C. Subrahmanyan Ayyar (Babuji),
Chandra's caring father, in 1917.

Lalitha and Chandra in Madras in 1936, shortly
after their wedding.

Edward Teller (left)
and Enrico Fermi
at the University
of Chicago in 1951.
AIP Emilio Segrè
Visual Archives

J. Robert Oppenheimer
(second row, in jacket
and tie), with Richard
Feynman (on his left)
and Enrico Fermi (first
row, right), at a seminar
at Los Alamos in 1944.
AIP Niels Bohr Library

Stirling Colgate in
Los Alamos in 2003.
Photograph by
Arthur I. Miller

Einstein and Eddington in Eddington's garden at Cambridge University in 1932. Eddington recalled once having invitations for dinner with Einstein in London and Manchester on the same evening. He commented, "This is doubtless explainable by the Principle of Indeterminacy; still I hope on this occasion ψ [Einstein's wave function] will have a more condensed distribution." *Winifred Eddington*

Chandra giving his lecture "The Series Paintings of Claude Monet and the Landscape of General Relativity" at Presidency College, Madras, in 1992. Lalitha is at his side. This is the same room where he had courted her more than sixty years earlier. *Courtesy Physics Department, Presidency College, Chennai*

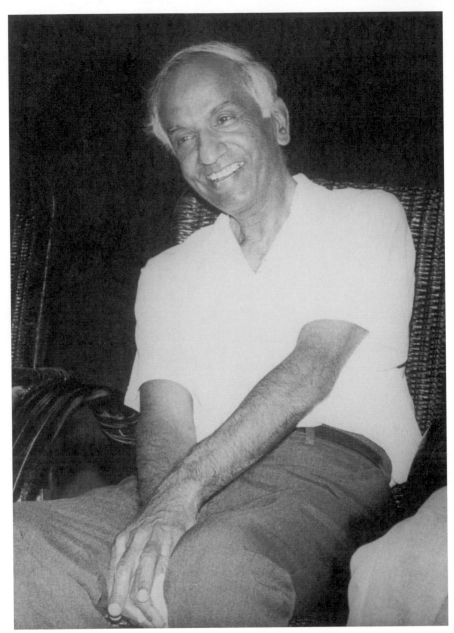

A relaxed Chandra at the Raman Institute, Bangalore, in 1982. *Courtesy G. Srinivasan*

five mad Englishmen," he wrote to his mother.[57] Eddington won, of course.

Schwarzschild was concerned only with the effects of a spherical object on the space and time around it. He did not discuss the fact that at a place very close to the center of the object, his solution gave the result of infinity. Later the place at which the solution becomes infinite was dubbed the Schwarzschild radius. It defines a region whose gravitational pull is so strong that nothing can escape, not even light; in other words, it takes an infinite amount of time for light to escape. Schwarzschild's solution becomes infinite once the object has dwindled enough to be bounded by its Schwarzschild radius. Every object has a Schwarzschild radius. If a star shrank until it was smaller than its Schwarzschild radius, its gravity would become so powerful that it would distort space around it so that nothing could escape. The same applies to the Sun, Earth, you, me, or a can of baked beans. The Sun's Schwarzschild radius is about 2 miles, as opposed to its actual radius of 432,000 miles. Human beings have a Schwarzschild radius about the size of a proton. The Schwarzschild radius of a proton, in turn, is unimaginably tiny. At the time, scientists considered all this theorizing pure science fiction. In reality, they thought, nothing could shrink to such minuscule proportions.

In 1926 Eddington had imagined what might happen to a star that shrank to the point that it distorted space around it. It would, he wrote, "produce so much curvature of [space] that space would close up around the star, leaving us outside (i.e. nowhere)."[58] He was describing what would happen to a star that shrank below its Schwarzschild radius: it would collapse for an infinity of time, becoming infinitely small and infinitely dense — in other words, a singularity.

Unlike their predecessors, however, Oppenheimer and Snyder took the idea of the Schwarzschild radius seriously and set about investigating whether it might apply to real stars. Their astonishing discovery was that under certain conditions a massive star could actually implode until it was smaller than its Schwarzschild radius, pulling space around it and disappearing from view. Scientists first used the term "Schwarzschild singularity" to refer to the Schwarzschild radius, because it takes an infinite amount of time for light to get out. But this turned out to be a misnomer. While the Schwarzschild radius defines the region, it is the imploding star that is the singularity, because it ends up infinitely small and infinitely dense.

Mysterious and unfathomable though this solution was, it was exactly what Chandra had described when he discovered what stars above a certain mass would do. His results had finally been substantiated.

Oppenheimer and Snyder's work threw up an intriguing puzzle, which the two were at a loss to explain. The viewpoint of an observer moving alongside a star that has shrunk almost to the size of its Schwarzschild radius and that of another observer viewing it from a distance seemed irreconcilable. The observer alongside sees matter sweeping inward faster and faster as it is tugged by the increasingly strong gravitational field near the horizon defined by its Schwarzschild radius. But when the star falls over the horizon, no more light can emerge from it. The distant observer, on the other hand, will report that the collapsing star "freezes" when it reaches the size of its Schwarzschild radius. Because of the viselike grip of the shrinking star's gravity, light takes longer and longer to escape, and eventually the star appears not to be moving any more, frozen in space and time. The "star thus tends to close itself off from any communication with a distant observer," they wrote.[59] They noted that the speed of the collapse might slow once they were able to take account of the behavior of highly compressed matter. But they could not see how to reconcile the two apparently conflicting views.

Ironically, in 1939, at the same time that Oppenheimer and Snyder were engaged in their groundbreaking work, Einstein himself was attempting to prove that "the 'Schwarzschild singularity' does not exist in physical reality [because] matter cannot be concentrated arbitrarily."[60] No object, he asserted, and certainly not a star, could ever be compressed to a size smaller than its Schwarzschild radius. Most uncharacteristically for Einstein, there were errors in his results. Like most others, he refused even to contemplate the idea that stars might collapse completely.

Dark clouds were gathering over Europe. Nazism and Communism were about to change the course of physics and astrophysics and the lives of its foremost figures in ways that could never have been imagined. Among those who fled Hitler and emigrated to the United States were Bethe, Bohr, Einstein, Fermi, and the eminent astrophysicist Martin Schwarzschild, Karl's son. Stalin's "gifts" to the United States included Gamow and Teller. For the first time science would be a major contributor to warfare. To his eternal chagrin, Oppenheimer found his vocation as the administrator of the biggest scientific enterprise of the twentieth century: the Manhattan Project, centered at Los Alamos. In 1942 he began full-scale recruiting for the project, whose purpose was to produce a weapon of such massive destructive power that it could obliterate an entire city — the atomic bomb. In his late thirties, Oppenheimer was one of the older scientists on the project; the average age was twenty-four. The Nobel laureate Luis Alvarez, who worked closely with Oppenheimer during the Manhattan Project, believed that if Oppenheimer had lived into the 1970s, when neutron stars were an established fact and the search for black holes was well under way, he would have received a Nobel Prize "for his contributions to astrophysics."[61]

Chandra was eager to play a role in defense work. Not yet a U.S. citizen in 1941, he initially had problems being cleared. His old friend from Cambridge, John von Neumann, was instrumental in resolving them. Chandra became a civilian consultant at the Army Ordnance Department's Ballistic Laboratory at the Aberdeen Proving Grounds in Maryland. There he worked on ballistics problems, some similar to those on which Milne was working in England.

Many scientists were about to disappear to the Southwest. Two years later, Chandra began receiving letters from P.O. Box 1663, Santa Fe, New Mexico. They were from someone he had not seen since the Washington meeting in 1938: Hans Bethe, now the head of the Theoretical Division at Los Alamos.

Bethe's first letter was dated March 20, 1944. "Johnny von Neumann has written to you on our behalf asking whether you would be interested in joining our project," he wrote.

> We are in great need of your help and we believe that you would be the best man to ask to take charge of certain calculations which have some loose connection with the work you have been doing at Aberdeen. We have no other person here who understands this type of problem with the exception of Johnny, who is here only a fraction of the time. You know that many of your friends are here.[62]

The bureaucratic red tape that had plagued Chandra when he arrived in London in 1930 also entangled him at Yerkes in 1944. Not until September 1944 did he receive clearance to go to Los Alamos. By then the Allied forces had fought their way off the Normandy beaches and were pouring across France toward Germany; there was talk that the war might be over by Christmas. Chandra was reluctant to move Lalitha away from their home. He was also deeply disturbed by the racial abuse he had suffered in Maryland: "Even at the Aberdeen Proving Grounds I used to encounter racial prejudice in many forms — in restaurants and places like that — and I was slightly afraid of driving down south," he recalled.[63] In the end he decided to decline Bethe's offer.

Chandra was well aware of what was going on at Los Alamos. He had kept abreast of developments in nuclear physics and knew of the "disappearance" of top researchers in the field. Bethe repeated his offer, agreeing that the war in Europe seemed to be over but wrote, in September 1944, that the "work here is likely to last at least until the end of the Pacific war and possibly longer."[64]

The first atomic bomb fell on Hiroshima on August 6, 1945; the second on Nagasaki three days later. In an interview in 1977 Chandra agreed that the first atomic bomb was unavoidable, but he felt that the "second did not seem to me necessary at all."[65] He went further. After first saying that the decision to drop an atomic bomb on Japan had no racist overtones, he added, "But it did occur to me that if the war in Germany had not been over, the bomb probably would not have been dropped on Germany." There was an awkward pause, broken by the interviewer, who said, "Well, back to your astrophysical work."[66]

When Chandra finally went to Los Alamos in the 1950s, it was to work on the next generation of nuclear weapons, whose explosive characteristics were remarkably similar to those of supernovae.

10. *Supernovae in the Heavens and on Earth*

L OOKING BACK on the turbulent times in Cambridge after January 11, 1935, Chandra wrote that he was astonished that he was "never completely crushed by these stalwarts. I decided that there was no good my fighting all the time, that I am right and that the others were all wrong. I would write a book. I would state my views. And I would leave the subject."[1] It took him less than a year to complete his first book on white dwarfs, *An Introduction to the Study of Stellar Structure,* though he got, he said, no "special exhilaration" from writing it.[2] The spark was gone, beaten out of him by Eddington's onslaught.

"This idea of permanently leaving a subject after writing a book did not occur in a natural way," he recalled. He "hated to relinquish the subject" but "had no choice." There was so much more to investigate about white dwarfs, such as what might happen if they rotated or pulsated. But he had had enough trouble: "Adhering to my point of view would only make more of a Don Quixote of me than I already appeared to be among the astronomical community."[3] He finally understood that he was just tilting at windmills. Eddington would never change his mind. So he gave up speculating on white dwarfs. Later he published *The Principles of Stellar Dynamics,* an elegant, highly sophisticated mathematical study dealing with the equilibrium of galaxies and solar systems. But he carefully avoided considering the structure of white dwarfs.[4]

In 1939 Gamow launched into a particularly intriguing area of speculation. His idea was to use Bethe's new nuclear physics, which analyzed how stars shine, to explore the turning point in the lives of stars, when they have burned all their fuel and are getting old.[5] He developed an evolutionary scenario which assumed that robust stars, those on the main sequence of the H-R diagram, stayed there, becoming brighter and moving up the main sequence until they had burned up all their hydrogen. Then, he postulated, they moved left — off the main sequence — contracting and becoming dimmer and eventually expiring as white dwarfs, except for those that exceeded Chandra's maximum mass, which would blow off mass in a supernova explosion and end up as neutron stars. Like Bethe, Gamow concluded that because heavier stars like red giants were not on the main sequence, they were obviously not fueled by hydrogen. He suggested that they were young stars, which shone because gravity was compressing the gas particles of which they were made. Eventually they would start to burn hydrogen, and would move onto the main sequence, like stars with a smaller mass, but would go on to explode into white dwarf fragments.

Chandra had a keen eye for lurking stability problems; his forte was identifying the exact point at which a star is likely to collapse. He spotted a flaw in Gamow's argument. The challenge was irresistible, and he decided to turn his attention to white dwarfs one last time.

During the billions of years when a star is in its prime of life, it burns the hydrogen in its core and converts it into helium, leaving behind a core of helium "ash." The hydrogen at the center burns first, because it is most concentrated there and the internal temperature is highest; this phase is called hydrogen-core burning. When the nuclei of two hydrogen atoms (protons) fuse, the fusion process produces deuterium, a variant of hydrogen with a single electron and therefore the same chemical constitution as hydrogen. The deuterium nucleus, with a neutron as well as a proton, is known as a deuteron. (Variants that have the same number of protons in their nuclei as the usual form of the element but differ in weight are called isotopes.) The deuterons then fuse with the protons (hydrogen nuclei), producing helium. Thus the original hydrogen core gradually turns to helium.

Gamow had cavalierly assumed that a star burns up *all* its hydrogen; Chandra realized that if so, the star must leave behind a heavier and heavier helium core as it moves in its evolutionary path across the H-R diagram. That scenario made him uneasy. He decided to look into the problem with a Brazilian postdoctoral student, Mario Schönberg, who was better informed than he about the nuclear physics of stars.

Chandra and Schönberg investigated whether this emerging helium core could remain stable, without collapsing or exploding, throughout the process of burning hydrogen. They came up with the surprising result that the helium core reached the maximum mass it could attain without collapsing when a mere 10 percent of the hydrogen fuel had been consumed.[6] That is known as the Chandrasekhar-Schönberg limit. But what happened next?

If the star's total mass was below the Chandrasekhar limit, it would expire as a white dwarf. If not, then — shades of Milne — it would have to eject enough mass to get below the upper limit, then pass through a series of degenerate cores, finally reaching a completely degenerate rock-hard state. One way the star could do this was by blowing up, that is, becoming a supernova. At this point Chandra and Schönberg did not consider the possibility that the star might collapse to nothingness. They assumed that somehow it had to become a white dwarf.

Their work showed that stars did not spend as long on the main sequence as everyone had thought. The next problem was to identify the intermediate stages before they began their demise, either as white dwarfs or neutron stars. To investigate this, astrophysicists began to use the H-R diagram to study stars in star clusters. Some regions of interstellar space called nebulae have enough gas particles that they can begin to swirl together and form stars. In these star nurseries, most of the stars are born at roughly the same time. As Strömgren discovered, a star's position on the H-R diagram is determined by its mass and hydrogen content, and a star's appearance changes as it burns up hydrogen. Thus, because the stars have different masses, we can use their positions on the H-R diagram to work out their ages and evolution. By the late 1940s, studies of star clusters revealed a "bridge" between the strip where giants are found and the main sequence. On the basis of theoretical discoveries such as Chandra and Schönberg's, astrophysicists were able to construct models of stars suggesting that after the Chandrasekhar-Schönberg limit has been reached, when 10 percent of a star's hydrogen has burned up, a new evolutionary process begins.[7]

Once the star has reached the Chandrasekhar-Schönberg limit, the helium core begins to cool. As radiation pressure decreases, gravity begins to dominate in the tug of war between the two opposing forces. This causes the helium core to contract, and the contraction causes it to heat up again, this time to such a degree that the shell of the original hydrogen core surrounding it starts to burn. This creates more helium, which falls onto the core, causing it to contract even more. As a result of this internal heating, the star's outer layer expands and the star becomes a red giant. Far from being young stars, as Gamow had suggested, red giants are actually stars in the final phase of life. They are ten to a hundred times larger than the Sun, about 100 million miles across (the radius of Earth's orbit around the Sun) and a hundred to a thousand times more luminous.

The Sun has enough fuel to burn its hydrogen core for another 10 billion years, as it has been doing for about 5 billion years; in another 5 billion years it will become a red giant, vaporizing the inner planets of the solar system, including Earth, and burning off the atmospheres of the rest. In red giants that have developed from stars no more than eight times the mass of the Sun, stellar winds begin to blow away the outer layers, which have cooled off and are loosely held, so that the mass decreases. Meanwhile the helium core continues to contract until eventually its temperature rises to 100 million degrees kelvin and it reignites. This stage is followed by successive paroxysms of ignition and reignition, creating a core of carbon and oxygen and blowing off great quantities of hot, glowing gases, which produce the brilliantly colored display we see as a planetary nebula.

The star's remaining carbon-oxygen core, covered with a thin surface layer of hydrogen and helium, is a white dwarf. Electron degeneracy pressure creates a rock-hard core, without enough mass to generate the gravity needed to make it collapse any further, so the burning caused by the core collapsing stops. It has burned up nearly all of its fuel, making it dim, but has undergone extreme contraction from the inward crush of gravity, making it hot. Its mass is less than the Chandrasekhar limit, which saves it from the fate of eternal collapse. The white dwarf cools down, the carbon and oxygen crystallize, and this mysterious object, which has opened the doors to all the wonders of the cosmos, expires as an actual diamond in the sky.[8]

During the war, Chandra later recalled, he could discuss astrophysics only with students, because the other scientists were working full-time on defense projects. Most of the students were women who could not be drafted and were free to continue their research.[9] What really interested him were shock waves and radiative transfer, the process by which radiation travels through the atmosphere of stars, where it is first absorbed by the atoms, then ejected or scattered.[10] In 1942 he gave a colloquium on these two related topics at the University of Chicago. The large audience included Eugene Wigner (another brilliant Hungarian émigré physicist and Dirac's brother-in-law) and Fermi. From 1944 on, he focused completely on radiative transfer.

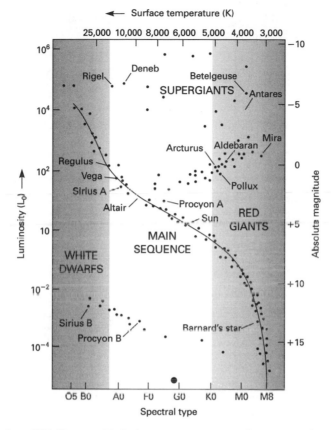

A modern H R diagram. Each dot represents a star whose spectral type and luminosity (given in units of the Sun's luminosity) have been measured.

This was the happiest time of Chandra's life: "I'm always nostalgic about that period . . . Before that, there was unhappiness connected with the Eddington controversy, and a certain diffidence whether I could make the grade in science."[11] He enjoyed rediscovering and improving on the results of the British school of high mathematical physics, which included such venerable nineteenth-century figures as Lord Rayleigh and George Stokes, who were heroes to Chandra. The two men had discovered many equations and effects that bear their names. By 1948 Chandra felt that he had reformulated the theory of radiative transfer into a "coherent mathematical picture with elegance and beauty."[12] He decided to wrap up the subject as it then stood and state the problems that still needed to be solved in a definitive volume that would be referred to for years to come. The book, another magnificent tome packed with equations, was *Radiative Transfer.*

For Chandra this was a satisfying and rewarding period in his teaching as well. At Cambridge his wardrobe had become extremely Anglicized; he carried this style over to Yerkes. In winter he wore a dark suit, in summer a light gray one (colleagues and students came to call the shade "Chandra gray") with a white shirt and conservative tie. Students remembered that he always spoke in complete sentences like a formal Englishman. But what they remembered best was his blackboard technique. With his long delicate fingers, he handled the chalk like a paintbrush, writing out equations in beautiful script. His lectures were so tightly organized and logical that his students later used their class notes as sources for their own teaching and research. He worked through the solutions of equations in great detail, rarely making a mistake. The only criticism ever made was that he emphasized mathematical methods rather than the basic physics in the equations. Colleagues joked that whatever the official title of Chandra's course, it should have been called "Mathematical Methods of . . ."[13] His principal criterion for selecting Ph.D. students was their skill in mathematics. They all considered him a superb teacher.

In 1945 Struve asked Chandra to be the associate managing editor of the *Astrophysical Journal;* in 1952 he took over as managing editor. He insisted on clarity of presentation. "A bad sentence cannot be corrected; it should never have been written," he often said.[14] For this job he began traveling to Chicago for one day a week, then two, to consult with the University of Chicago Press, the journal's publisher. He usually took several students with him in his car, and the trips became occasions for stimulating discussions. "We will meet at 6.00 A.M., plus or minus fifteen minutes," he would tell them, adding playfully, "I can be plus or minus. You should be plus."[15] Usually he spent Thursdays and Fridays on the Chicago campus.

By the late 1940s, Chandra had an international reputation. Around this time he began to initiate serious contact with the physicists at the main campus in Chicago. The astronomy department there had organized a series of courses for students who hoped to research their theses at Yerkes. Struve asked Chandra to teach the first course, entitled Topics in the Theory of Stellar Atmospheres (Astronomy 301), thus giving students their first chance to see this mysterious figure who spent his time in the wilds of Wisconsin. The course has become legendary thanks to an apocryphal story dreamed up by the president of the university, John T. Wilson, when he introduced Chandra before the 1975 Ryerson Lecture. Chandra liked the story so much that he never set the record straight.

Twice a week during the term of 1948 and 1949, so the story went, Chandra made the arduous three-hour drive from Yerkes to the Chicago campus to teach a course on astrophysics to just two students, Tsung-dao Lee and Chen Ning Yang. Yang was an instructor in the physics department (he had just completed his Ph.D. under Teller), Lee a graduate student of Fermi's, working on the internal structure of white dwarfs. It was Fermi who had suggested that Lee should consult with Chandra. The course, Wilson claimed, was the only physics class in history in which every student went on to win a Nobel Prize. Both Lee and Yang won it in 1957 for the overthrow of parity, their discovery that nature can in fact distinguish between left and right.[16] The professor would have to wait another twenty-six years.

The real story was ferreted out by Donald E. Osterbrock, an eminent astrophysicist turned historian of the subject and a former student of Chandra's.[17] In fact several students were registered for the course. Lee and Yang sat in, as did Fermi and other members of the physics department. As the term went on, the numbers dwindled. Lee and Yang stuck it out, along with six registered students, Osterbrock among them. One staff member who lasted almost to the end was Marcel Schein, a Belgian-born cosmic ray physicist. He looked very much like a well-known professional wrestler of the time called the Angel, so Chandra's students referred to Schein as "the Belgian Angel." He always sat in the front row and invariably went to sleep with his head hanging over the back of his chair, snoring loudly. Chandra would stand right over him with a look of disgust while he lectured, but Schein never woke up. In one class, however, Chandra really did teach only Lee and Yang. That winter the Midwest was hit by a severe blizzard that all but shut the area down. True to form and against everyone's advice to stay in Williams Bay, Chandra heroically made the trip by train and foot into Chicago, then hiked to the university. Only Lee and Yang showed up for that class, having been told by Fermi that the professor was on his way.

For Chandra the postwar years were ones of mixed happiness. Like an author in search of a plot, he was at a loss for what to do after completing his work on radiative transfer. He did not recover his stride until 1952, when he settled on his next topic, hydrodynamic and hydromagnetic stabilities, the study of flow; the terms refer to the transition from the normal flow of a fluid to highly turbulent, chaotic flow, as in whirlpools, and to the conditions for generating magnetic fields that maintain their shape. The mathematics involved in their study is extremely complex. Chandra probably chose this course because of his contacts with physicists on the Chicago campus. By now he had moved far away from his early focus on white dwarfs. Besides his editorial duties in Chicago, he attended colloquia in the physics department, which was packed with brilliant physicists, and also at the Institute for Nuclear Studies, headed by the charismatic Enrico Fermi.

During the war, Fermi had become curious about astrophysics and wanted to learn more. After hearing Chandra's lectures, he suggested that they collaborate on a problem. Fermi's method for learning something new was to talk to the experts rather than delve deeply into the relevant literature, then combine his new knowledge with his own deep understanding of physics. To Chandra, Fermi seemed "like a master musician who, when presented with a new piece of music, would play on sight with great conviction."[18] The difference in styles between Chandra and Fermi went beyond their approach to physics problems. In his immaculate suit and tie, Chandra was formal and aloof, while Fermi was outgoing, with sleeves rolled up and no tie; he often lunched on hot dogs with his students in a local diner. Short and stocky, he was a smiling, popular man with an Italian's ease of manner.

Fermi made it all look easy. He had a natural grasp of the big picture: he could work out almost any problem from basic equations. Research scientists have bookcases filled with advanced texts; Fermi had only one book in his office — a table of integrals (a handbook for mathematical calculations). In 1953 Chandra and Fermi produced two papers on the magnetic field in the arms of spiral galaxies. It was a brilliant collaboration. Chandra's forte was patiently seeking appropriate and elegant mathematical solutions. Fermi was just the opposite; he had no interest in elegance of mathematical presentation. Chandra was entranced with Fermi's command of physics and his "use of whatever means got him the answer."[19] Their first paper was written primarily by Fermi and contained few equations. The second, written by Chandra, was mostly equations and very few words.

Little by little, however, the atmosphere at Yerkes began to go sour for Chandra. From the very beginning, he had harbored a particular resentment against Struve. In later years he would look back on his time at Yerkes with the conviction that he had been unfairly treated. He learned that Kuiper and Strömgren, who had arrived at the same time, had been hired as assistant professors while he had entered as a mere research associate. They had been promoted and received salary raises and tenure the following year, while Struve appointed Chandra assistant professor with no increase in salary. It was six years before he was promoted to associate professor, and he was not made a full professor until 1944.

It turned out that Struve had a good reason for recruiting Chandra at a lower level. To build Yerkes into a first-rate astronomy department, he was recruiting brilliant young men regardless of nationality. But in introducing Chandra into the department, Struve was contending with entrenched racial prejudice. The University of Chicago was effectively segregated, a state of affairs supported by many powerful staff members, notably Henry G. Gale, the dean of physical sciences, and Gilbert Ames Bliss, a famous mathematician and head of the mathematics department. Both were strongly opposed to the appointment of someone who was "Indian and black."[20] Astronomy was linked to the mathematics department, which made Struve's job all the harder. Struve went over their heads to Robert Maynard Hutchins, the president of the university, who supported Struve's strategy of hiring young talent. This infuriated Gale, and he made Struve promise never to allow Chandra on the Chicago campus. Struve ignored him. Gale retired in 1940, but by then other professors had made a point of encouraging Chandra to spend more time in Chicago. Kuiper and Strömgren were much better known to American astronomers than Chandra, and there was nothing to hinder their progress. It was natural for Struve to hire them as assistant professors. In Chandra's case, Struve had to work around Gale and Bliss. It caused less upheaval to appoint him at a more junior level.

Russell and Shapley had warned Chandra about the racial tensions in Chicago and had advised him to accept the Harvard position. Shapley feared that Chandra's political views would be unacceptable in Chicago's rather conservative political atmosphere. Knowing about Chandra's visit to the Soviet Union in 1934, he even told Struve (quite falsely) that Chandra was a Communist. Many people interpreted Chandra's strong reactions to the racism he encountered in England as radical politics. But Hutchins, a man of extremely liberal political views, made it clear to Struve that he was interested only in whether Chandra was the best person for the position. Tall and handsome,

Hutchins was a born leader with great charisma. He had become president in 1929, when he was just thirty. Young though he was, he had already proved himself a dynamic administrator as dean of the law school at Yale. He served as president until 1945, and was then chancellor until his resignation in 1951. He was directly responsible for the university's meteoric rise.

Chandra did not learn of Hutchins's courageous stance until the 1960s, at a time when there was a violent reaction against desegregation. In a public lecture, Hutchins claimed that even the hallowed halls of high academia had had their share of racist incidents, and he cited two incidents that had occurred at the university while he was president. One was in the medical school; the other was the opposition by the head of a certain science department to the appointment of a theoretical astronomer because he was Indian and black. The speech made waves across the nation.

Chandra and Hutchins maintained close contact over the years. When Hutchins announced his retirement, Chandra dashed off a note to him saying that his resignation was a "disaster for the University."[21] Hutchins replied with warmth, "My leaving is not a disaster for the University. You wait and you'll see. I am very proud that I helped bring you here."[22] At his retirement ball, students flocked around him to shake hands. Such popularity was unusual for a university president then, and it still is today.

In 1946 Chandra was offered a professorship at Princeton to succeed Russell, at double his Chicago salary, and he accepted with alacrity. Hutchins persuaded him to change his mind. While it would be a great honor to succeed someone like Russell, he argued, would it not be more advantageous to initiate his own research program instead of continuing someone else's? To illustrate his point, Hutchins asked Chandra to name Lord Kelvin's successor at Glasgow University. Chandra could not. Hutchins then offered Chandra a distinguished service professorship at a salary that matched Princeton's offer. Chandra accepted the honor and the sorely needed salary boost. Six years later, he was promoted again, to Morton D. Hull Distinguished Service Professor.

In 1951 Hutchins invited Strömgren, who had returned to Copenhagen before the war, to take over as director of Yerkes. Strömgren believed curriculum reform to be long overdue at Yerkes. For well over a decade Chandra had taught the theoretical courses himself. He would usually teach a course for three years running, after which his enthusiasm would wane. He would then write a book on the subject and move on to a new research topic and new courses. Rather than producing well-rounded research students, this teaching method turned out students who could work with Chandra on whatever his current research project happened to be. It was not a satisfactory system.

So Strömgren appointed a committee, made up mainly of younger members of staff, to redraft the curriculum. Chandra saw the committee as a cabal. In his absence they voted to throw out his curriculum. As far as he was concerned, it was yet another byzantine plot against him, akin to Eddington's in 1935.

The following year, to everyone's relief, Fermi invited Chandra to leave Yerkes and join the physics department in Chicago. After that Chandra rarely taught at Yerkes. His research interests had begun to diverge sharply from astrophysics. At first he stayed on the campus when he visited Chicago. Then in 1959, the Chandrasekhars rented a pied-à-terre near the campus. After being an unofficial member for twelve years, Chandra was officially appointed to the physics department in 1964.[23] That year he and Lalitha bought an apartment in the city. Chandra looked upon his experiences at Yerkes as another humiliation, but at least now he was associating with people such as Fermi.

Fermi was eager for Chandra to participate in research at the Institute for Nuclear Studies, which was linked to but separate from the physics department, and from 1952 on, Chandra was an unofficial member there. The institute was perpetually buzzing with excitement. Staff there held joint appointments, initially in chemistry and physics. Soon the circle widened to include astrophysics, astronomy, geophysics, and mathematics, making it the greatest concentration of research talent in the United States. The core figures from Los Alamos had all relocated to Chicago. Besides Fermi, who had received his Nobel Prize in 1938, there were three other past or future Nobel Prize winners: Maria Goeppert-Mayer, Harold C. Urey, and Walter F. Libby.[24] After Fermi's death in 1954, the institute was renamed the Enrico Fermi Institute. Around the same time, Chandra was formally appointed to this hybrid organization.

Roger Hildebrand, then a young staff member, remembered the fifties as Chicago's golden years. Throughout the institute, there was an "enthusiasm, a love of science, no snobbery; the whole spectrum of physics was discussed, whatever was interesting." The most popular meetings were the Thursday afternoon seminars, for staff only, run by the German-born physicist Gregor Wentzel, who had arrived at the university in 1948 at the age of fifty. Wentzel's credentials were impeccable. After completing his Ph.D. in Munich under Sommerfeld, he had worked with Heisenberg and Pauli, producing important results in elementary particle physics. He and Chandra became so close that Chandra even let him smoke his huge cigars in his presence. Each of the luminaries had a special chair in the institute's colloquium room. If, God forbid, the wrong person sat in one by accident, the "owner" would stand in front of him, uncomfortably close, and stare. It was a daunting experience for a young staff member. When there was a lull in the discussion, Wentzel might turn to one of the junior lecturers, seated at the back of the room, and say, "Well, Roger, what are you thinking about these days?" Hildebrand recalled that "you'd damn well better have something good to say or you could be demolished!"[25]

As it happened, Chandra's new area of interest, hydrodynamic and hydromagnetic stabilities, in addition to his reputation in the field of radiative transfer, made him more attractive than ever to the new powers that had emerged at Los Alamos, such as his old friend Edward Teller, who had left Chicago in 1949 to return to defense work. The new project was to build a bomb so powerful that it would make the atomic bombs dropped on Japan look like mere firecrackers. The weapon was to be a hydrogen bomb, and its development would depend heavily on understanding the physics of stars.

A detonating atomic bomb is like an exploding star, a hydrogen bomb even more so. In each case radiation has to find a way out. If it escapes too easily, the star or bomb will burn up too quickly; if it is too constrained, the explosion will occur prematurely. When scientists design a nuclear bomb, the aim is to hold in the radiation just long enough that nearly all of the nuclear fuel is cooked, then to let it escape as rapidly as possible. Thus the study of what makes radiation flow in a smooth, stable way and of the resistance of the medium through which it flows is critical to both bombs and stars. Similarly, hydrodynamic and hydromagnetic stabilities, essentially the study of flow, are relevant to exploring the behavior of shock waves and how they can be used to compress matter — a vital area of knowledge in the making of bombs. Chandra's expertise in all these fields was essential.

Back in 1934 Fermi, then in Rome, had thought of bombarding nonradioactive elements with neutrons to see if he could produce radioactive elements. Charged particles are repelled by similarly charged ones and therefore have difficulty penetrating a nucleus. Until the discovery of the neutron, alpha particles (the nuclei of helium atoms) were the projectiles used in collision experiments such as these. The disadvantage was that alpha particles, which consist of two protons and two neutrons, have a positive charge. A heavy nucleus, however, is made up of many protons and therefore has a very large positive charge, which the alpha particle must overcome in order to penetrate it. But the neutron has no electric charge, making it the perfect "bullet" to bombard atoms and smash right into the nucleus.

A logical extension of Fermi's experiments was to explore what happened when neutrons struck a naturally radioactive substance such as uranium. To his amazement, Fermi found that the bombardment produced elements heavier than uranium, the heaviest element known to occur naturally on Earth. These new elements are highly radioactive and break up into stable, lighter elements over periods of time ranging from minutes to millions of years; they are therefore no longer naturally present on Earth.[26] In 1938 Fermi received the Nobel Prize for his idea of using neutrons to produce radioactive elements and for discovering elements heavier than uranium. Uniquely in twentieth-century physics, he did outstanding research in both theory and experiment.

Then in 1938 — as the world was on the brink of war — Fermi's experiments led to an earth-shattering discovery. Reexamining his data, some German scientists realized that he had not simply created a new element, as he had thought; he had actually split the nucleus of the uranium atom.[27] Once a neutron has smashed its way into the uranium nucleus, pandemonium ensues, as the nucleus's 146 neutrons and 92 protons try to eject the intruder. The whole nucleus begins to vibrate like a drop of water. First it stretches into a dumbbell shape, pinched in the middle, then splits like an amoeba reproducing. This is nuclear fission, and it happens in a tiny fraction of a second. When the nucleus splits, it creates the elements barium and krypton. The total mass of the new particles, however, is slightly less than that of the original neutron plus the uranium atom. This "missing mass" is converted into energy as predicted in Einstein's equation $E = mc^2$. But the minuscule amount of missing mass multiplied by the velocity of light squared (186,000 × 186,000) results in the release of a colossal amount of energy. The two new elements, barium and krypton, are radioactive and release tremendous radioactivity. The fission also produces at least two neutrons, which can then smash into other uranium nuclei, leading to an uncontrolled chain reaction and creating a nuclear explosion. For this to happen there has to be a critical mass of uranium, about 10 kilograms; if there is less, there will be no explosion, only a fizzle. The nuclei also have to be packed together tightly enough to provide a sufficiently large number of targets for each neutron.

When Bohr heard that the nucleus had been split, he immediately understood the horrific potential of the energy generated. Perhaps he remembered the haunted face of Alfred Nobel, who discovered dynamite, which he hoped would be used only for peaceful purposes. As had happened with dynamite, someone was bound to think up a lethal use for nuclear fission very soon.

In September 1941, Fermi was in his office at Columbia University in New York, mulling over the possibility of an atomic bomb — a fission bomb. With him was Teller, who believed that the fission bomb would be limited in power because the vital critical mass placed an upper limit on the amount of fissionable material that could be piled on. The more powerful the device, the quicker it blows itself apart. In fact, the maximum explosive power of an atomic bomb is around a megaton, the equivalent of a million tons of TNT. Fermi suggested that a fission bomb might be used to raise the temperature enough to start the process of fusing hydrogen atoms to make a hydrogen bomb — a fusion bomb. Such a bomb would be modeled on a star, for it is the fusion of hydrogen atoms that produces the energy that makes a star shine. While fission blows apart, fusion glues together: two nuclei are fused to make a heavier one. Unlike a fission bomb, a fusion bomb needs no critical mass and has no upper limit to its power. Like a stoked fire, it burns ever more brightly as more fuel is added. It has unlimited explosive power.

At first Teller was incredulous at the notion that an atomic bomb could act as a trigger for a fusion bomb. But a year later, when the atomic bomb project was well under way, Teller returned to Fermi's suggestion. Initial calculations indicated that an atomic bomb might generate a temperature of 10 million degrees kelvin. From his work on stars, Teller knew that this was high enough to ignite hydrogen. Compared with a fission bomb, the fuel for a thermonuclear (fusion) bomb cost a pittance. Deuterium could be extracted from water for a few cents a gram. In contrast, extracting the isotope of uranium that can most easily be induced to fission, and which accounts for less than

1 percent of the element naturally, was immensely expensive.[28] Bethe's first reaction was that the fusion bomb, or "super," was "a terrible thing."[29] For Teller, it became an obsession. "Something changed in him after he joined the Los Alamos Project," recalled his old friend Gamow.[30]

From the moment that Teller arrived at Los Alamos in April 1943, he caused problems. To start with, he was offended when Oppenheimer appointed Bethe, not him, as head of the theoretical division. Teller's group was directed not to worry about the super but to create a fission bomb in which the fissile uranium was squeezed together in a process called implosion. Tolman, the expert on the general theory of relativity at Caltech, had suggested the previous year that an amount of fissile substance just below the critical mass could be made to explode if it was sufficiently compressed for the neutrons in it to set off a chain reaction. To create an implosion — blowing things together rather than apart — scientists put chemical explosives around the inside of a hollowed-out sphere containing a sphere of fissionable material at the center and set them to explode simultaneously. The resulting series of concentric shock waves moves inward, crushing the fissionable material until it is dense enough to spark a chain reaction. But Teller's heart was in the super, not the fission bomb. He sulked and did not put his heart into his work. Bethe became increasingly annoyed. Oppenheimer did not want to lose someone as innovative as Teller. In 1944 he finally allowed him to leave Bethe's theoretical division and work on the super with his group.

For the atomic bomb, scientists could perform tabletop experiments to determine the feasibility of a particular design. But the hydrogen bomb needed an atomic bomb to ignite it; all Teller and his group could do was to make simulations based on mathematical models. But working out the ideal configuration of atomic bomb and fusion material that would set off a thermonuclear reaction required horrendously complicated calculations, beyond the capacity of IBM's mechanical calculating machines at the time.

In 1945, however, the ENIAC (Electronic Numerical Integrator And Computer), the world's fastest electronic computing machine, which had just been built at the University of Pennsylvania, went on line. It could handle the calculations for the super. ENIAC was soon outmoded, and in 1949 it was replaced by the MANIAC (Mathematical Analyzer Numerical Integrator And Computer). Besides the one at Los

Alamos, another MANIAC was built at the Institute for Advanced Study at Princeton, at the instigation of the physicist John Archibald Wheeler. A veteran of the Manhattan Project, Wheeler was disappointed with the progress at Los Alamos on the super. He wrote to the head of the physics department at Princeton that "thermonuclear research involves strong Princeton specialties — nuclear and atomic theory, ideas from astrophysics, and hydrodynamics."[31] (Hydrodynamics was what Chandra was working on at Chicago.) Thus Princeton became the second center for the hydrogen-bomb project.

Despite his extreme right-wing anti-Communist views and salt-of-the-earth demeanor, not unlike that of a midwestern farm equipment salesman, Wheeler had one of the most way-out minds in physics. He had spent most of the war in Hanford, Washington, studying basic problems of reactor physics essential for the production of plutonium, an alternative fissile material to uranium.[32] During the 1930s, Wheeler and Bohr had worked together closely on the new field of nuclear physics. When Bohr arrived in New York City in January 1939, Wheeler met him. Bohr immediately filled him in on the latest development — nuclear fission. The two set about trying to understand the full significance and workings of this momentous new phenomenon. Wheeler agreed with Teller that it was vital to develop the next phase of nuclear weaponry, the super. But Oppenheimer, who was reluctant to sanction a bomb even more powerful than the atomic bomb, was the primary obstacle to its development. As chairman of the General Advisory Committee (GAC) set up to advise the newly created Atomic Energy Commission (AEC), Oppenheimer had a great deal of say in the matter. Teller was growing increasingly antagonistic toward him.

Teller wanted to go immediately for a high-megaton bomb that would dwarf the atomic bombs that had been dropped on Hiroshima and Nagasaki. The Hiroshima bomb, dubbed Little Boy, had been equivalent to 20,000 tons of TNT. The ENIAC had been built for the specific purpose of testing Teller's "classical super," which consisted of a fission bomb attached to one end of a pipe filled with liquid deuterium. His theory was that the heat from the exploding fission bomb would travel down the pipe, triggering fusion reactions in the deuterium. He estimated that the classical super packed around 10 megatons of explosive power. Another of Teller's designs was a hemispherical bomb made of concentric shells of deuterium and fissionable material. He called it the Alarm Clock, because it would wake up the world. But

Teller's preliminary calculations showed that the bomb would not be very powerful, so he did not take it any further.

Then, on August 29, 1949, the Soviet Union exploded Joe 1, almost an exact copy of Fat Man, the bomb dropped on Nagasaki, which used the implosion mechanism. The design had reached the Soviets courtesy of Klaus Fuchs, a thirty-eight-year-old German émigré scientist who had arrived in Los Alamos in 1942. There he was privy to the most highly classified information — which he passed to the Soviets.[33] Ironically, Fuchs had joined the implosion group as Teller's replacement. President Harry S. Truman made the news of the Soviet bomb public on September 23 with the dramatically terse announcement that "we have evidence that within recent weeks an atomic explosion occurred in the USSR." The U.S. monopoly over nuclear weapons was over. Banner newspaper headlines appeared immediately. There was near panic in the country. What should be the American response? The official debate was carried out in the highest secrecy. Part of the scientific community agreed with the AEC that the Soviets knew that the next step was the super and were determined to build it first. Suddenly everything had changed. Sorting out the design problems for the super was now top priority.

Most of the GAC, chaired by Oppenheimer, cautioned against entering a potentially disastrous arms race. But powerful and vocal scientists, notably Teller, supported the super. In alliance with key congressmen and senators, they carried the day. The argument that tipped the scales was the fact that Fuchs had also been privy to Teller's designs for the classical super and the Alarm Clock. On January 31, 1950, Truman announced to the world that the United States had embarked on a project to build a hydrogen bomb. The nuclear arms race was on. Teller asked Oppenheimer to work on the super, but he bluntly refused. He had concluded that any advances the Soviets could make on Fuchs's information would be "marvelous indeed," because simulations indicated that Teller's classical super was unworkable.[34] The bomb designers had reached an impasse, which Oppenheimer interpreted to mean that the super was simply not feasible.

Then in December 1950, the mathematician Stanislaw Ulam had a flash of inspiration. He had worked on the super since 1943, when he was assigned to Teller's renegade group at Los Alamos. Dark, urbane, always impeccably groomed and dressed, Ulam — known as Stan — radiated an air of carefree confidence. Born in 1909 in Lemburg, Poland (now Lvov, Ukraine), he was a mathematical prodigy. He learned his trade in the remarkable school of prewar Polish mathematics and completed his Ph.D. under the world-renowned Stefan Banach. In 1934, sensing the approaching menace of Nazism, Ulam had decided to start making contacts elsewhere. Traveling around Europe, he stopped off in Cambridge, where he met and impressed Eddington and Chandra, among others. His take on their relationship was that Chandra collaborated with Eddington, "for whom he had mixed feelings of admiration and rivalry."[35] Their paths were to cross again at Los Alamos in the 1950s, when Chandra was a consultant on the theory of turbulence and other hydrodynamical problems.[36]

Staring out the window of his living room in Los Alamos, Ulam suddenly realized that the key to creating the super was implosion, the very same phenomenon that Chandra had predicted in white dwarfs.[37] Ulam suggested using a fission bomb to trigger a hydrodynamic shock wave, which would be carried by a flow of neutrons. To shape the shock wave properly, he suggested physically separating the fission bomb (or primary) from the fusion fuel (or secondary). Ulam described his idea to Teller late in January 1951. Almost immediately, Teller saw that Ulam had hit upon something quite new. A fission bomb creates not only a flood of neutrons but also high-energy radiation (x-rays). Teller was well aware that it would be very difficult to focus the neutron shock wave so as to implode the fusion material. The radiation, however, could be focused; moreover, the x-rays would reach the fusion material before the neutrons because light travels faster than any particle with mass. The solution was to use radiation to induce implosion.

What emerged from their collaboration became known as the Teller-Ulam configuration, the basic design for all fusion bombs. The details are still classified, but the general idea is that the fission bomb generates x-rays, which are somehow used to implode the fusion material and begin a thermonuclear reaction (fusion of hydrogen). The result is a massive explosion, very similar to that in a star squeezed by gravity. The design was clearly feasible, and everyone wanted to move ahead quickly, for the Soviets were probably not far behind. Even Oppenheimer, who had been the most outspoken of the scientists opposed to the production of hydrogen bombs, conceded that the Teller-Ulam configuration was "technically sweet."[38]

By 1952 the bomb was completed. It was tested on November 1, on the island of Elugelab in the Eniwetok Atoll, 3,000 miles west of Hawaii.

The bomb, referred to as Mike, had a yield of 10 megatons, bigger than anyone had expected. Elugelab disappeared completely. The fireball alone was more than three miles across, big enough to destroy New York City, frying most of its inhabitants and blasting the rest to bits. In a billionth of a second the Mike fireball created every natural element in the universe as well as artificial ones. But one problem remained. Mike was not a deliverable weapon; no means had yet been found to solidify the fusion components, which had to be packed into a huge refrigerator. The entire bomb weighed 65 tons. Two years later the hydrogen bomb was "weaponized," that is, redesigned so that it could be transported by plane. The new weaponized bomb, called Bravo, was detonated in March 1954. The explosion was more than twice as big as expected — 15 megatons, creating a fireball four miles across.

On the other side of the world, in the Soviet Union, physicists had been working around the clock since the end of the war to develop an atomic bomb, spurred on by the knowledge that they were under attack for being ideologically impure. "Leave [the physicists] in peace. We can always shoot them later," Josef Stalin was quoted as saying to the feared head of the secret police, Lavrenti Beria, director of the gulags and the instigator of purges that killed hundreds of thousands of Soviet citizens.[39] Stalin understood that scientists would be indispensable in any future conflict with the West. He demanded immediate results, which is why the Soviets' initial atomic bomb was a copy of the plutonium bomb, Fat Man, whose design had been handed over by Klaus Fuchs. Failure was not an option, especially as Beria was overseeing the bomb project personally. The Soviets then tried to produce a hydrogen bomb, modeled on Teller's super, at a top-secret site called the Installation, also known as Arzamas-16 or "Los Arzamas."[40] In reality it was the town of Sarov, 240 miles east of Moscow.

The two most important Soviet scientists involved in the effort were Andrei Sakharov and Yakov Zel'dovich. Like Gamow and Teller, they were physical opposites. Sakharov, tall, thin, invariably dressed in black, and terribly introverted, was working on his Ph.D. when his studies were interrupted by World War II; he spent the war years as a factory worker. In 1947, at twenty-seven, he finally completed his Ph.D. in nuclear physics. He was immediately assigned to a team looking into the possibility of developing nuclear weapons, led by Zel'dovich.[41]

Zel'dovich was almost the exact opposite of Sakharov. A powerfully built man with a rapid stride, a swarthy complexion, and close-cropped hair on a bullet head, he carried himself with chest stuck out and head held high. A very confident and assertive character, he appeared much taller than he actually was. Zel'dovich was a human dynamo who thrived on physical challenges. In the middle of a discussion of astrophysics in his office at home, he would suddenly suggest to colleagues that some exercise was in order. Gymnasiums were not common in Moscow, so Zel'dovich devised exercises such as playing catch with a medicine ball in the stairwell of his housing complex. Bold and brash, he was the life and soul of any party, always on the dance floor with any woman he could find.

In 1929, at the age of fifteen, Zel'dovich graduated from high school and began studying for a career as a laboratory technician. In 1931 he went on a school visit to the famous Physical-Technical Institute in Leningrad. The professors were struck by the young man's penetrating questions and invited him to work there in his free time. Over the next ten years, while still a student, he lectured at the institute on the latest developments in quantum physics in such depth and breadth that he was invited to become a member of the chemical physics department. Three years after his initial visit, he completed his Ph.D. It was an amazing career path for someone who was effectively self-taught. Zel'dovich's work in shock waves, gas dynamics, and their application to explosions was, and still is, highly regarded.

At the time of the German invasion of the Soviet Union in June 1941, Zel'dovich was studying chain reactions in uranium. Along with his fellow scientists, he moved back to working on conventional explosives. He helped invent the famous Katyusha missiles, which terrified the Germans. In 1943 the Soviet government decided to embark on its own nuclear weapons program and drafted Zel'dovich into a small elite team. At the Installation, Zel'dovich headed one of several competing design groups. Part of the research was delegated to universities, including Moscow State University, where Landau was working. At first, Sakharov's team was delegated to check and refine the calculations produced by Zel'dovich's group, and they resented this subsidiary role. "Our job is to kiss Zel'dovich's ass," one of Sakharov's colleagues put it — in private, of course.[42]

While the first Soviet atomic bomb had been developed largely thanks to Fuchs, the first successful hydrogen bomb, similar to Teller's Alarm Clock, was developed by Sakharov. It was exploded on August

12, 1953, at the testing ground at Semipalatinsk in Siberia. It yielded 0.4 megaton, which made it only ten times bigger than the atomic bombs used in World War II. The Soviet scientists immersed themselves deeply in the search for design innovations that would produce a higher-yield bomb to rival Mike.[43]

During those years, Zel'dovich branched out into particle physics and astrophysics. Working with him, Sakharov read all he could on gas dynamics and astrophysics. They had all realized that "the physics of stars and the physics of a nuclear explosion have much in common."[44] Sakharov was familiar with Landau's paper on collapsed stars written in 1932, when Landau rediscovered Chandra's upper limit for the mass of a stable white dwarf, and his 1938 paper that looked into the minimum mass necessary for a neutron core to form within a star. Zel'dovich had first understood the role of explosions in astrophysics after hearing Chandra lecture in Moscow in 1934 on his upper limit and on the possibility that stars might collapse or explode. Wheeler, too, recalled that scientists working on implosion began to look toward the "astronomer's techniques."[45] The groups at Los Alamos and "Los Arzamas" agreed that "the high temperatures generated in thermonuclear reactions led to the development of the physics of high pressures and high temperatures," very much like reactions in stars.[46]

In the spring of 1954, Sakharov and Zel'dovich finally cracked the Teller-Ulam configuration. With the help of the Soviets' own high-speed computers, improvements in bomb design followed quickly. On November 22, 1955, the two tested a bomb with a yield of 1.6 megatons that was capable of being delivered. They went on to produce larger bombs, and on October 30, 1961, detonated a 50-megaton bomb, dubbed the Czar bomb, which to this day remains the largest nuclear weapon ever exploded.[47]

In their temperaments, Sakharov and Zel'dovich mirrored Ulam and Teller. Both Ulam and Sakharov were interested in computational aspects of physics; Teller and Zel'dovich were more intuitive and tried to avoid detailed calculations. "Zel'dovich could see the solution of many problems without long calculations," recalled his former student and collaborator Sergey Blinnikov.[48] Landau suffered a tragic car accident in 1962 which seriously impaired his mental ability, and from then on Zel'dovich was the undisputed leader in Soviet theoretical physics. He wrote dozens of books and papers. When Stephen Hawking, one of the most important figures in astrophysics today, met Zel'dovich, he was amazed to discover that he was just one man. He had assumed that the name was a pseudonym for a group of authors who had worked collectively — how else to explain such phenomenal productivity? In recognition of his achievements in the Russian nuclear program, Zel'dovich was made a laureate of Lenin and Hero of Socialist Labor (the highest civilian award) three times, an extraordinary achievement.

The scientists who took part in wartime research never forgot the lessons they learned from it. Take, for example, Robert F. Christy, whose contributions to the design of the bomb dropped on Nagasaki were so great that it was often referred to as the "Christy gadget."[49] During a sabbatical from Caltech in 1960 at Princeton, Christy decided to "learn something about the stars." After all, he reasoned, the "mathematical approach was very similar to what we had been working on in Los Alamos during the war. And I thought it was interesting, in a way, that the theory used to make atomic implosion bombs was the same theory I could apply to certain kinds of variable stars. It's interesting to see how things relate to each other." He had always been impressed with Eddington's work and decided to look further into the mechanism of Cepheid variable stars. He was delighted when the RAS awarded him the Eddington Medal for his results. "Eddington himself I always considered a physicist," he wrote, but in fact "he's a great astrophysicist."[50]

Wheeler too was fully aware of the intimate connection between bombs and stars. Early in the 1950s, at Princeton, armed with his state-of-the-art MANIAC computer, he set out to resolve what he considered nothing less than *the* fundamental problem of physics — "the fate of great masses of matter."[51] He began by exploring what happens to very cold matter, by which he meant the burned-out core of a star in which the nuclei have given up the last of their energy. His method of studying cold matter was to formulate an equation describing its state that took account of densities ranging from very low to very high, including, whenever necessary, the principles of general relativity and what was then known about nuclear forces. Wheeler had worked on the atomic and hydrogen bombs and had done nuclear physics research with Bohr in the late 1930s. This background allowed him to deal with the way matter behaves over broad ranges of temperature and pressure. Two graduate students, B. Kent Harrison and Masami Wakano, worked with him. They used the MANIAC to input data into their equation of state from just about every known star, working through several per hour, a process that would have been hopelessly time-consuming before the computer was invented. Chandra had made laborious calculations on a mechanical calculator, as had Serber and Volkoff in their work with Oppenheimer.

The MANIAC computer enabled Wheeler and his team to explore in detail the evolution of a large number of stars. They came up with some truly astonishing results. It was clear from their super-equation for the state of cold matter that while a star with a mass less than the Chandrasekhar limit would collapse to a white dwarf, a burned-out relic with a larger mass would keep on collapsing and become a neutron star. There was no alternative, no intermediate stage.

Harrison, Wakano, and Wheeler provided the first detailed glimpse of the fate of a star more than eight times the mass of the Sun. Their results (brought up to date) shed new light on the life of stars. What they found is that the high gravity of such a massive star makes its interior hotter than the center of the Sun, so it takes less time — about 10 million years — for the hydrogen to burn up. Once the hydrogen fuel is spent, the star begins to die. First the helium core cools and begins to collapse. Then the crush of gravity raises the temperature again, to the point that helium begins to burn. After another million years, the helium too is spent, leaving a carbon core surrounded by layers of hydrogen and helium ash. That core in turn starts to cool and then collapse under the gravitational pressure of the mass around it. Again the temperature rises, the core reignites, and, after six hundred years, burns up, leaving a neon core. The same process repeats again and again, more and more rapidly; the heavier elements burn more quickly because they are more stable, so the same amount of fuel generates less energy. Neon burns off in a year, oxygen in six months, and silicon in a day.

Silicon is the last element in the core to burn. What is left then is iron, the element with the most stable nucleus. Iron will undergo fusion or fission only if energy is supplied; it cannot generate any energy by itself. The formation of an iron core signals the beginning of the end for a star that started out eighteen times the mass of the Sun. At this point the temperature of the iron core is a million degrees kelvin, and the pressure is 10 million grams per cubic centimeter. The star is now an onionlike structure made up of layers of different elements.[52] The layer of silicon ash is 4,000 miles across, twice the size of the Moon and 50 million times as massive. Inside it the iron core is 1,000 miles across. This is a tiny fraction of the volume of the whole star, which has ballooned to 20 million miles across, exerting an incredible pressure on the core.

The interior of the star is so hot that all the atoms have been stripped of their electrons. The core now consists of iron nuclei and electrons moving close to the speed of light — relativistic electrons. But the immense pressure of the 20-million-mile onionlike structure around it squeezes the mixture of iron nuclei and relativistic electrons closer and closer together. Eventually the electrons smash into the iron nuclei and combine with the nuclear protons to produce neutrons and neutrinos. This creates a host of nuclei of elements heavier than iron that contain more neutrons than usual — "neutron-rich" nuclei. Many things now begin to happen. The total effect is to reduce the number of electrons — lessening the degeneracy pressure that caused the iron core to be as hard as rock — and drive the mass of the core above the Chandrasekhar limit. The core, becoming unstable under the enormous inward pull of gravity, collapses to a neutron star.[53]

Harrison, Wakano, and Wheeler's analysis also suggested something else — that it was not out of the question that a star with a large enough mass might continue to collapse forever, until it was "nowhere," as Eddington would have put it. Was there a force that could prevent the core of a star from collapsing right through the white-dwarf and neutron-star stages into nothingness? That is what Eddington and Milne had sought in their day, and right into the 1960s almost everyone continued to hope to find such a force. Wheeler and his team could not find it. So what did the core then collapse to? Wheeler suspected that at such extremely high temperatures and pressures the behavior of matter might be governed by a whole new set of physical laws. The answer, he suggested, might be found at the "untamed frontier between elementary particle physics and general relativity," in the realm of a quantum theory of gravity.[54] But in fact the definitive solution would come from an unexpected quarter.

11. How the Unthinkable Became Thinkable

WHEELER'S RESULTS seemed to reveal that a star far larger than the Sun could ultimately collapse to nothing — a notion so outrageous that he simply refused to believe it himself. The complicated equations he and his team had processed through the MANIAC computer showed that a very massive star would shrink until it was unimaginably tiny and dense. Its gravity would be so strong that it would pull space around itself like a shroud and be swallowed up by it. But to Wheeler this was an unacceptable conclusion. Quite apart from the absurdity of it, there was no observational evidence whatsoever to support it.

At the Solvay Conference on physics in 1958, Wheeler argued that there must be a way for stars to shed mass to get below the maximum possible mass for a neutron star, which he and his team had calculated to be about twice the mass of the Sun. It was like a rerun of the sessions at the Royal Astronomical Society in the early 1930s, with Wheeler insisting — armed with no evidence, just his own preconceived notion of the way things had to be — that something had to put a brake on a star's collapse.

Oppenheimer took on the role that Chandra had played in those earlier debates. He for one was prepared to consider that Wheeler's results could be taken at face value. He had an answer that cut through Wheeler's byzantine arguments: "Would not the simplest assumption about the fate of a star of more than the critical mass be this, that it undergoes continued gravitational contraction and ultimately cuts itself off more and more from the rest of the universe?"[1] But Oppenheimer, who had written only a few papers based on purely theoretical concepts, was virtually a lone voice, and he carried little weight in the astro-physics community. Wheeler, conversely, was running a large research program that included applying nuclear physics to the study of realistic stars. At this stage, for all their theoretical work, no one had actually observed any way in which stars might die other than by becoming white dwarfs. But a discovery soon to be made would finally persuade Wheeler that Oppenheimer was on the right track — and would bring Chandra and his theory back into the center of the debate.

It all came about when a man named Stirling Colgate had a startling insight. At the time he was America's premier diagnostician of thermo-nuclear weapons — hydrogen bombs — at Livermore National Labo-ratory in California. He realized that flashes of light from supernovae might set off the detection devices in American satellites spying on the Soviet Union and start World War III, even though supernovae were some 100,000 trillion miles away and had exploded more than 100,000 years before. Lean and rugged and with a perpetual suntan, Colgate has a raffish, devil-may-care style that belies his age (now touching eighty) and an ongoing serious dedication to scientific research. His in-terests are wildly varied, his expertise spans experimental and theoreti-cal physics, and he has a feel for machinery that is right out of Robert Pirsig's *Zen and the Art of Motorcycle Maintenance.*

Colgate's introduction to nuclear weapons occurred, dramatically, in the place where it would all happen, long before it did. He was a stu-dent at the Los Alamos Ranch School when, in December 1942, a dele-gation arrived. It included an impressive number of military officers to-gether with two civilians, one in a porkpie hat, the other in a fedora. Immediately after the visit, the school made a shocking announce-ment: it was to close immediately. Colgate and the two other seniors would graduate on the spot, ahead of schedule. Colgate and his preco-cious friends quickly worked out what was going on. The two civilians, "Mr. Jones and Mr. Smith," were scientists. In fact, they were Robert Oppenheimer and Ernest O. Lawrence, a nuclear physicist from Berke-ley and a 1939 Nobel laureate. Oppenheimer personally delivered the commencement address to the three-man graduating class, then or-dered several buildings bulldozed and made plans to build laboratories and offices. Oppenheimer, Colgate remembers, did not seem at all un-comfortable with the high-handed takeover, though Lawrence was a lit-tle abashed. From that moment Colgate developed a serious mistrust of Oppenheimer, which he maintained throughout his later dealings with him, though he also sympathized with Oppenheimer's regret and guilt as a result of his role as the leader of the atomic bomb project.

Colgate, a scion of one of America's wealthiest families, has always gone out of his way to be his own man. Two years after that fateful meeting in Los Alamos, he enlisted in the merchant marine, which was, he recalled, "perfect for me."[2] He had to establish himself among the older experienced seamen, for whom being a whiz kid in science meant little; what counted was being able to get the job done. On August 6, 1945, the captain summoned the crew to the mess hall and told them that America had dropped an atomic bomb on Hiroshima, adding that he would "appreciate it if Mr. Colgate would tell us what it means." Colgate knew the answer — though he points out that at the time his conjectures were classified information, especially his explanation that each fission reaction would release two or more neutrons, producing an uncontrolled chain reaction and leading to an apocalyptic explosion.

When he was discharged in 1946, Colgate returned to Cornell University, where he had already spent a year studying electrical engineering. "Physics," he decided, was "the thing for me." Edwin Salpeter, a professor there, remembers how impressed everyone was with this extraordinary young man. Colgate completed a B.S. and a Ph.D. in nuclear physics, then took a position as a postdoctoral fellow at Berkeley.

In 1952 he moved to Livermore National Laboratory, recently created by Teller with the encouragement of the United States Air Force to provide competition for Los Alamos, which Teller felt was lagging behind in weapons research. Teller made it amply clear that Livermore's sole mission was to research and develop a hydrogen bomb. Aware of Colgate's reputation, Teller suggested that he take on the important task of making diagnostic measurements for the forthcoming tests. "Well, of course," he replied. It was a natural progression for him. "I was always enamored with explosives, and eventually I graduated to dynamite and then nuclear bombs," he explains nonchalantly.

The diagnostic work was divided into two parts. One was to analyze what the bomb yielded by studying the radioactive products of the explosion, scooped from the atmosphere by purpose-built aircraft. Colgate's responsibility was "fast diagnostics" — measuring the range of energy of the neutrons and high-energy gamma rays bursting out of the explosion. Scientists hoped to deduce from this information how well the bomb's mechanisms functioned and precisely how quickly events progressed from fission to fusion during the Teller-Ulam process, before all the components had been vaporized.

New diagnostic instrumentation had to be designed as quickly as possible. The first hydrogen bomb, Mike, was already scheduled to be tested that same year, 1952, on the island of Elugelab. Colgate often had to shuttle back and forth between Livermore and Los Alamos. He would fly to El Paso, Texas, take another plane to Albuquerque, New Mexico, then drive to Los Alamos. On one of these trips, he decided that some R and R was in order. From El Paso he crossed the border into Mexico and strolled into a bar in Juarez. There he got into conversation with an ex-Marine who happened to be a graduate student in astrophysics. Colgate's new pal was from the Deep South, and this was the 1950s, long before political correctness had become de rigueur. In a long slow drawl, the ex-Marine declared that he worked for one of the blackest people you ever would see. He had been brought up to believe that such people could not be too bright, but this guy was just the smartest person he had ever met. "And, if you can believe it," he concluded, "he has an absolutely crazy name, too: Chandrasekhar!" Colgate had come across Chandra in the context of his work on radiative transfer but had never known that he had any connection with astrophysics.

Back at the bar, Colgate recalls, scientific bonding led to innumerable rounds of drinks. The next morning the two men were so hung over that Colgate still can't remember how they got from Juarez to El Paso. They were almost thrown off the El Paso to Albuquerque flight. From there, Colgate's new friend went on to Chicago, and Colgate arrived at Los Alamos in a terrible state. As he tells it, the first person he ran into there was Chandra, working as a consultant on the upcoming Mike test. Without thinking, Colgate walked up to him and blurted out the whole story. Suddenly he felt "sad and terribly embarrassed." But Chandra just roared with laughter. Unfortunately, charming though the tale is, the only person who can remember Chandra being at Los Alamos at the time is Colgate himself, so perhaps the tale has become embellished over the years. But it does show the impact Chandra had on people and the kind of myths that grew up around him. It might even be true. This was how Colgate remembers being introduced to Chandra. Their paths were to cross again in the astrophysics world a decade later.

Since Colgate did not belong to the "glamorous" category of bomb design, and his work was top secret, he is not usually mentioned in books on the development of the hydrogen bomb. Nevertheless, two years later, at the test of Bravo, the first deliverable thermonuclear bomb, the twenty-nine-year-old scientist was in charge of a staff of thousands.

The Bravo test was a great success in terms of both its yield and the diagnostic methods used to assess it. Colgate had won his spurs. Teller gave him carte blanche to choose his next research project, and he opted to look into thermonuclear fusion and plasma physics, both peaceful uses of nuclear energy. But one can never escape one's past. In 1959, on the advice of the senior scientists at Livermore and Los Alamos, the State Department recruited him as the scientific consultant on nuclear test ban negotiations in Geneva. Both sides needed to "agree on a detection system so we don't all lob nukes at one another." Colgate proposed accomplishing this with satellites designed to detect nuclear testing.

Then he remembered some research he had done in 1956 with his colleague Montgomery Johnson. At the time, the U.S. government had suggested adding another battlefield to air, land, and sea by making outer space a fourth dimension for warfare. Colgate and Johnson had been recruited to investigate the blow-off from a hydrogen bomb exploded in space. They had conducted simulations suggesting that the result would be "monster" amounts of x-rays and gamma rays, just as if it were a supernova.[3] A hydrogen bomb, it seemed, was remarkably similar to a supernova. But the problem was that if a satellite was sent up to detect nuclear testing, a supernova might set off the detection device, thus accidentally precipitating a cataclysmic war. "It was just natural to bring up supernovae in Geneva if we wanted to get a good spy satellite," recalls Colgate. "The Russians obliged and they too made a big thing about gamma-ray bursts" — highly energetic radiation from outer space, which could come from a supernova. Colgate, as a very junior member at this august gathering of senior Soviet scientists, felt inadequate in his knowledge of supernovae, but "as I learned in later years, so were they."

By 1959, research on x-ray and gamma-ray emissions from space had become quite sophisticated. X-rays cannot penetrate Earth's atmosphere, so experiments were carried out using equipment sent up on rockets, at first on captured German V-2 missiles. Then in 1961, satellites detected gamma rays emitted by the 50-megaton Czar bomb, which the Soviets had just exploded in violation of the Soviet-American moratorium on nuclear weapons. The Americans were apoplectic. Clearly the Soviets had agreed to the moratorium as a ruse, to buy time to prepare for their next series of tests while the United States was honoring it. Livermore and Los Alamos went into overtime. Understanding supernovae became top priority.

Teller urged Colgate to follow up on this line of inquiry, but in fact Colgate had already started to move over to astrophysics. As soon as he perceived the connections between "supernovae and nuclear weapons," he went to see William ("Willy") Fowler, Caltech's expert on nuclear reactions in stars. They struck a deal. For a couple of days each week, Colgate gave lectures on explosions — his area of expertise — at Caltech, and in exchange he boned up on astrophysics. Both the staff and the graduate students enjoyed shooting down his budding ideas on supernovae. After hours studying, Colgate would spend his afternoons and evenings at nearby Venice Beach.

Colgate carried out his initial research on supernovae with Johnson. Their work was based on the classic 1957 paper by the husband-and-wife team of Margaret and Geoffrey Burbridge at the University of California, San Diego, together with Willy Fowler and the Cambridge astrophysicist Fred Hoyle. The Burbridges, Fowler, and Hoyle — known in the trade as B^2FH — argued persuasively that as stars evolve they produce heavier and heavier elements. When a star much more massive than the Sun has completed its evolution, it has an iron core at the center with a density of 10 million grams per cubic centimeter. The natural next question for astrophysicists was, What makes a super-heavy star such as this become a supernova and what is left behind? Everyone agreed that Zwicky's hypothesis from the 1930s was almost certainly right: there had to be something like a neutron star left after a cataclysmic supernova explosion.

After all its silicon has burned up, an aging star is a huge ball of layer upon layer of burned-out nuclear ash surrounding an unburnable iron core. At this point there can no longer be any nuclear reactions which will give the star energy, so it begins to cool off. Crushed by the enormous weight of the onion-like layers above it, the core begins to implode. By now it is a mixture of neutron-rich nuclei, electrons, and protons at a temperature of about 5 billion degrees kelvin, caused by the extreme compression. At that temperature the high-energy radiation (x-rays and gamma rays) in the star is powerful enough to smash the neutron-rich nuclei into helium nuclei (alpha particles), protons, and neutrons. It is the fusion of these light nuclei that produces hydrogen bombs and makes stars shine for billions of years.

Suddenly, in the blink of an eye, all this is undone. All these reactions absorb energy rather than provide it, so the only way the star can get any more energy is by shrinking under the massive inward shove of its own gravity. As a result the core begins to heat up again. The high-energy radiation then turns on the alpha particles, ripping them apart into their constituent protons and neutrons. More and more electrons collide with the protons in the heavier nuclei that remain intact and turn themselves into neutrons and neutrinos.[4] This depletes the supply of electrons and thus weakens the electron degeneracy pressure, which can no longer support the core and keep it rock hard. Without electron degeneracy pressure the entire star collapses, but not all at once.

At a speed of 36,000 miles per second, the iron core and its shell of silicon ash shrivel in a fraction of a second from a ball more or less the size of Earth to an incredibly dense sphere a mere 12 miles across. But this sphere is only a third of the total mass. This collapse happens so quickly that the onion layers of carbon, oxygen, neon, helium, and hydrogen are left behind — a shell without a center. As the core shrinks at lightning speed, burned-out nuclear ash from the onionlike structure around it falls after it. The core reaches its limit of compression, then stiffens and rebounds like a coiled spring. This "core bounce" creates a shock wave that undulates outward at 6,000 miles per second, plowing through the ash still plummeting from the onion layers above the core. When scientists create a model of a supernova explosion, one problem is to make sure that the shock front does not stall — that it is not stopped dead by the mass of material plunging in. If the wave rolls on, it creates an explosion that produces the exact yield of chemical elements one sees in an actual supernova explosion.

The inside of the collapsed core is like a Cubist painting, an *Alice in Wonderland* scenario with no distinction between inner and outer. Everything merges into everything else, with exotic elementary particles, perhaps even free quarks, springing up, as well as huge numbers of neutrons and some protons and electrons. This is a neutron star.[5]

For a neutron star to be left after the star has imploded, the core has to have a strong enough repellent force to counterbalance the attraction between neutrons and protons (which are naturally drawn together by the nuclear force to create nuclei). Without a strong repellent force, the attraction between neutrons and protons, together with the force of the star's gravity, will be enough to make the star go on collapsing, dwindling ever smaller until it disappears into nothingness. The star also has to blow off enough mass to get below the maximum mass for a stable neutron star, which Oppenheimer and Volkoff calculated as seven tenths the mass of the Sun. (With developments in nuclear physics enabling them to formulate realistic models of supernovae, astrophysicists currently calculate the mass above which a neutron star becomes unstable to be two or three times the mass of the Sun.) If the burned-out core of a star is greater than this maximum mass, the options of shriveling to a neutron star or ending up as a white dwarf are both closed, for the maximum mass for a stable neutron star is greater than the Chandrasekhar limit for a white dwarf (1.4 times the mass of the Sun).

Colgate and Johnson's research was a first attempt to understand the mechanism of a supernova explosion. In practical terms, what they were trying to find out was how to blow up a star. They assumed that the shock wave from the core bounce would smash into nuclear ash plummeting inward, pulled by gravity. They postulated that the shock wave would turn this matter around and heat it up, thus causing the supernova explosion. But this notion turned out to be wrong. Another member of the team, Richard White, produced computer simulations showing that the shock wave from the core bounce is not strong enough to accomplish all this. Instead of the star imploding in a supernova explosion, then shriveling to nothing, there is just a fizzle.

White studied physics at Pomona College in California, where he also took an evening course in the fledgling subject of computer programming. On graduating in 1956, he joined Livermore National Laboratory. At twenty-two, he was the first person there with any formal training in computers, which were still in their infancy. Programmers and theoretical physicists at Livermore simply learned about them on the job. Working with Colgate at Livermore, White later remembered, was "akin to having a close personal relationship with a whirlwind."[6]

Colgate and White decided to take a long hard look at supernovae by creating models of stars on the verge of collapse. A very ambitious program, it entailed constructing a mathematical description of the gas in the star, an equation of state that built in nuclear forces adjusted to prevent the star from collapsing completely, as well as the mixtures of chemical elements that made up the star. The equation of state they were looking for was far more complicated than the perfect gas law or Chandra's equation for an "ideal" white dwarf, both of which ignored interactions between particles of gas, and also far more sophisticated than the one used by Wheeler's team. The whole equation of state had to be encapsulated in computer software that could model real stars. White began with a computer program that combined software used to design bombs with the most up-to-date equation of state for a star from astrophysics. At that time, only Livermore and Los Alamos had computers fast enough to run such complicated calculations.

Working at a weapons laboratory, Colgate and White knew nothing about the supernova research that Zel'dovich and his group were carrying out. This was an era when Soviet scientists attended international meetings accompanied by "minders" — apparatchiks or even KGB agents with dubious scientific credentials. This is not to say there was no scientific communication between East and West at these events, but it was restrained. Zel'dovich, of course, was not allowed out of the Soviet bloc. On one rare occasion, according to an oft-told story, the organizer of the astrophysics section of a meeting of the American Physical Society decided to make a goodwill gesture by inviting a member of the Soviet delegation, chosen at random, to give a scientific lecture. Unwittingly he chose a KGB agent.

At first, says Colgate, he "didn't think he was doing anything particularly original." As far as he knew, he and White were just creating models of scenarios that other people had already suggested, like those in the 1957 B²FH paper, but that had not been run through a computer. After Colgate and Johnson hypothesized that a supernova explosion is sparked by the shock front from core bounce smashing into infalling nuclear ash from the collapsing star and igniting it, Fowler and Hoyle began to fill in the details. But Colgate and White, working together, were able to show that Fowler and Hoyle had made a mistake in analyzing the physics of the explosion.[7]

The challenge was to work out how core bounce could create an explosion. At this point, Colgate had a flash of inspiration: "Why not look into something that everyone else had neglected?" At the moment when the Earth-size core of the star is crushed to less than 12 miles across, the electrons in the core are smashing together with the protons in hydrogen atoms and heavier nuclei such as iron, and turning themselves not only into neutrons but also into trillions upon trillions upon trillions of neutrinos. After the bounce, all of these particles fly outward like a sonic boom at more than 6,000 miles per second. Perhaps, Colgate thought, neutrinos were the clue. Perhaps they heated the nuclear ash plummeting inward so much that they powered an explosion.

"That did it." Now Colgate was eager to learn about neutrinos and tie them into his and White's computer simulations. He set off for Caltech again to meet with Christy. He had been impressed with Christy's work on Cepheid variables, which made use of atomic bomb design and for which he had been awarded the Eddington Medal. Christy, in turn, was intrigued by Colgate's proposals. Colgate felt comfortable enough to air what he thought was an outlandish suggestion: "Can neutrinos be treated as if they were a gas of radiation?" Christy replied, "Of course. Just look at what Chandra did for an electron gas." He was referring to Chandra's discovery of electron degeneracy pressure — that at the density of a white dwarf, a gas of electrons develops an outward pressure that prevents white dwarfs smaller than 1.4 times the mass of the Sun from collapsing. Like electrons, protons, and neutrons, neutrinos can also develop a degeneracy pressure. Aha, Colgate thought. "A vital piece of the puzzle had been neglected, namely, the huge degeneracy pressure expected from the flood of neutrinos. Excitement within the astrophysics community ran high. Colleagues generously offered advice and time. Science was really working at every step, no one was jealous." Amazing results were emerging.

"The way I thought about it then was, you talk about certifying a nuke, I keep talking about how could you possibly ever certify a supernova?" says Colgate, with a glint in his eye. By "certifying" a hydrogen bomb, he meant that he had worked out the theory of the bomb and understood how it exploded. But that was still far from true for supernovae.

By inputting models of stars into their computer and just letting it run, Colgate and White discovered far more than they had set out to find. It turned out that they had proved numerically what most astrophysicists, even Wheeler himself, had refused to believe, despite the evidence of his own and Chandra's mathematics — that stars really could undergo an endless collapse.

Up until the time they ran their first computer simulation, even the most cutting-edge astrophysical research had been based only on simplified models of stars. Those investigations were valuable, but everyone knew that the models were far removed from real stars, in which many different factors could prevent collapse or implosion. "Prior to

the Colgate-White research, conventional wisdom held that singularities were untenable," White recalled; astrophysicists refused even to
consider that a real star might begin a process of infinite collapse.[8]
They took for granted that stars would always shed enough mass to
shrink below the Chandrasekhar limit, so that when their fuel burned
out they would quietly expire as white dwarfs. The possibility of complete collapse did not even need to be considered; it was a mere mathematical conceit.

At that time, Wheeler often traveled to Livermore and was in regular
contact with Colgate and White, as well as Chandra, just as their results
were pouring from their computers. Colgate recalls:

> Wheeler, and Chandra as well, were incredibly satisfied to see that when
> suitable numbers and details referring to the constitution of matter were
> taken into account — that is, equations of state and nuclear reactions —
> it is clear that the core really can collapse. It made it very clear to Johnny
> [Wheeler] and to Chandra that collapse was inevitable and then the
> beauty is that we can't really put our finger on what happens next.

As Wheeler exclaimed excitedly to Colgate, "The big deal, Stirling, is
that it makes a reality of these things!"[9] But to Colgate it was not such a
big deal; he had acquired his experience doing diagnostics on some of
the largest hydrogen bombs ever produced by the United States, working on both theoretical and experimental physics. His work had always
involved first working out the theory and then putting it into practice.
He was convinced that concepts like complete collapse or relativistic
degeneracy were practical realities. For him it had never been mere
theory:

> The sense of the reality of the nuclear physics, of the radiation, of the hy-
> drodynamics seemed to me absolutely beyond question. I had under-
> stood all the phenomena first, then designed the experiments, then saw
> that the results of the experiments confirmed in detail the design physics
> and the calculations. It was not that I was nonchalant, it was instead the
> conviction that physics works at a very very deep level.

As Kip Thorne, now an eminent relativity theorist at Caltech, recalls,
Wheeler was so excited about Colgate and White's work that after one
visit he shelved the topic of the day in his relativity class at Princeton,
where Thorne was a student, and gave an impromptu lecture on what

he had just heard.[10] For Wheeler, it was a turning point in his research on the fate of the stars. It had finally been proved, conclusively enough to overcome even his most deep-rooted objections, that if a neutron core could not eject enough mass to get below the maximum mass to make it stable, it would have to collapse indefinitely, to a point of infinite density.

Colgate and White wrote up their work and in 1962 submitted it to *Reviews of Modern Physics*. The editors "could not figure out what to do with it within the context of physics, because it was really astrophysics," recalls Colgate. "After two years of muddling around," the paper was passed on to Chandra, the editor of the *Astrophysical Journal*. The paper was a mess. Despite his sessions boning up on astrophysics at Caltech, Colgate was writing about astrophysics as if it were physics. "We were rediscovering astrophysics," remembers Colgate. "Chandra took one look at it and said, 'This has got to be published. But, look here, Stirling, there's some astrophysics you need to learn.'" Chandra must have realized immediately how important this research was in verifying his life's work. The paper finally appeared in 1966. It had taken four years to publish one of the most important papers in astrophysics. Colgate is careful to point out that "Chandra carefully edited it and in places rewrote it." Among his alterations was to add a reference to a paper of his own alongside one of Eddington's.[11]

The essence of Colgate's discovery was that neutrinos provided the key to understanding how stars go supernova. Neutrinos might be the catalyst that would allow the shock wave from the rebounding core to continue outward through the star without being stalled by nuclear ash falling toward it. The core produces huge amounts of neutrinos, which are carried along by the outgoing shock wave. They heat up the matter plummeting in from the onionlike layers, which then turns around and sweeps outward, blowing off the star's outer layers and exposing the neutron star at its core.

However, fascinating as this scenario was, neutron stars remained purely hypothetical. Astrophysicists were skeptical that they really existed. Then in 1967, two astronomers at Cambridge, Jocelyn Bell and Anthony Hewish, were studying radio signals emitted by certain stars. They noticed that some were pulsing at surprisingly regular intervals, within a range of 0.25 and 2.0 seconds, like the beam of a lighthouse sweeping over the sky. What could they be? To those with fertile imaginations it sounded very much like a message being semaphored from a civilization on another planet. The pulses were too close together to come from a pulsating star like a Cepheid variable, which changes its size over a period of weeks. Nor could they be a signal from a particularly hot spot on a rotating white dwarf star; if it was spinning that fast it would be ripped apart.[12] Then, the following year, pulsing radio waves were identified in the Crab Nebula, a remnant of a supernova.

It was a remarkable moment for astronomers. Whatever was emitting these waves, it was not a white dwarf. Until then astrophysicists had assumed that white dwarfs were the end of the line. For all the talk about neutron stars, they were still in the realm of hypothesis, while the idea of stars collapsing into nothingness had been mathematically verified but still seemed like wild speculation. People doubted that either phenomenon could really exist. Now, however, it seemed that only one thing could be emitting the pulsing radio waves: a neutron star. This was the first concrete observational evidence that a star could die in some other way than as a white dwarf.

The discovery made everything fall into place. As the core of a huge collapsed star, a neutron star was bound to spin much faster than the original giant star, just as an ice skater spins faster and faster as she draws in her arms. Similarly, the magnetic field of a neutron star can be some 10 billion times that of the giant star it had been before its collapse. These spinning neutron stars became known as "pulsars," from the original notion that something was pulsing.[13] Since their discovery, more than a thousand pulsars have been found, and it is estimated that there are a million in our galaxy alone.[14] Bell and Hewish had discovered what Zwicky and Baade first predicted in 1933, what Oppenheimer and Volkoff had provided a theoretical basis for in 1938, and what had been almost ignored ever since. Neutron stars existed.

Although the Colgate-White model would be revised and refined in the coming decades as computing power increased, the basic idea remained sound. However, there was another, even more dramatic way in which a massive star might die. In the years after Chandra discovered his maximum mass, astrophysicists had circled around this possibility. Colgate and White's computations, which so excited Wheeler, had revealed that, in mathematical terms at least, it was a certainty. But even when the mathematics seemed to point to it as inevitable, astrophysicists refused to believe that a star could dwindle not just to a white dwarf or a neutron star but to a state of unimaginable nothingness — a never-ending collapse beyond smallness to a colossal density and a gravity so intense that it seemed to violate the laws of space and time. It was not until the early 1970s that the impossible finally began to seem possible.*

* See Appendix B for current developments in supernovae.

PART III

What Happens When Stars Die

12. The Jaws of Darkness

S TIRLING COLGATE and Richard White's computer simulations were the first step in persuading scientists that stars really could collapse completely. Mathematicians, meanwhile, had begun to accumulate results that helped to clear up the seemingly discordant views of the two observers in the puzzle proposed by Oppenheimer and Snyder back in 1939. To recap, a collapsing star is about to reach the size of its Schwarzschild radius. A nearby observer sees it shrinking faster and faster, almost at the speed of light, as it falls through its event horizon — the surface defined by its Schwarzschild radius.[1] To a distant observer, however, the collapsing star appears to freeze. This turned out to be an illusion: the star's gravity becomes so strong as it disappears behind the event horizon that light can no longer escape.[2] Solving the paradox required finding mathematical equations that would enable the two observers to compare their views of what happens to matter trapped in a region of space and time so greatly contorted by gravity. It turned out that Eddington had first worked out how to do this way back in 1924, while he was examining Schwarzschild's solution to Einstein's theory of general relativity, without realizing that it could be applied to collapsing stars.[3]

But the real stumbling block was the concept of a star many times bigger than the Sun shrinking to an infinitesimally tiny point. Physicists were prepared to accept any number of scenarios, however strange, illogical, contradictory or downright impossible they might at first appear, as long as the mathematics bore them out. Astrophysicists, though far more hidebound, were slowly beginning to accept that a star could disappear into nothingness.

In 1965, a year before Colgate and White completed their computer simulations, a thirty-four-year-old English mathematician, Roger Penrose, at Birkbeck College, London, had the brilliant idea of applying topology to stars. Topology is a branch of mathematics that studies the properties of objects and their surfaces that remain unchanged when the objects are distorted. A standard example is a coffee cup and a doughnut. They have different shapes but the same topology, because they each have a hole — the coffee cup's is in the handle. They can be transformed one into the other without tearing their surface. Penrose used topology to study the surface of the event horizon, to try to work out a connection between the shape of an imploding object and the singularity it can form. He came up with a watertight proof, framed in the most abstruse and rigorous mathematical detail, that whenever a star disappears through its event horizon, it is bound to continue collapsing until it becomes a singularity. This was the argument that clinched the case for complete collapse in purely theoretical terms.

The next question was whether such an infinitely small and infinitely dense star could be seen. When general relativity is applied to the question of the collapse of stars, a singularity is the logical conclusion. But at a singularity the laws of physics break down and the course of the universe becomes unpredictable, which is highly disturbing to scientists. In 1969 Penrose put forward the "cosmic censorship conjecture" — that there can be no naked singularities.[4] By definition, singularities and the bizarre physics they produce are hidden from view behind the event horizon, from which nothing can escape. But general relativity does not take quantum effects into account. It could be that a new physics — a quantum theory of gravity — is required to describe what goes on beyond the event horizon; there might not be a singularity to worry about after all.

As it happened, Penrose's brilliant topological idea was not pure mathematical speculation. He had been inspired by an extraordinary discovery made by Maarten Schmidt, a thirty-four-year-old Dutch-born astronomer at the California Institute of Technology, two years before Penrose published his paper. Schmidt had been deciphering data on the composition of light as it is broken up by a spectrograph into its different frequencies, which appear as spectral lines on a photographic plate.

Scientists have learned practically everything we know about the stars — from their chemical makeup to their movements — by analyzing their spectra. A prime example is the Doppler effect, which reveals how fast a star is moving in relation to Earth. The Austrian scientist Christian Doppler explained this phenomenon, which causes the pitch of an ambulance siren to change as it moves away from us: the distance between successive peaks in the sound wave from the siren increases by the distance the ambulance has traveled between emitting the two peaks, and our ears interpret the longer sound wave as a lower pitch.

In 1842 Doppler predicted, on the basis of the wave theory of light, that light would behave in the same way — that the color of a luminous body such as a star will vary depending on its speed relative to the observer. Twenty years later, astronomers confirmed that prediction. To measure a Doppler shift in light, scientists compare the wavelength of a spectral line from a particular element in the laboratory with the wavelength of the same spectral line in a star. Thus they can calculate the speed of the star in relation to Earth. A star moving toward us (a star orbiting another in a binary system, for example) appears bluer than it would otherwise because its spectral lines are shifted toward shorter wavelengths, at the blue end of the spectrum. This effect is known as a blueshift. Stars moving away from Earth appear redder because their spectral lines are shifted toward longer wavelengths, at the red end of the spectrum. This optical redshift is different from the gravitational redshift discussed earlier, which is a consequence of the effect of gravity on light.

In 1929 the forty-year-old American astronomer Edwin Powell Hubble used the optical redshift to make a momentous discovery that contributed greatly to our understanding of the universe. For almost a decade he had been working to classify galaxies in terms of their shapes and distances from Earth, using the 100-inch Hooker Telescope at Mount Wilson Observatory near Pasadena, at that time the biggest and best telescope in the world. As a result of painstaking measurements, he discovered that galaxies are redshifted — in other words, moving away from us — at a speed that depends on their distance. The farther away the galaxy, the faster it is moving. What is now known as the Hubble law states that the speed at which a galaxy is receding from us is the product of its distance from Earth times the Hubble constant. A wealth of observations has set the Hubble constant at 70 kilometers (42 miles) per second per 19 million trillion miles (a megaparsec). In other words, for each 19 million trillion miles a galaxy is away from us, the speed at which it is receding increases by 70 kilometers per second.

Hubble's discovery was the first piece of observational evidence that the universe is expanding, and thus was evidence for the Big Bang theory of the universe's creation, which astrophysicists began to formulate in the late 1960s. According to this theory, some 13 billion years ago space exploded out of a tiny, superdense, superhot state — a singularity in space and time — and has been expanding ever since, like the surface of a balloon. The different galaxies are like spots on the balloon's surface; they move farther and farther apart as the universe expands. In other words, the redshift is the result, not of the relative motion of the galaxies and Earth, but of the expansion of the universe itself. It is a "cosmological redshift."[5]

Shorter-wavelength radiation — gamma rays, x-rays, ultraviolet radiation, and often infrared radiation — cannot penetrate Earth's atmosphere, so astronomers use telescopes carried on orbiting satellites to observe it. Radio waves, however, at the opposite end of the spectrum, have long wavelengths and low energy and frequencies. They pass through Earth's atmosphere, so astronomers can study them from the ground. Radio astronomy began almost by accident in 1932, when Karl Jansky, working at the Bell Telephone Laboratory, was trying to identify some hissing static that was interfering with transatlantic telephone calls. The static turned out to be radio waves coming from the central portion of the Milky Way that were far stronger than any from the Sun. Initially radio astronomers used a concave dish to focus radio waves into the center, where they were scooped up by an antenna. But after World War II, in which radar played a vital part, radio astronomy became radically more sophisticated, with highly trained operators using much-improved instruments.[6] Radio astronomy was responsible for one of the most important discoveries of all.

When charged particles accelerate, they emit radiation. High-speed electrons spiraling through a galaxy's magnetic field emit long-wavelength radiation, that is, radio waves. Every galaxy emits radio waves, but some are more powerful emitters than others and emit radio and light waves of equal intensity. That is highly unusual; most galaxies mainly give out light. The most intense radio waves from distant galaxies are a billion times stronger than the radio waves from the Milky Way. By the late 1940s, astronomers had detected three different sources of exceptionally strong radio-wave emissions, far more powerful than any that had ever been detected from a galaxy. They pinned the emissions down to certain constellations, but no one had any idea what they could be.[7]

Walter Baade (Zwicky's one-time collaborator) and the fifty-six-year-old German-born astrophysicist Rudolf Minkowski decided to try to locate the optical counterparts of these mysterious sources.[8] In 1954, using the largest optical telescope in the world, at the Palomar Observatory on Mount Palomar in California, they found an object that they knew could not be a star, more or less at one of the sources of strong radio-wave emissions, in the constellation Cygnus, outside our galaxy. They called it Cygnus A. Its redshift indicated that it was speeding away from us at 10,200 miles per second — 5 percent of the velocity of light. Applying the Hubble law showed that it was 4,500 million trillion miles away.[9] Sirius, the brightest star that can be seen with the naked eye in the night sky, is 52 trillion miles away, while the Sun is a mere 93 million miles from us. If Cygnus A were closer and radio waves could be seen, it would have a radio "luminosity" 10 million times that of the entire Milky Way. It is in fact the strongest source of radio waves in the northern sky.

Studying their photographic plates, Baade and Minkowski suggested that the source of this barrage of radio waves might be two galaxies colliding. But the galaxies would have to be colliding again and again, with enormous violence, to maintain such a huge output of radio waves — an unlikely idea.[10] Around that time, astronomers using radio telescopes at Jodrell Bank in England noticed that the radio emissions came from two widely separated "lobes." It appeared that Cygnus A was made up of an extremely high energy, central "starlike" object with a radio lobe on either side.

By 1963 astronomers had made two other oddball sightings using radio telescopes. These were catalogued as 3C 48 and 3C 273.[11] As with Cygnus A, astronomers using optical telescopes found optical counterparts. But these were far more mysterious even than Cygnus A. Certain of the lines in their spectra made absolutely no sense: they did not represent any known elements.

Might this mean some entirely new elements in the universe? Maarten Schmidt and another astronomer at Caltech, Jesse L. Greenstein, a contemporary and former colleague of Chandra's at Yerkes, often discussed this possibility with Willy Fowler, Caltech's expert on the chemistry of stars. Usually scientists discuss their work over coffee, but with

Fowler it was martinis. He kept saying that this "new element business" made no sense.[12] It was Schmidt who proposed that the mysterious spectral lines in 3C 273 could be lines from the light emitted by hydrogen atoms but with their wavelengths redshifted by an astounding 16 percent. If that was the case, it would mean that 3C 273 was 12 billion trillion miles away from us, even more distant than Cygnus A, and receding at 25,000 miles per second, 15 percent of the speed of light.[13] Schmidt put forward his theory rather tentatively, saying only that explaining this phenomenon in extragalactic terms seemed the most "direct and least objectionable."[14]

Within days, Greenstein and Thomas A. Matthews at Caltech had worked out that the apparently indecipherable spectral lines of 3C 48 were redshifted by 37 percent, meaning that the source was receding from us at 46,500 miles per second, 25 percent of the velocity of light, and, applying the Hubble law, that it was 24 billion trillion miles away.[15] In other words, it would take light from these sources more than a billion years to reach us. Looking at them was like glimpsing the universe over a billion years ago.

Given that they were so far away and yet could still be photographed, 3C 48 and 3C 273 also had to be enormously bright. 3C 273 turned out to be a thousand times brighter than its entire galaxy, which it completely obscured. It was as bright as 25 trillion Suns. This starlike object, which was right in the nucleus of a fast-moving galaxy, was a hundred times brighter than the luminous galaxies surrounding radio sources such as Cygnus A. Schmidt also noticed a wisp of a jet, as he put it, extending from the object toward one of the lobes. A close connection between the jet and the starlike object, he mused, would be "intriguing."[16]

These extraordinarily bright starlike objects have as much energy as would be produced by applying Einstein's equation, $E = mc^2$, to the mass of a billion Suns — a galaxy's worth of stars. Astronomers measured variations in their output, and by 1965 had discovered that 3C 273 generated its energy — produced its "light," as it were — within a region 1 trillion miles across. The radiation from objects such as 3C 273 differed dramatically from that of ordinary stars. It consisted almost entirely of very high-energy gamma radiation and x-rays, emitted from a material hotter than 100,000 degrees kelvin.

One possible explanation was gravitational redshift, in which intense gravity can increase the wavelength of light, making it look redder.

Greenstein and Schmidt examined this hypothesis and concluded that it was very unlikely that the redshifts for 3C 48 and 3C 273 could be produced by gravitational means. To produce such extremely high gravity, the objects would have to be so hyperdense that the laws of physics, at least as they were known at the time, would be strained.[17] They also came up with a name for these starlike objects, dubbing them "quasi-stellar objects," which they shortened to "quasars." But they did not completely discount the gravity hypothesis. Perhaps a gravitational redshift might take place under some other condition.

There was no agreement at all among astrophysicists on the source of the quasars' enormous power. Perhaps, exaggerating Baade and Minkowski's original suggestion, the energy was generated by several galaxies colliding. But such a collision would require an impossibly high efficiency rate for the conversion of mass into energy. Or could it be that 100 million closely packed stars on the verge of going supernova, each thirty to fifty times the mass of the Sun, suddenly exploded in a chain reaction? In 1967, Stirling Colgate formulated a model along these lines and demonstrated that, incredible though such a celestial fireworks display might sound, it agreed fairly well with the data. The idea was ingenious, but it was fiendishly difficult to work out the details.[18]

Hoyle and Fowler, meanwhile, had been working on a simpler but very dramatic model.[19] Since the radio waves seemed to emanate from the center of the galaxy, perhaps a quasar was some sort of supermassive star. This star, formed from the dense cloud of gas at the center of a galaxy in the usual way, might have become so massive that its outward radiation pressure exceeded the pressure of gravity pushing inward. At this point it would have to stop taking in gas or it would explode. Fowler and Hoyle suggested that these supermassive stars must be 100 million times the mass of the Sun. After each period of burning, the star would collapse under the pressure of its own gravity. Perhaps, then, supermassive stars undergoing gravitational collapse were the source of the quasars' prodigious energy. These two ideas — Schmidt's theory of quasars and Hoyle and Fowler's suggestion that their energy might come from a massive star that had collapsed to about the size of its Schwarzschild radius — inspired Penrose to look more deeply into the mathematics of a star's collapse.

One day, walking across the Caltech campus, Fowler bumped into his colleague Richard Feynman, a brilliant forty-five-year-old physicist. Two years later, in 1965, Feynman was to share a Nobel Prize for figuring out a cure for the ills of quantum electrodynamics, a cure that had eluded Dirac, Heisenberg, and Pauli (the theory had predicted, among other things, that the electron had an infinitely large mass). Feynman cut straight to the chase. "Willy, you know those supermassive objects that you and Fred have been working on are unstable," he said. "They'll collapse due to general relativity." Hoyle and Fowler had been using Newton's theory of gravity in their model. But the general theory of relativity gives gas molecules a far stronger gravitational attraction than Newton's theory does. This meant that as a supermassive star contracted while letting off energy in a cosmic belch, it would be unable to stop contracting and would go on collapsing indefinitely. Fowler ran back to his office. He "immediately put general relativity in [and] found out right away that Dick was right, and the damn things would collapse."[20] He tried to save his model by varying the amount of nuclear fuel burned and making the star more realistic by making it rotate, "but it really wasn't enough."[21]

By 1960 Chandra had turned his attention to relativity theory, a field he had long avoided. He did so after a long period of studying hydrodynamic and hydromagnetic stability and writing another of his famous tomes (published in 1961), titled, not surprisingly, *Hydrodynamic and Hydromagnetic Stability*. Chandra was not happy with the way his work in this area had turned out. The calculations had been long and hard, and he felt that he had wasted ten years working on a problem that was ultimately inconsequential. He decided to turn to "more contemplative matters." But the decision to turn to relativity was more than that — it was "something to do with vanity." For years he had wanted to work on pure theoretical physics, like Dirac. "There's not a single table in any of his papers," Chandra remembered thinking.[22] But he was hesitant. Could he come up with significant ideas? He confided in his friend and colleague in the physics department, Gregor Wentzel, who urged him to take the leap. After all, he said, "They can't fire you."[23]

In 1962, to update himself on current developments, Chandra attended a meeting in Warsaw on the general theory of relativity. There he discussed astrophysics with Zel'dovich, whom he had not seen since his trip to Russia twenty-eight years earlier. He came away feeling that there was a place for someone with extensive experience in astrophysics who could formulate problems "whose solution might be useful."[24]

Around that time Willy Fowler's predicament about the stability of supermassive stars was being discussed. When a physicist of Feynman's stature comments on a topic on the edge of science, the news spreads immediately. Chandra was familiar with Fowler's computer calculations in which stars 100 million times the mass of the Sun burn up their nuclear fuel, then collapse. He had also heard about Feynman's intuitive response and Fowler's desperate attempts to rescue his theory. The topic struck a chord in Chandra, taking him back to his days in Cambridge thirty years earlier, and he set to work. The task was like the one that Eddington had set himself in 1916: to explore the stability of stars going through periods of bulging followed by compression. Chandra simply applied the general theory of relativity and thus, he claimed, obtained results Eddington could have found decades earlier.

In 1964, a year after Feynman's flash of intuition, Chandra published his first papers on the general theory of relativity.[25] He proved with mathematical certainty that if a star 100 million times the mass of the Sun began to pulsate, it would become unstable. Once it had pulsed down to the estimated size of a quasar, 100 billion miles across, it would have to collapse completely and would therefore cease to exist. Thus Hoyle and Fowler's postulated supermassive star could not be the source of the quasars' energy. Chandra was delighted with this result. He was, he declared, "pretty convinced" that singularities — stars of infinite smallness and infinite density — really could exist and that enormous stars, above the upper limit he had established for the mass of a white dwarf, could shrivel to nothingness and disappear into a niche in space and time.[26]

That December all the major players in quasar research, including Chandra, gathered for the Second Texas Symposium for Relativistic Astrophysics at the University of Texas in Austin. Information and ideas were flooding in. In session after session, Fowler and his coworkers "presented another talk with yet another theory to fit the rapidly expanding database. Some of their theories put the quasars at the edge of the solar system, others at the edge of the universe."[27] One day redshifts were gravitational; the next, cosmological. The scientists proposed different theories for different data at different times of day.

Fast jets took the delegates as far as Dallas, but from there they had to travel to Austin on propeller-driven planes — the "milk run." Heading back to Dallas on one such flight after the meeting, Richard White found himself among such luminaries as Margaret and Geoffrey Burbridge, Chandra, Fowler, Wheeler, and Schmidt. Soon after takeoff, when the seat-belt sign went off, Fowler stood up, cupped his hand over his brow in a dramatic gesture, scanned the passengers in their cramped seats, and announced in his booming voice, "If this plane were to crash, we could get a new start on this quasar problem."[28]

The mystery of quasars and what made them so bright had still not been solved. Then in 1964, Edwin Salpeter at Cornell University and Zel'dovich and his co-workers in the Soviet Union proposed a scenario in which an object more than a million times the mass of the Sun might move through an interstellar gas cloud, scooping up gas particles. Eventually this supermassive star would undergo gravitational collapse through its event horizon, collapsing until it was an unimaginably tiny, dense point. Then it would drag space and time in around it, continuing to move along, scooping up gas particles like a cosmic vacuum cleaner. Accelerating to speeds close to that of light as they were sucked over the event horizon, particles near the edge would crash into each other, heat up, and emit enormous amounts of radiation in the form of x-rays. Salpeter and Zel'dovich had discovered a way in which far more radiation could be generated than that produced in a nuclear reaction — as much as the energy radiated by an entire galaxy.

All the various lines of research in different fields were converging. A year later, in 1965, Penrose produced his elegant topological proof that if a star disappears through its event horizon, it is bound to keep disappearing forever. Then in 1966, Colgate and White used computer simulations to investigate the actual mechanisms by which stars could collapse. It was John Wheeler who, in 1967, found the perfect and deeply memorable term to describe the region of space into which a collapsed star falls: a "black hole."[29] At the center of a quasar was a black hole, with particles on the verge of tumbling over its event horizon emitting enormous amounts of radiation. In other words, quasars were powered by black holes. No one could doubt any longer that black holes existed. The tempo of research quickened.

In 1969, Donald Lynden-Bell, an English astrophysicist at the Royal Greenwich Observatory in Sussex, took Salpeter and Zel'dovich's model and applied it to the center of a galaxy. He suggested that a massive star that completely collapsed through its event horizon would become such a strong source of gravity that it would drag everything in its vicinity into a disk that would spin around it. From this idea it suddenly became obvious that black holes too must be spinning. As massive stars blow off mass and grow smaller, they spin faster and faster. If that was the case, then black holes were not merely receptacles for dead stars — they had structure. As it happened, the mathematics of a spinning black hole had already been proposed in 1963 by a twenty-nine-year-old mathematician from New Zealand, Roy Kerr, who had done a Ph.D. at Cambridge and was on the staff of the University of Texas in Austin.

Kerr applied Einstein's general theory of relativity to space and time around a spinning spherical object such as a star.[30] His solution, which provides a way of measuring distances and times in the space distorted by the object, is referred to as the Kerr metric, or measure. The metric we use in everyday life is based on Euclidean geometry, the geometry we learn in school, which assumes that space is flat and three-dimensional.[31] The Kerr metric describes the structure of the curved four-dimensional geometry required by the general theory of relativity, that is, the three dimensions of space plus time. Schwarzschild's solution of Einstein's general theory of relativity — the Schwarzschild metric — described space and time around a nonspinning spherical object; as Oppenheimer and Snyder realized in 1939, it is actually the metric for the region of space containing a nonspinning collapsed star. Thus spinning black holes are referred to as Kerr black holes, and nonspinning ones as Schwarzschild black holes. Kerr's metric becomes Schwarzschild's when the black hole's rotation is set to zero.

The power of Kerr's solution is that it applies to every possible black hole. Its beauty lies in its simplicity, in that its description of every black hole requires only two parameters, the black hole's mass and its spin. The mass of a black hole is the mass of the star and the material that has fallen into it, which can be measured straightforwardly if the black hole is in a binary system; its spin is the number of times it rotates per second, a number that can be in the thousands if it results from the collapse of a huge star.

But how was such a simple classification possible? After all, each spinning, collapsing star had different surface characteristics. Yet after the star collapsed through its event horizon, the event horizon always turned out to be perfectly smooth, like water after a pebble has fallen into it. Every physical characteristic of the star that fell through, momentarily deforming space and time, vanished except for its mass and spin. These were left behind, a bit like the Cheshire Cat's smile. This astounding discovery was made in the 1960s, primarily by Zel'dovich and his group. Once again, Wheeler found memorable words for it: "black holes have no hair."[32] No matter what sort of star fell through its event horizon, it left no trace of its previous idiosyncrasies. Penrose's cosmic censorship hypothesis went on to place a limit on the mass and spin of a black hole. If the black hole was spinning faster than the limit, the rim would "roll down" to reveal the singularity within, which was unacceptable according to the laws of physics.

Chandra was eager to join the party. In the autumn of 1971, he took research leave to spend three months with one of the most active groups in the field, headed by Kip Thorne, then thirty-one and professor of theoretical physics at Caltech. Thorne had been involved in black hole physics right from the start and was now at the leading edge of research. He had studied under Wheeler and was one of the authors of *Gravitational Theory and Gravitational Collapse.* Six feet tall, rail-thin and wiry, with long hair and a beard, Thorne exudes the assurance of an expert, moving with ease from the complex mathematical theory of black holes to highly technical aspects of the equipment used to detect them. He imparts to his students a "we're all in it together" sense of teamwork.

Thorne remembers Chandra affectionately. Always immaculately dressed in his conservative "Chandra gray" suit, he would eagerly join Thorne and his scruffy graduate students for lunch in the student cafeteria (the "Greasy"), rather than the much more formal and elegant Athenaeum, Caltech's famous faculty club, where eminent guests such as Einstein stayed. Chandra was always in his office bright and early with his door open, ready to talk to any astronomer who passed by.

Two of Thorne's graduate students, William H. Press and Saul Teukolsky, drew Chandra's attention to Kerr's result. He was stunned by the power of its mathematics:

> In my entire scientific life, extending over forty-five years, the most shattering experience has been the realisation that [Kerr's] exact solution of Einstein's equations of general relativity provides the *absolutely exact representation* of untold numbers of massive black holes that populate the universe. This "shuddering before the beautiful," this incredible fact that a discovery motivated by a search after the beautiful in mathematics should find its exact replica in Nature, persuades me to say that beauty is that to which the human mind responds at its deepest and most profound.[33]

Teukolsky went on to look for a way to calculate how spinning black holes interact with their surroundings when, for example, they are hit by electromagnetic waves or by gravitational waves.[34] The latter are the ripples in space and time caused by the movement of objects; they are predicted by general relativity but have yet to be observed directly.[35] Cutting through the complications of Teukolsky's work, Chandra sensed the glimmer of a connection between the structure of black holes, electromagnetism, and gravitational waves. Teukolsky recalled Chandra's intense determination to understand this connection in as complete a way as possible. The resulting search involved Chandra in some of the most complex calculations he had ever done.

By then the perception of black holes had been transformed. Far from being rejected as ugly, ungainly objects that spoiled a harmonious universe, they had come to be seen as the "most perfect macroscopic objects there are."[36] This was an utter reversal of the situation on January 11, 1935. As Chandra put it when he began his research, quoting Eddington's famous dismissal with mingled bitterness and irony, "We are today considering seriously situations that were brushed aside as *reductio ad absurdum* not so long ago."[37]

On August 5, 1971, the spaceship *Apollo 15* was about to start its long journey home from the Moon. This was the first manned scientific expedition into outer space. Colonel David R. Scott and his crew had already walked on the Moon and circled Earth outside their spaceship, carrying out repairs. But as well as being pioneers of exploration, they were required to carry out a variety of scientific experiments. That day the crew members were tracing the sources of x-rays passing through space. They had identified rapid irregular bursts of x-rays coming from Cygnus X-1, an x-ray star 20,000 trillion miles away that was the companion star of the blue supergiant HDE 226868, in the constellation Cygnus.[38] Their results would be incorporated into observations made since the previous year in a scientific spinoff of the cold war.[39]

After the Soviets tested the massive Czar bomb in 1961, the United States launched surveillance satellites that would detect x-rays from Soviet hydrogen bomb tests. Circling Earth, these satellites began to pick up all sorts of surprising signals and information, such as intense x-rays that seemed to be coming from far outside our galaxy. On December 12, 1970, NASA launched SAS-1 (the first of three Small Astronomical Satellites) from the San Marco Platform in Kenya to continue steady surveillance of these x-rays. The satellite — renamed *Uhuru*, the Swahili word for "freedom" — homed in on the source of the x-rays, which turned out to be Cygnus X-1. But whereas *Uhuru* could study this source for only two minutes at a time, the *Apollo 15* crew could monitor Cygnus X-1 for an hour.[40]

What first drew scientists' attention to Cygnus X-1 was the extreme variability of its x-ray bursts. Some continued for months, others were as short as a thousandth of a second. Because an object flickers for exactly the length of time it takes light to travel across it and because the speed of light is constant, scientists can work out the size of an object from the longest duration of light it emits. They calculated that the x-rays were coming from a compact object about 1,800 miles across at most — less than a quarter the size of Earth. What tiny object could be generating so much power? From comparisons with other blue supergiants, astrophysicists had estimated that HDE 226868 was about thirty times the mass of the Sun. As Cygnus X-1 was its binary companion, they took account of the shape of the orbit of the two and the mass of HDE 226868 to deduce that it was at least seven times the mass of the Sun. That made it too big to be either a white dwarf (maximum mass, 1.4 times the mass of the Sun) or a neutron star (maximum mass, 2 to 3 times the mass of the Sun). With growing excitement they realized that Cygnus X-1 had almost all the necessary characteristics of a black hole.

Today most astrophysicists agree that Cygnus X-1 is indeed a black hole, relentlessly devouring its companion star. The "mouth" of the black hole itself, its event horizon, is just 26 miles across. Deep inside is the collapsing star, continuing its eternal demise. The matter that the black hole sucks from its blue supergiant companion forms an "accretion disk," a broad flat disk of matter revolving around the event horizon like a vast compact disk in a CD player. The x-rays come not from the black hole — no radiation can escape from there — but from the gas in the accretion disk. The inner parts of the disk rotate faster than the outer ones, and as the inner particles rub against their slower neighbors, they heat up from the friction and give off x-rays. Matter from the accretion disk is continually being dragged into the black hole the way water swirls down a drain, from the inside outward. By measuring the spectrum of the x-rays, astrophysicists calculate that they come from a gaseous accretion disk some 4 million miles across. The hottest part — the portion emitting the x-rays — is some 300 miles across.

Today we take this scenario for granted. But for decades research was impeded because astrophysicists simply refused to believe that stars could disappear.

Chandra was aware that black hole research would not be an easy field to enter, for it was already full of brilliant and experienced young men. Also, by 1975, when he started writing papers, it seemed that the most exciting discoveries had been made and the golden years of black hole physics were over. All that remained was to clarify and systematize the details. But that was exactly what Chandra liked best. Instead of juggling numbers, he would be working in pure mathematics, dealing with symbols without "a single table" of figures, as in Dirac's work.

Chandra began by streamlining the subject and creating a direct and simple derivation of Kerr's metric.[41] It was quite an accomplishment. In the opinion of experts, including Kerr himself, it could not have been done without horrendously complex calculations. The next step was to establish a method of analyzing what might happen when a rotating black hole is hit by electromagnetic or gravitational waves and how that black hole might affect a particle moving near it. Scientists usually begin by taking only small changes into account and setting aside larger ones. Chandra took the reverse path and "just bull-dozed in," as he had once described Bethe's approach to physics problems.[42]

In the 1980s, further mathematical developments of Kerr's solution and better data from observations of quasars finally enabled astrophysicists to come up with a unified description valid for all sources of radio waves, from weaker radio galaxies to quasars. It was a stunning marriage of general relativity and theoretical and observational astrophysics. A quasar, it seems, is powered by a supermassive spinning black hole, accompanied by its accretion disk, which is also rotating. (The black hole powering the quasar 3C 273 is a billion times the mass of the Sun.) As the disk spins, it compresses the gases that make it up. This compression, together with the friction of the gas particles, generates extremely high temperatures. The highest energy radiation comes from the innermost part of the accretion disk, which spins fastest and is thus hottest. The whole disk is about a trillion miles across. As particles are sucked into the black hole in 3C 273, which is nearly a billion miles across, it gives off gamma rays and x-rays.

In recent years, astronomers have been able to make many observations of quasars using radio, x-ray, and infrared devices. There are, it appears, two lobes of gas on either side of an active galaxy, or quasar, at whose center is a supermassive black hole surrounded by an accretion disk. Shooting out of the inner part of the disk are two back-to-back streams, narrow jets of electrons traveling at speeds close to that of light and emitting radio waves. These jets can extend over vast reaches of space — 100,000 trillion miles, five to ten times bigger than the entire galaxy in which the quasar is located. As the jets ram into the lobes and excite the atoms in the gas, the lobes in turn emit radio waves. The luminosity of the biggest quasars obscures the stars in the host galaxy, so that there appears to be just a single star, one quasi-stellar object.[43]

Quasars are found only in older galaxies, far, far away in both space and time. Over time, the accretion disk around a spinning black hole uses up all its gas particles, and the quasar ceases to emit light — it shuts down. We can only conclude that there must be millions of black holes out there. Thanks to these developments, beginning in the 1960s, Chandra's discovery of 1930 was finally vindicated and Eddington proved wrong.

13. *Shuddering Before the Beautiful*

ON FEBRUARY 6, 1960, Babuji died suddenly of a heart attack. He was seventy-five. Chandra had not seen his father since 1951, and that had been on his first visit home in fifteen years. Babuji had always regretted that his brilliant son never returned to live in India, which led to periods of great strain in their relationship. The problem had come to a head in 1953, when Lalitha and Chandra were pondering whether or not to become naturalized American citizens. It was a momentous decision for them, because it would mean giving up their Indian citizenship. There were practical reasons for it, such as Chandra's difficulties in leaving and returning to the United States on an Indian passport. On a deeper level, having noticed the poor research conditions in India during his visit and been reminded of the savage academic rivalries there, he had become convinced that it would be very hard to return to the country of his birth.

America was also becoming more liberal. In 1952 Adlai Stevenson, a brilliant and enlightened Democrat, campaigned for the presidency on a platform that included ridding the country of McCarthyism. The Chandrasekhars, caught up in the liberal fervor, threw themselves into Democratic party activities in Williams Bay. Then in December the Immigration and Naturalization Act was passed, allowing a small quota of Asians to become U.S. citizens. After much deliberation, the Chandrasekhars decided to go through the arduous procedure, and in October 1953 they became Americans.

For Babuji the news was a slap in the face, both for himself and for India. He accused Chandra of turning his back on his homeland and his Hindu heritage. Deeply hurt, the Chandrasekhars pointed out that they had lived in the United States for seventeen years and intended to remain there. They had discussed the matter with the Fermis, the Kuipers, and the Schwarzschilds, all of whom had chosen to become citizens. Chandra and Lalitha intended to keep close ties with India, just as their friends did with their native countries. Nothing helped — Babuji was deeply offended. Chandra had no communication with him for several years.

And now he was dead. For Chandra, the news must have come as a shock. No doubt he thought back to his lonely days at Cambridge, when Babuji had lent a sympathetic ear. Perhaps he remembered too the emotional evening in London when Babuji was about to return to India, when Chandra prayed to God to make him a worthy son. As soon as he could, a year after Babuji's death, he made the long journey to India. At Chandra Vilas he sought out all the letters he had sent to his father, going back to the time when he was at Presidency College.

Chandra was now concentrating his efforts on general relativity. But the field was darkened for him because one of the dominant figures in relativity studies had been his demon, Eddington. Then one Sunday morning in September 1974, when he was sixty-three, he felt a terrible tightening in his chest. Lalitha rushed him to the hospital, where he had an angina attack. Chandra recalled the doctor's assessment of his condition: "On a scale of 5, 5 being fatal, mine was 3½."[1] Had he not been in the intensive care unit at that moment, he would have died. Forcefully reminded of his own mortality, he thought back to Babuji's death. Three years later, Chandra had another heart attack and underwent an extremely risky triple bypass operation. The medical staff suspected there would be postoperative problems, and indeed they came to a head early the next morning. Excessive bleeding resulted in a dangerous buildup of blood in his lungs, necessitating an emergency procedure. Assuming that her husband was recovering safely, Lalitha had already gone home; Chandra had to be roused to sign the permission slip himself. A lengthy convalescence followed. "It was extremely painful," he remembered with anguish.[2]

He did not fully recover his health until 1980. His work on black holes kept him going, but the years of gloom and disappointment were taking their toll. Chandra had always kept an informal diary of the problems he was working on and his scientific progress. Now his notes took on a more painfully introspective tone.

On January 20, 1976, he copied out some lines from Gustav Mahler's

Das Lied von der Erde ("The Song of the Earth"), which must have seemed to resonate with his own deepest feelings:

> My fortune was not kind to me in this world!
> Whither I go? I go, I wander in the mountains,
> I seek rest for my lonely heart![3]

That November he was agonizing over his research into black holes. He was trying to disentangle some lengthy equations, but they did not yield.[4] Dispirited, he wrote:

> I feel utterly discouraged, despondent, and lonely. Life has lost its savour. And I am desperately tired. I wish that I could overcome the present impasse in my work: I have been up against a stone wall for nearly three months . . . How I long for the time when the book will have been written; and I with a sigh of relief leave this university!; and go into exile like the saints of old. I wish, I wish but so many vain things continue to haunt me.[5]

Chandra was beginning to assemble his results into a book that would be titled *The Mathematical Theory of Black Holes*. But he felt tormented over whether writing the book was worth the effort. Had he lost his passion for research? By then he had received every important scientific award except the highest of all, a Nobel Prize. His work on white dwarfs had been recognized in 1974, when the American Physical Society presented him with the Dannie Heinemann Award for excellence in mathematical physics, always his "first love." He had also received the prestigious Gold Medal of the Royal Astronomical Society, and the Bruce Medal, given by the Astronomical Society of the Pacific. But he complained that these were neither fulfilling nor encouraging "in the kind of work that I considered most relevant" — his discovery of the upper limit to the mass of a white dwarf, which was mentioned briefly in the award citations.[6]

Four years later, Chandra's mood continued to be dark. Just before his seventieth birthday, on October 18, 1980, he wrote: "Why has the optimism of the forties, and even the remnants of the sixties, dissipated? How is it that my outlook has become dark and pessimistic: with no hope, no joy?"[7] The next day he meticulously wrote a couple of lines in his neat handwriting on the same page:

> Oh!
> Let me not imagine vain things
> Let my consolation be my book.

He found it impossible to hide his deep sadness. Once in the 1980s, a physics department meeting finished earlier than expected. The professors relaxed and, one by one, began to reminisce. When it was Chandra's turn, his eyes filled with tears. He felt, he confessed, only regrets about his life. Perhaps it had all been in vain. "I am afraid that by giving to the pursuit of science the highest priority, one necessarily distorts one's personal life," he said in an interview in 1977. "That includes my marriage, in the sense that life has been very hard for my wife . . . I had, in fact, no time for anything else [besides science]."[8] He often referred to Lalitha as his "noble wife," as indeed she was. She had consciously made the decision to give up her own career in physics to make life as comfortable as possible for her husband, so that he could concentrate entirely on his research. She took care of all the household tasks except shopping, which he enjoyed. As she says, rather sadly, gazing wistfully into the distance, "so many things could have been different."[9] She had dreamed of working with her husband and keeping up to date with developments in science. But that had proved impossible.

"What is the point of it all? I feel so cheated," Chandra scribbled in his diary.[10] The only remedy was to carry on working obsessively, even though he questioned whether the work ultimately had any meaning. Work became his salvation, helping him through the bouts of severe depression that followed his heart attack. He seemed to believe, like Picasso, that as long as he kept busy he would not die. Then late in 1982, when he finished his book on black holes, the futility of existence overwhelmed him once again. He recorded the emotional ups and downs of the previous half century in a long diary entry:

> During the 40s, 50s, and 60s, I did have occasional spells of depression and gloom. But the overriding anxiety to be scientifically productive subdued those periods; besides, certain hopes I had were realized. And on the whole, I maintained a feeling of wholesome well being. But since 1969, I have worked and lived with essentially no encouragement; and whatever hopes I may have had, were shattered. And nothing assuaged the enveloping gloom. Still the persistence to be productive kept me going. The only bright interludes were brief, few, and far between — as during May 1976 when I separated Dirac's equation and February 1978 when I decoupled the Newman-Penrose equations governing the perturbations of the Reissner-Nördstrom black hole. I escaped the gloom of the 70's during 1980–82 by my preoccupation with the book. I cannot imag-

ine what those years might have been had I not had the book constantly weighing on me. Now that the anodyne that kept me going during the past three years has been exhausted, I am at a complete loss; and I do not have even the desire to be productive. The book seems to have taken revenge and destroyed my motivations. The future is bleak and there is no prospect even for a moment of joy. I am against a dark impenetrable wall; and I foresee no release.[11]

The weight of memory began to take its toll. It was not just the usual malaise that every writer feels on completing a book. Something else was preying on his mind. Perhaps through his study of black holes he had hoped to fulfill his dream of becoming a theoretical physicist, like Dirac. But his work had been inconclusive and not well received, and the honors had come late in life.

In *The Mathematical Theory of Black Holes*, Chandra often states that he is seeking the "simplest solution." But in the end both style and simplicity eluded him. That is especially evident in the monumentally complex calculations by which he solved the problem of what happens to a black hole hit by gravitational waves. It tormented him that he was unable to arrive at the solution by simpler means. Perhaps in the future, he wrote, the "complexity will be unraveled by deeper insights." In the meantime, "the analysis has led us into a realm of the rococo: splendors, joyful, and immensely ornate."[12] For details of the calculations, he refers the reader to the six hundred folio-sized pages and six notebooks that he put on deposit in the Joseph Regenstein Library at the University of Chicago. They are awe-inspiring to behold.

His contemporaries complained that Chandra's excessive use of mathematics tended to obscure the underlying physics. But that was his way of solving the sorts of problems he liked to work on. No doubt Eddington would have gleefully ripped apart *The Mathematical Theory of Black Holes*, as he had Chandra's second book, on stellar dynamics. Colleagues joked that Chandra would rise at the crack of dawn and write two hundred equations by seven A.M. At a colloquium, he displayed equations so long they did not fit on a single transparency. Nevertheless, as he reminded everyone, "You may think I have used a hammer to crack eggs; but I have cracked eggs!"[13] In describing *The Mathematical Theory of Black Holes*, the eminent British astrophysicist and Astronomer Royal, Sir Martin Rees, quoted the nineteenth-century polymath William Whewell's description of Newton's presentation of his theory of motion: "We feel as when we are in an ancient armory where the weapons are of gigantic size; . . . we marvel what manner of men they were who could use as weapons that we can scarcely lift as a burden."[14] That was Chandra's style, a classical style very different from the elegant, modern, yet powerful topological approach introduced into black-hole literature by Roger Penrose, whom Chandra regarded as "of course, at the pinnacle."[15]

On October 19, 1983, Chandra' seventy-third birthday, it was announced that he had been awarded a Nobel Prize. Finally, he had received the greatest honor science has to offer. Congratulations poured in. But Chandra's own response was ambivalent. At the ceremony in Stockholm, the citation specified that the award was in recognition of "one of Chandrasekhar's most well known contributions, his study of the structure of white dwarfs," in other words, the Chandrasekhar limit.[16] Chandra was aggrieved. As far as he was concerned, the body of work he had produced over his whole life had been disregarded. In an autobiographical statement he listed the seven periods of his career and the epic publications that marked each: stellar structure (1929–1939), stellar dynamics (1938–1943), radiative transfer (1943–1950), hydrodynamic and hydromagnetic stability (1952–1961), equilibrium of ellipsoidal figures (1961–1968), general relativity and relativistic astrophysics (1962–1971), and the mathematical theory of black holes (1974–1983). "However, the work which appears to be singled out in the citation for the award of the Nobel Prize is included in the following papers," he continued.[17] Almost all were from the 1930s, when he was in his youth. The award statement, he argued, distorted his life's work because all it acknowledged was his early insight into white dwarfs. His friends tried to convince him that in reality the award was to honor a lifetime of research.[18] And at last his discovery of the maximum mass for white dwarfs had been fully recognized. But he was still dissatisfied.

Chandra was awarded the Nobel jointly with Willy Fowler. Fowler was immensely popular across the astrophysics profession, while Chandra was not, although of course he was highly respected. On the day the prize was announced, a symposium on nuclear astrophysics was in progress at Yerkes. Fowler gave a stirring yet self-effacing speech in response to the momentous news. David Schramm, a prominent astrophysicist at the University, phoned Chandra to congratulate him. Lalitha answered. The Nobel was long overdue, she said. Her exact words were "It's about time." Since Chandra had launched his career at Yerkes, suggested Schramm, perhaps he would like to come out and join the celebration. After all, Willy was there. Chandra politely declined.[19]

Peter Vandervoort, a former student of Chandra's, was a colleague at the University of Chicago. To mark the occasion, he gave a lecture to his graduate class in astrophysics on Chandra's maximum mass. To his surprise, when he slid up the front layer of blackboards, he found a derivation of it already written there. Vandervoort told Chandra of this coincidence, to which Chandra replied ungraciously, "I bet five years ago no one in the department could have done that derivation."[20]

Soon after the award was announced, Chandra was interviewed. It would have been different, he said, if his work had been recognized in the 1930s and the damaging controversy with Eddington had never happened. Instead, he had received no encouragement to go any further. He recalled attending several of Russell's lectures on stellar structure and hearing no reference to his work. Then he recalled the shabby treatment he had received at Yerkes when he arrived there.

The burning desire to have one's work recognized with a Nobel Prize can have a damaging effect on a scientist's life. Some who have done important work listen for the phone to ring in October, only to wait and wait. Only three scientists can win a Nobel in any one field.[21] Some become bitter because they feel they have been unjustly overlooked. By the time Max Born received a Nobel Prize in 1954, he had become paranoid about it. Chandra recalled being in the audience with Born in 1933 at a ceremony at the Cavendish Laboratory in Cambridge, to celebrate the fact that Dirac and Heisenberg had received Nobels. When the guests of honor filed in and everyone stood up to applaud, Born's eyes filled with tears. "I should be there, I should be there," he said.[22]

Some Nobel Prize winners are so overwhelmed by the lucrative speaking engagements and invitations to move into higher administrative positions that they give up their research careers. For Chandra it was just the opposite: the prize was an incentive to step into new fields of research. He decided to look into what happens when gravitational waves collide with each other and with stars, making the stars they hit change shape and "breathe" (nonradial oscillation). The collision then generates new outgoing (scattered) gravitational waves, in a process very much like waves in the ocean hitting a rock and rippling back to intersect with the incoming waves. Chandra's *Selected Publications* lists over twenty-seven articles on these subjects. His productivity never flagged.

Valeria Ferrari, now a professor of physics at the University of Rome, worked alongside Chandra in his research on colliding gravitational waves. They met in Rome in the summer of 1983. She impressed him sufficiently that he invited her to spend some time in Chicago, perhaps later that year. Excited about the invitation, she carefully prepared for this new project. Three days after the Nobel Prize had been announced, she was at his office, ready to start work. Chandra, still reeling from the sudden publicity, was stunned to see her. Ferrari's telexes had not arrived, a situation reminiscent of Chandra's first unheralded appearance in Cambridge as a student. She still recalls his graciousness amid all that was going on in his life and, most of all, his "humility of starting learning something new, digging into some difficult problem and trying to unravel its structure and order. It was certainly a quest for harmony and symmetry that led Chandra to think that the theory of colliding waves could be constructed along the same lines as the theory of black holes."[23] It was an enormously difficult task, yet a year later their first paper appeared. Chandra always referred to it as "the Valeria paper."[24]

In the mid-1980s, well into his seventies, Chandra set to work on yet another project. He had been invited to speak at a conference to celebrate the three-hundredth anniversary of the publication of Isaac Newton's *Philosophiae naturalis principia mathematica* (usually known as the *Principia*) in June 1987. As an expert on the general theory of relativity, he agreed to speak on Newton and on gravitational theory after Newton. With his usual seriousness, he began to delve into Newton's magisterial work. He was profoundly moved by the elegance of Newton's presentation and of his results, which had laid the foundations of modern science and brought about the Age of Reason. Chandra entitled his lecture "The Aesthetic Base of the General Theory of Relativity." But he was not satisfied with it and did not submit a text for the conference proceedings.[25] He described what he tried to do as follows: "I decided to try to read [Newton's *Principia*], but that I estimated would take a year, so I took 15 propositions, tried to prove them on my own, and then compare them with Newton's proofs. I was astounded at his originality."[26] Even on a historical subject, he still took a scientific approach.[27]

He was hooked. Newton became the love of this last phase of his life. Though now an old man, he planned his work for the future:

> There are two things that I should like to do before I die:
>
> 1. To write to my satisfaction and send to Press my *Lobegesang* [song of praise] on Newton.
> 2. To be able to give a carefully prepared Lecture on a suitable occasion on "The serial paintings of Claude Monet and the landscape of general relativity." [*But I shall write the essay regardless — even if the occasion does not arise.][28]

For the next two years he devoted himself to these two projects. Eight years after the Cambridge conference, *Newton's Principia for the Common Reader* was finally published. Uncharacteristically, Chandra wrote it hastily, as if racing against time, sending each chapter off to the publisher as it was completed. He also wrote jottings to himself, entitled "How I Came to Writing My Book on the *Principia*." He noted that he intended his method to be very different from that of the historians of science who dominated Newton studies. While they, as he put it, took "the easy way of talking in large, soft and vague terms," he chose to analyze Newton's work using the heavy machinery of twentieth-century mathematics.[29] If anyone could do this, it was he. At this point he was probably the world expert on Newtonian mechanics.

Chandra stated his approach in straightforward terms. He intended his book to be assessed "as an undertaking by a practicing scientist to read and comprehend the intellectual achievement that the *Principia* is."[30] Historians of science might argue that this approach would obscure Newton's thinking. He was, after all, a man of the seventeenth century, whose principal interests were biblical chronology and alchemy. When asked why he had chosen to ignore practically all the historical work on Newton, Chandra replied, "Why shouldn't I? Was Newton wrong?"[31] The reviews of *Newton's Principia for the Common Reader* spoke highly of its technical merits. Robert Westfall, one of the high priests of Newtonian studies, described it as "required reading," although he added that "historians of science will find it a disturbing book."[32] Chandra never saw this review, which was published in 1996.

He also completed his essay on Monet and the landscape of general relativity, as he had promised himself. Perhaps he felt that his life was drawing to a close. He had things he wanted to say, even though they were in areas he had never touched on before, such as aesthetics. He had the opportunity to present his thoughts on Monet as a lecture several times, first at the Inter-University Centre for Astronomy and Astrophysics in Pune, India, in 1992, then at Presidency College the same year, and in his last colloquium at Chicago's department of physics on October 6, 1994. His colleagues there were taken completely by surprise. What on earth would Chandra come up with next? How could abstruse equations of general relativity that took up an entire page be beautiful? Chandra had his own ideas.

He had already written several articles on the relationship between art and science, collected in a book with the elegant title *Truth and Beauty: Aesthetics and Motivations in Science,* published in 1987. In it he explored grand themes against a background of personal reflections. Chandra loved reading biographies of famous scientists. Seeking out the differences and similarities in the creativity of artists, writers, and scientists, he found that many artists and writers continue to develop over the years and create ever better works, even blossoming in old age. But most highly creative scientists do not continue to be so creative. At forty-two, for example, Newton gave up serious scientific research; after the age of thirty-five, James Clerk Maxwell, the unifier of electricity, magnetism, and light, never did anything to match his earlier achievements; and Einstein spun no more great theories after the age of thirty-six. Chandra compared these men with Lord Rayleigh who, throughout his long career, sought to bring logical consistency, clarity, and mathematical rigor to subjects ranging from electricity and magnetism to light, heat, and sound. But from this analysis he came to what must have been a rather bitter realization — that his own most important discovery had been his first.[33] That, indeed, had been the opinion of the Nobel committee. What about all his other contributions, which were as solid as those of Lord Rayleigh?

When all was said and done, in Chandra's view, scientists pursued their research not because it was useful or would be understood by more than a few people. He always bore in mind the words of G. H. Hardy: "What is important is to have added something to knowledge."[34] Hardy had once advised him in Cambridge:

If anybody asks you why you are an astronomer and what use is astronomy, then to such a philistine the correct answer is, 'Astronomy may not be useful, indeed my work may not be of "importance" in any sense, but I am convinced that my work in astronomy is at least the most important part of my personality. That I am married, have a salary, am brown in color, loved by others are of no significance except to myself. But my creative work has a value to others (however few these others may be) which has a significance apart from and therefore of the greatest value in myself.'"[35]

For Chandra, the quintessential loner who never considered himself "part of the astronomical establishment," it was an appropriate expression of purpose.[36]

Chandra's Monet essay is deep and original. He had shaped it so as to convince colleagues of the importance of his discoveries in black-hole physics via another avenue, the interplay of art and science. Kerr's conclusive proof that every black hole can be characterized by just its mass and spin had caused many things to fall into place for Chandra. Perhaps part of the reason for his soul-searching was that he had taken the leap into theoretical physics so late in his life.

From his earliest letters home, it is clear that Chandra loved art and literature. He believed that an appreciation of the arts "might help one to do science better."[37] In his essay on Monet, Chandra takes scientists to task for writing about the beauty of Einstein's general theory of relativity, "though they are mostly silent or vague on the elements that constitute its beauty."[38] He demanded that they define their terms. He asked himself "how one may evaluate scientific theories as works of art in the manner of literary or art criticism," and offered two criteria. The first was from the seventeenth-century English philosopher Francis Bacon: "There is no excellent beauty that hath no strangeness in the proportion!"[39] The general theory of relativity demonstrates "strangeness" in that it relates phenomena that were previously assumed to be completely unrelated, namely the structure of space and the matter in it. Large masses and very dense objects, such as white dwarfs and neutron stars, distort the space around them. Stephen Hawking brought more surprises and more strangeness when he introduced quantum concepts into black-hole physics and revealed that black holes will eventually evaporate, though over trillions and trillions of years, longer than the expected lifetime of our universe.[40]

Chandra's second criterion was from Heisenberg, his boyhood hero: "Beauty is the proper conformity of the parts to one another and to the whole."[41] In Chandra's opinion, Heisenberg's criterion was applicable to "both the arts and the sciences."[42] In his essay on Monet and general relativity, he expounded on Kerr's solution of Einstein's general theory of relativity, which he saw as an important example of the beauty to be found in scientific forms. Kerr's solution began as a mathematical exercise in the analysis of space and time around a rotating sphere. But further investigation revealed the astounding consequences of assuming the sphere to be a spinning star inside a spinning black hole. That black holes could be identified just by their mass and spin stunned everyone. Chandra's comment was "It seems to me that there are a number of instances in which what the human mind perceives as beautiful has counterparts in nature. [This] is to me in many ways a very sobering thought."[43]

Monet was Chandra's favorite artist. Looking at the series of paintings of Rouen Cathedral with his mathematician's eye, Chandra saw the immutable objects behind the surface glitter of light and color. In those paintings the immutable object is the cathedral, which exists as a thing in itself regardless of the light illuminating it. "For me a landscape does not exist in its own right: it is the surroundings that bring it to life," he wrote, quoting Monet.[44] What brings the landscape of general relativity to life? According to the theory, that landscape is spacetime, which is determined by its geometry and described by the complicated equations that shape it. Just as painters have in their mind's eye colors, contours, and symmetries, the scientist uses geometry and equations to describe the reality behind appearances. Those mathematical forms play the same part as light and color for the artist.

Already deep into his black-hole research, Chandra had been inspired by a paper by Penrose that explored the problem of colliding gravitational waves.[45] Surprisingly, the mathematics indicated that after colliding, the waves bounce apart, then collapse and form a singularity that is mathematically the same as a black hole. As he had that day on the Arabian Sea more than fifty years before, Chandra noticed something everyone else had missed. His deep knowledge of the mathematical complexity of general relativity "led him to think that the theory of colliding waves could be constructed on the same lines as the theory of black holes"[46] — particularly Kerr's theory. He could bypass the difficult mathematical equations dealing with gravity waves because the structure of spacetime after a collision of gravity waves is the same as around a rotating black hole.

Monet's series of paintings struck Chandra as an excellent way of describing the unexpected symmetry between black holes and colliding gravitational waves. He presented the fearfully complex equations describing these two phenomena as two "*tableaux*" showing how the Kerr metric describes a black hole and how gravitational waves collide. Superficially they are as different as two of Monet's paintings of Rouen Cathedral. From there he went on to tease out the underlying immutable object, but without the complex mathematical details of his scientific papers.

With dazzling mathematical sleight of hand he showed that both sets of equations are generated by the single equation $\mathscr{E} = p\eta + iq\mu$, representing the distortion of spacetime resulting from colliding gravitational waves as well as the distortion around a rotating black hole. "It is a remarkable fact," he wrote, "that the entire set of equations of both tableaux can be reduced to the same equation."[47] The meaning of the symbols aside, suffice it to say that $\mathscr{E}$ was Chandra's "immutable object," the reality behind the glittering surface.[48] It ties together two apparently distinct parts of astrophysics; the properties of black holes and the scattering of gravitational waves become different aspects of the same general theory, encapsulated in the mathematical quantity $\mathscr{E}$. Here was a precise way of talking about the beauty of general relativity. Chandra was extremely excited about this unexpected simplification. In his Monet essay he quoted a wider description by Heisenberg of what constitutes beauty in science: "If nature leads to mathematical forms of great simplicity and beauty — to forms that no one had previously encountered — we cannot help thinking that they are true and that they reveal genuine features of Nature."[49]

Some of Chandra's colleagues did not share his enthusiasm. Valeria Ferrari recalled, "I think that he was mainly disappointed because people did not fully appreciate and understand the deep physical insights that had been achieved in unifying in a coherent view these theories [black holes and colliding gravitational waves], a result that he considered a major achievement in his work."[50]

In his earlier book *The Mathematical Theory of Black Holes*, Chandra reported a discussion he had had with the sculptor Henry Moore about whether sculptures should be viewed from close up or far away. Moore replied that they should be seen from every distance because their beauty is revealed on every possible scale. Like a Michelangelo statue, to Chandra "the mathematical perfection of black holes of Nature is, similarly, revealed at every level by some strangeness in the proportion in conformity of the parts to one another and to the whole."[51] Theirs is a frightening yet beautiful strangeness.

Eddington wrote of falling into a black hole as falling into "nowhere." But is it really "nowhere," at least as we understand that term? At the edge of the singularity — the star that has shrunk to an infinitesimally small, infinitely dense point — the laws of classical physics fail, as does the general theory of relativity: they yield singularities — infinities — that create untenable contradictions. But quantum physics, a more realistic description of the universe, can handle the infinities at which classical physics balks. Deep inside a black hole, the laws of quantum gravity take over. Physicists conjecture that in such an extreme region, space and time are unglued from each other and there is no causal order of events, no before or after. The laws of quantum physics signal the onset of strangeness and ambiguity. Space loses definite form, leaving a fluctuating "quantum foam," like an amorphous mass of soap suds. It is a world with no certainty, in which probability reigns.

It is thrilling to try to imagine what further surprises are waiting in the wings. In mathematical terms the singularity deep inside this bottomless gravitational pit is uncannily similar to the singularity that generated the Big Bang. Might it spawn "baby universes" that will evolve into universes capable of supporting life? Again and again, mathematics has turned out to be a key to unlocking nature's deepest secrets, as Chandra perceived — and marveled at — toward the end of his life.

Chandra formally retired from the University of Chicago in 1980. Until 1985 he continued to hold the Morton D. Hull Distinguished Service Professorship as a full-time, paid, postretirement appointment. He then took up a nonsalaried position, funded by a research grant, so that he could invite his favorite young collaborators, like Ferrari, to work with him. On August 21, 1995, he was awakened by severe chest pains. He did not want to disturb Lalitha, so he quietly dressed and slipped out of their bedroom. Probably, as he always did, he glanced quickly at the mirror on their dresser, which held a postcard-sized reproduction of a painting by Monet of a little girl. "That is you," he often told Lalitha. He took the elevator downstairs, got into his car, and drove to the university clinic. As he was entering, he collapsed. He had had a massive heart attack. He died later that day, with Lalitha at his side.

Later she sprinkled some of his ashes on the university grounds. Some she threw into the air, some she sprinkled into Lake Michigan, and some she took back to India, "so he is here with us since he is everywhere."[52]

14. *Into a Black Hole*

FOUR YEARS after Chandra's death, a little after midnight on July 23, 1999, the space shuttle *Columbia* lifted off from Cape Canaveral. It was carrying the latest state-of-the-art x-ray telescope — the Chandra X-ray Observatory.[1] Lalitha was guest of honor at the launch. Would Chandra have been pleased? Lalitha thinks not: "I can imagine him coming home and telling me, if he should be alive today, 'They built an x-ray satellite and named it after me. So what?'" Perhaps the publicity would have violated his intense sense of privacy, as he felt his Nobel Prize had. As Lalitha puts it, "All he cared for was his work."[2] He would have been more interested in what the satellite revealed about the universe than in its name.

Appropriately, the mission of the Chandra X-ray Observatory has been to seek out black holes. It has been extremely successful, clearing up mysteries that have troubled astronomers and astrophysicists for years. One of the more intriguing concerned the galaxy NGC 6240.

NGC 6240 is in the constellation Ophiuchus, which is 2,400 million trillion miles away (that is 25 trillion times the distance from Earth to the Sun; we see NGC 6240 as it was 400 million years ago). Previous x-ray observatories had identified a mysterious source of x-rays right in the center of the galaxy. Then a combination of radio, optical, and infrared observations picked up a second source. The Chandra X-ray Observatory, with its super-sharp resolution, was able to identify that the x-rays were being emitted by two active galactic nuclei powered by supermassive black holes.[3] From this, astrophysicists deduced that NGC 6240 is the product of two galaxies colliding, each with its own active black hole. It is a so-called starburst galaxy — a star nursery in which stars are forming, evolving, and sometimes exploding in the aftermath of a galaxy merger estimated to have happened some 30 million years ago. The two supermassive black holes are 18,000 trillion miles apart, and each has an enormous gravitational field. They will gradually drift closer and closer, form a binary system, spiral around each other, and finally merge to form an even larger black hole. This picture lends support to the hypothesis that supermassive black holes at galactic centers are formed by the merging of other black holes.

The merge will take about 400 million years. It is a three-part process. The two black holes spiral around each other, then merge, then click together in the final note of this cosmic tango, a phase known as ringdown. This final "ping" comes when the opening in spacetime (the event horizon) snaps out of its deformed shape caused by the two black holes combining. Astrophysicists predict that the result will be a single spinning black hole with a smooth opening — since black holes "have no hair" — leaving no evidence that there were originally two. The fingerprint of the spiraling and the merger will be imprinted in the gravitational waves and can be extracted using the mathematics of the general theory of relativity. The strongest part of the signal will probably come from the merge and the ringdown. Depending on the character of the gravitational waves, it should be possible to test whether the final black hole, the product of the merge, is actually as predicted by Kerr's theory of spinning black holes. Astrophysicists estimate that several cataclysmic merges of this sort take place every year in the observable universe, which is just as well, because we will have to wait 400 million years before we can detect the NGC 6240 merge from Earth.

In the last few years, scientists have designed ground-level instruments to detect the gravitational waves rippling out from ringdown and other violent events in the cosmos. One is the Laser Interferometer Gravitational Wave Observatory (LIGO), in which one laser beam is split in two by a beam splitter. There are LIGOs in Livingston, Louisiana; Hanford, Washington, and Cascina, Italy. Each beam races back and forth along a path on one of two arms, each two miles long, forming an L. Where the two beams recombine they form a characteristic interference pattern that can be accurately monitored. Any change in the distance that the laser light travels will disturb that pattern. LIGO is so sensitive that it can detect a change of one part in a trillion billion in one of the laser beams. Ripples in the fabric of spacetime caused by a cosmic event such as the spiraling of two black holes or two neutron stars toward each other, scientists hope, will alter the length of at least one of the two paths, disturbing the interference pattern and enabling astrophysicists to detect gravitational waves.

A very exciting project being developed jointly by NASA and the European Space Agency (ESA) is an orbiting detector, the Laser Interferometer Space Antenna (LISA), to be launched sometime around 2011. It will consist of three spacecraft equipped with laser devices, located at the three points of an equilateral triangle, 3 million miles apart. The three vehicles will direct laser beams at one another. Out in space LISA will be far more sensitive to the long wavelengths of low frequency gravitational waves. On the ground there is too much background noise from events such as earth tremors to make such measurements, and the LIGO arms are too short to detect long wavelengths. LISA will be able to detect binary stars that orbit each other for periods of minutes and hours, whereas LIGO is sensitive to much more rapid phenomena, such as the final spiraling in of binary black holes.[4]

Other projects are under way. One is a program that will be run on a supercomputer at the Lawrence Berkeley National Laboratory. The plan is to calculate the sort of gravitational waves that two spinning Kerr black holes with no particular alignment of the axes they are rotating on — no "north-south alignment" — are likely to generate at the moment when they merge. The gravitational waves created at ringdown have momentum, like breaking waves at the water's edge, which can knock you down. As the two black holes collide and merge, the gravitational waves emitted by the newly formed black hole as it snaps into its smooth shape should cause it to recoil, perhaps even knocking it out of the center of its host galaxy. The core of the galaxy where this takes place should be scarred by this cosmic cataclysm and exhibit an unusual radiation profile. Sir Martin Rees has suggested something even more spectacular — a recoil so violent that the black hole breaks free of its galaxy and hurtles through space as if from a catapult.[5]

Amazingly, the Chandra's companion observatory, the Hubble Space Telescope, has observed a black hole streaking across our galaxy at 70 miles per second. Catalogued as GRO J1655-40, it is the first known example of a black hole in runaway motion imparted by a kick from the supernova explosion of a massive star. It can be tracked because it is traveling with a companion star that it is gradually gobbling up. Astronomers believe that the binary companion is a survivor of the same supernova explosion that gave birth to the black hole. The companion star orbits GRO J1655-40 every 2.6 days. Fortunately GRO J1655-40 is a safe distance from us, around 36,000 to 54,000 trillion miles away.

The Chandra Observatory has also homed in on the jets emanating from quasars. Observations of radio waves show that the inner portion of a jet shoots out from around the black hole at the center of the quasar at half the speed of light. The Chandra Observatory has picked up x-rays generated farther from the center at the point where the jet plows through gas in the galaxy and comes to a halt. Astronomers hope that images taken by Chandra will help them understand how these jets affect the regions around them.

The Chandra Observatory has even "heard" the sound of a black hole. This cosmic performance took place in the supermassive black hole in the Perseus galaxy cluster, some 1,500 trillion trillion miles away. More than 53 hours of observation using special image-processing techniques revealed wavelike features around the black hole that appear to be sound waves. Scientists think that these result from explosive activity near the black hole caused by large amounts of gas falling into it from smaller galaxies that it is cannibalizing. The sound is pitched at a steady B flat, but we cannot hear it because it is a low grumble, fifty-seven octaves below middle C. To give some idea of just how low this is, a piano keyboard spans a mere seven octaves. The hum of a black hole is over a million billion times deeper than can be heard by the human ear and is the deepest note ever heard in the universe.

One of the most exciting discoveries of all is right in the middle of our own galaxy, the Milky Way, a mere 168,000 trillion miles away. Through observations of radio waves, infrared waves, and x-rays, astronomers have penetrated the gloom of thick interstellar dust that surrounds the galaxy's core. We have known for some time that right at the center there is a very strong source of radio waves and light, far too strong to come from a star, a pulsar, or an exploded fragment of a supernova. Astrophysicists hypothesized that every galaxy has a black hole at its center and suggested that this was a supermassive black hole feeding off its accretion disk. They named it Sagittarius A* ("Sagittarius A star," or Sgr A* for short).

In 1996 astronomers at the Max Planck Institute for Extraterrestrial Physics in Garching, Germany, started to construct a precise three-dimensional map of the orbits of star clusters rotating around the center of the galaxy. In October 2002, observations began to come in. From data collected over ten years using the highest-resolution equipment available, including the European Southern Observatory's Very Large Telescope at Cerro Paranal, Chile, they were able to confirm that the star clusters were moving around a nonstellar source of radiation, and they recorded a dark object at the exact location of the compact x-ray source Sgr A*.[6] The star closest to Sgr A*, known as S2, orbits it once every 15.2 years at a distance of 10 billion miles, and from this information astrophysicists have deduced that Sgr A* is 3 to 4 million times the mass of the sun and has an event horizon 14 million miles across — smaller than Earth's orbit around the Sun. It spins on its axis once every 11 minutes.

Something that massive and compact could only be a supermassive black hole — right in the middle of our own galaxy. That is the only hypothesis that fits the facts, providing the most clear-cut evidence yet for a black hole. Simplicity and beauty reign, as Chandra believed.

Astronomers are continuing their study of Sgr A*, collecting data they hope will reveal how a supermassive black hole in the center of a galaxy is born and how it evolves. Even when it flares up, Sgr A* emits x-rays of rather low intensity, which indicates that it is a starved black hole, perhaps created by a recent supernova explosion that cleared away most of the interstellar gas around it. Thus it transpires that black holes are not cosmic vacuum cleaners, voraciously and incessantly gobbling up galaxies.

Chandra would undoubtedly have been fascinated by the deep-sky survey shown in the insert photograph, a mosaic of thirty separate Chandra Observatory images taken during July 2001, showing an oblong section of the core of the Milky Way, some 2,400 by 5,400 trillion miles. If you look hard you can see hundreds of white dwarfs, neutron stars, and the flaring x-rays that mark the location of black holes, all bathed in an incandescent fog of gas, many millions of degrees kelvin in temperature. Sgr A* is to the right of the bright patch at the center. The escaping gas is rich in elements forged in the furnaces of supernovae, which will be deposited in galactic suburbs and perhaps elsewhere in our universe.

All black holes must have spun at one time. As the star inside collapses, the black hole spins faster and faster, and the whirling accretion disk adds even more spin. But this theory had never been verified for an actual black hole. Then, in September 2003, the Chandra Observatory detected strong evidence. As a black hole hurtles through space, it sucks in gas from the companion star that whirls around it, building up its accretion disk. If it has no companion, it draws in surrounding interstellar gas to form the disk. As the gas is dragged toward the black hole, it heats up to tens of millions of degrees kelvin. Iron atoms in this gas produce distinctive x-ray signals that reveal the orbits of particles around the black hole. How close a particle's orbit is to the black hole depends on the curvature of spacetime near the hole. If the black hole is spinning, it drags spacetime along with it, twisting it up into a funnel whose event horizon is smaller than if it were not spinning. Thus a particle orbits closer to a spinning black hole than to one that is not spinning.

The Chandra Observatory can detect this smaller orbit because the x-rays emitted by the particle orbiting the spinning black hole have a longer wavelength and different characteristics from those orbiting a nonrotating black hole. There is also a stronger gravitational redshift because the particle is closer to the event horizon. Astrophysicists studying x-rays from iron atoms orbiting black holes have found that some iron atoms orbit deep within the gravitational well, as close as 20 miles from the event horizon. Particles orbiting Cygnus X-1, however, get no closer than 100 miles from its event horizon, from which astrophysicists infer that Cygnus X-1 is a nonspinning black hole. These conclusions depend on complicated mathematical models of accretion disks. Tempting as these models are because of their beauty, there is at present no litmus test to tell us whether or not a black hole is rotating, although evidence strongly suggests that some are and that their properties are exactly as proposed by Kerr's theory.

In some cases, the companion star has expired before the black hole has had time to acquire much spin, and thus it will eventually run down. Cygnus X-1 may have been one of these, although astrophysicists have yet to agree that it is definitely not rotating. And why do some black holes spin faster than others? To make it more complicated still, a black hole's spin can result from the spinning collapsing star inside, from the spin of its companion star, and from the chaotic flow of hot gas from the accretion disk; also, the spins may not always have been in the same direction.

The Chandra Observatory has witnessed all sorts of extraordinary cosmic events. In one of the most lopsided gravitational battles ever observed, a star with about the mass of the Sun was accidentally knocked off course by another star and ended up in the neighborhood of the supermassive black hole at the center of a galaxy catalogued as RX

J1242–11, in the constellation Virgo. With a mass 100 million times greater than the Sun's, the black hole ripped the intruder to shreds, creating one of the most powerful x-ray bursts ever detected in that galaxy. Chandra picked up this catastrophic destruction on March 9, 2001. Its data were combined with those from the European Space Agency's XMM-Newton and earlier observations from the German-British-American x-ray observatory, ROSAT.[7] Such a phenomenon had been predicted since the earliest days of black hole theory, but it had never before been confirmed.

Black holes have opened the minds of scientists to incredible speculations and provided all of us with a breathtaking perspective on the universe of which we are a tiny part. Will we one day be able to produce black holes in the laboratory? Physicists are already making concrete plans to try. They will do this not by crushing matter to its Schwarzschild radius — the technology does not yet exist for that — but by using colliders to accelerate elementary particles to velocities near that of light, then smash them into one another. The collision will produce a tremendous concentration of energy, extremely hot, yet far smaller than a nucleus, thereby recreating conditions a fraction of a second after the Big Bang, when the temperature was about 10,000 trillion degrees kelvin and black holes were being created. The experiment is scheduled to be carried out in 2007, using protons and anti-protons, at the Large Hadron Collider at the European Organization for Nuclear Research, near Geneva. Physicists are already talking about "black hole factories" in which they hope to produce one black hole every second.[8] The black holes will be a million times smaller than a nucleus (10^{-13} centimeters), with a mass similar to that of a proton (10^{-24} grams). Stephen Hawking predicts that each one will evaporate in a fraction of a second, leaving behind a telltale flash of Hawking radiation, from which physicists hope to pry information about the formation of black holes as well as the structure of space at its tiniest level.[9]

According to the mathematics of Hawking's theory, the shape of the radiation spectrum of the Hawking radiation depends on the number of dimensions of the space through which it undulates. Thus Hawking radiation may also reveal information on the character of space around a black hole. Theories of the early universe suggest that at that time space had more than the four dimensions we now experience. The other dimensions did not expand into the world as we perceive it but remain "rolled up," making their presence felt only on a scale way beyond microscopic, a tenth of a trillionth of a trillionth the size of an atom (a length of 10^{-33} centimeters), where scientists expect quantum gravity to take over from general relativity.[10] This is the "Planck length," the smallest length conceivable in physics. It is constructed from three fundamental constants: the gravitational constant, G, representing the large scale of nature; Planck's constant, representing the atomic scale; and the velocity of light. To get some idea of its smallness, if an atom were magnified until it was the size of the entire universe, the Planck length would be around four feet.

The Planck length may be unimaginably small, but it is not zero — it defines a region of space with a volume of 10^{-99} cubic centimeters. Incredibly, scientists can explore nature at this minuscule scale in the laboratory. When we use radiation to investigate the structure of an object, we choose a wavelength about the same size as the structure. Radar uses wavelengths at least as large as an airplane. We need very small wavelengths to investigate atomic properties, and wavelengths trillionths smaller for looking into nature at the Planck length. Physicists can produce radiation with such wavelengths — and therefore with enormously high frequency and energy — using the Large Hadron Collider. They hope to succeed in reproducing conditions in the early universe, and, from the Hawking radiation given off by the black holes created in the collider, to learn something about the additional dimensions that are believed to be there. Perhaps deep inside a black hole — at the level where quantum gravity takes over — there is no endlessly shrinking star. Instead, perhaps extra dimensions come into play. Surprises are in store.

Scientists are also daring to explore fantastic scenarios that until recently had been dismissed as science fiction — such as whether black holes might be used as portals for travel to distant galaxies. The most daunting obstacle is the incredibly strong gravity of the eternally collapsing star, guaranteed to reduce any rash space travelers and their craft to atoms or less. To overcome this, some scientists have come up with the cunning idea of finding a black hole with a second "benign" singularity not too deep inside; they are currently working on mathematical models of this.[11] They also hypothesize that some supermassive black holes may have a hybrid structure, with areas in which the gravity is less intense. It might be possible to navigate a spacecraft toward these sections, thus ensuring the intrepid astronauts a greater chance of survival.

As a spacecraft plunges into such a black hole, distortions in the fabric of space and time around the benign singularity or the low gravity area jostle it violently, as if it were hitting a series of speed bumps. The craft accelerates with jarring speed, then bumps to a halt. When the astronauts look out of their porthole, they see a glaringly strange yet tranquil unearthly terrain. Checking their instruments, they discover that they have landed on a planet in a galaxy a million trillion miles away from Earth. They have traveled this vast distance in just a few minutes by their watches. While light coursing along the surface of spacetime takes a million years to make the journey, the folds in spacetime near the singularity have bumped the astronauts into a shortcut, like traveling from the North Pole to the South by going straight through Earth instead of around it.

The shortcut is actually a mathematical solution to Einstein's general theory of relativity, called a "wormhole." It is a rip in spacetime that permits the astronauts to tunnel through instead of traveling across the surface. A wormhole can pop up anywhere and close just as suddenly because of the force of gravity in the "throat" linking the two distant regions, which is so powerful that it prevents even light from passing through. One way to avoid being caught by the closing wormhole would be to find or develop an exotic material capable of pushing against the crush of gravity — a material with negative gravity, which is a phenomenon predicted by quantum physics — that would hold open the throat of the wormhole long enough for a traveler to get through. Perhaps in the distant future a superadvanced civilization will work out how to do this.

It is virtually impossible to find a wormhole, but as we have seen, we do know the locations of some black holes. Sagittarius A*, the nearest to Earth, is a good candidate for space travel, though unfortunately it is 150,000 trillion miles away. No doubt scientists will eventually crack the problem of placing humans in a state of suspended animation in order to survive interstellar travel, and perhaps one day we really will be able to send astronauts into a black hole and from there bounce them into a wormhole, launching them on a journey through time and space.[12]

Black hole research has involved taking the mathematics of general relativity to its logical conclusion, thereby producing results that for decades seemed totally absurd and divorced from reality. There are yet more deductions to be made from special relativity, the implications of which have yet to be taken seriously. It is now known for certain that — as predicted by special relativity — a clock in motion marks time more slowly than a clock at rest.[13] In the future, astronauts might be able to journey to a distant star, traveling at a speed close to that of light. According to their calendars, the astronauts would have traveled — and aged — only a few years. But on Earth a hige amount of time would have passed. Their families would have long since died, and generations would have come and gone. When they returned to Earth, these astronauts — whether they liked it or not — would find themselves in the future. Wormholes are another mathematical manifestation of relativity. Scientists have only begun to look into the many bizarre implications they may offer, such as reversing time. If our astronauts could make use of antigravity material to hold wormholes open, they might be able to travel to a star many trillions of miles away or even arrive before they had left.[14]

All these scenarios may seem mere mathematical conceits — but so did Chandra's flash of inspiration on the Arabian Sea that sunny August afternoon in 1930. No one believed him either for many years. Yet it was Chandra's proof that white dwarf stars can shrink to an infinitely tiny point of infinite density that paved the way for the discovery of black holes — mysterious chasms in the fabric of space and time, millions of miles across, vast enough to gobble up a star. It was a concept that forever changed the course of astrophysics and profoundly transformed our view of the universe.

Chandra was a man who dealt in complexities, in a world of intricate calculations, both numeric and symbolic. He helped reveal some of the secrets of the universe, but there were mysteries in his own life that he never really resolved. He was confident of his own brilliance, yet permanently bitter at never having received the recognition he thought was his due. After making his great discovery, he became more conservative in his approach to his work, focusing on making fields such as hydrodynamics and radiative transfer more rigorous and consistent rather than innovating. Until close to the end of his career, he was afraid to reach too high. Then he finally saw that he could achieve what he had always wanted — to be a theoretical physicist, albeit a very mathematically inclined one. He knew deep down that only by giving up astrophysics could he achieve his life's goal. Chandra often liked to refer to Eddington's image of Daedalus and Icarus. Scientists in the Daedalus mold play it safe with their theories, applying them only when they are confident of success. Icarus types are more adventurous, stretching their theories to the breaking point. Chandra yearned to be an Icarus, someone who liked to take chances. But by the time he did, the wonder years of black hole research had ended. Maybe he was thinking of this when, many years later, he transcribed some lines from one of Rabindranath Tagore's Cantos: "I forget, I ever forget, that I have no wings to fly, that I am bound on this spot evermore."[15] Yet he managed to achieve something exciting, interesting, and deep, incorporating notions from art and science to communicate his views, using the fluid prose and expansive knowledge that till then he had reserved for his personal correspondence.

In India the young Chandra had published complex papers, achieved the highest grades at university, and been very popular. He went from success to success, meeting and befriending world-class scientists such as Werner Heisenberg and Arnold Sommerfeld. India's greatest scientists, C. V. Raman and Meghnad Saha, foresaw a brilliant future for him. When he arrived in England, bursting with high hopes, he expected to be welcomed as the new Ramanujan. He brought with him a discovery about nothing less than the fate of the stars. Instead he found himself not only a stranger in a strange land but pitted against the most powerful astrophysicist in the world, one of the most important scientists of his day. Rather than comparing himself with other students, Chandra set himself up against the likes of Dirac, Fowler, and Rutherford, although he was always in awe of Eddington. Eddington, from whose books he had learned his trade, was his intellectual hero.

Whatever the complexity of factors driving Eddington's urge to undermine the young Indian scientist, for Chandra it became an obsessive quest to prove that he was as good an astrophysicist as Eddington. By the late 1930s his international reputation was soaring. He could easily have moved on and put the humiliation of January 11, 1935, behind him. But like the Ancient Mariner, he never ceased to recall the events of that day.

In the 1980s, Chandra began to make frequent visits to India, laying the groundwork for Lalitha to return permanently after his death. But she decided to stay on in the flat they had lived in for so many years, surrounded by memories. Chandra's study remains exactly as it was on the day he died. It is dominated by a massive wooden desk, from which he could look through the large window and see the campus. In the corner to the right of the desk is the chair where Lalitha would sit gazing adoringly at him as he worked. Sometimes he would glance at her and ask if everything was all right, whether she wanted anything. "Just love me," she would say, and they would embrace. For fifty-nine years they were soulmates. Most Indians are bad listeners, says Lalitha, but she was a good one; she opened Chandra's eyes and his mind. She knew him through and through. He was pleased and surprised at her insights about him. "You have made a discovery about me," he often said.[16]

Chandra loved to drive. Once, on a long journey along the Ohio Turnpike, Lalitha suggested turning on the radio to listen to some music. The drive the day before had been boring and tiring. But the radio was not working. So she suggested going through the alphabet, choosing a subject to talk about for each letter. They never got past A for astronomy. When she asked Chandra to tell some stories about astronomers, he said immediately, "Why don't I dictate?"

"That will be a fine idea," Lalitha replied. "About whom?"

Eddington's name came up instantly.

"The words came steadily while I wrote them down," remembers Lalitha. "Time passed quickly."[17]

Lalitha has her own take on Chandra's relationship with Eddington. As she sees it, while Jeans and Milne allowed their lives to be ruined by Eddington, Chandra's was a life of steel, forged in an Eddingtonian furnace. "Chandra had the character to overcome the shock of his experience with Eddington, and turn to other fields of investigation," she wrote:

> What Milne and Jeans did was the easy way: they kept their hurt alive and fretted over it. What Chandra did was the harder way: He set the whole matter aside, and told himself, "No, it will not help to brood over this. I must push it aside and proceed to do other things if I am to survive." And not only did he analyse the situation and come to the above decision: he was also able to act on it and was able to discover a whole new range of subjects he could explore, and in true gratitude be able to say, "I owe all this to Eddington!"[18]

Chandra seldom discussed his mental turmoil with friends or, surprisingly, even with Lalitha. One person he did try to confide in was his close friend of the early days at Cambridge, William McCrea. But while McCrea listened patiently to Chandra, he refused to defend him in public. In an obituary of his old friend for *The Observatory*, McCrea reviewed Chandra's achievements but dismissed the suggestion that the events of January 11, 1935, had had any negative impact on Chandra. McCrea based his remarks on his memory of events sixty years before; the two lost touch after Chandra emigrated to America. He wrote, "These days in the literature it is sometimes alleged that some astrophysicists at the time concealed their agreement with Chandra lest they should offend Eddington; this is nonsense historically since Eddington was even then regarding himself as the odd man out."[19] That is most unlikely, of course. In 1936, when Shapley asked astronomers to list their peers by rank, Eddington was number one on every list. He certainly did not consider himself the odd man out. McCrea went on, "Personally I cannot recollect details of contacts I had with Chandra over the rest of 1935, but had he ever betrayed any stress resulting from the meeting I am fairly certain I should have been aware of it — I never was."[20] Having had no further contact with him over the rest of 1935, McCrea was not the bosom pal he portrays himself to be. Although he offered to write to Eddington in 1935, he never did. Instead he chose to sit on the fence so as not to offend Eddington.

Chandra's close friend and colleague at the University of Chicago, James W. Cronin, who received a Nobel Prize in 1980, responded to McCrea's obituary. Cronin was one person Chandra had confided in, during their many relaxed walks around the picturesque woodlands of Williams Bay. "It was absolutely clear that this confrontation had a lifelong effect," he wrote in a letter to *The Observatory*. "It is true that he never expressed a personal bitterness towards Eddington, but to imply that the incident had little effect completely contradicts the many conversations we had on this subject."[21] McCrea replied that Cronin had been duped by the "mythology about Eddington which has come into being over more than half a century since the RAS meeting." No evidence had ever been unearthed that "Eddington did anything whatever that was in any way detrimental to Chandrasekhar's advancement."[22] Chandra, of course, had never made any such claim.

Cronin and others at Chicago felt that Chandra was never happy or content with his life.[23] The melancholy he had suffered as a boy in India was exacerbated at Cambridge when he tried to measure up to established scientists instead of comparing himself with his fellow students. He was obsessed with work, never congratulated himself on doing a good job, and was predisposed to gloom and fits of depression; his mental state only worsened after that January day, which led to an explosion of self-doubt. Sometimes he could not control his distress. It would surface in outbursts of emotion in front of colleagues and when he spoke tearfully to Lalitha, regretting the toll his work had taken on his personal life. By the 1980s he had received every prize he could possibly have won, far more recognition than Eddington received. Yet Eddington's shadow continued to hang over him.

Though Chandra loved reading the biographies of great physicists, he was sometimes critical of the lifestyles they revealed. Reading Heisenberg's reminiscences of his years of glory, Chandra commented that he doubted if he himself had had any such years. He was disturbed to read about the many women in Erwin Schrödinger's life. As far as he was concerned, clean living was an essential component of scientific achievement.[24]

Lalitha, an accomplished singer in the Indian tradition, often sang for her husband. Right after their marriage, when she entered Chandra Vilas for the first time as Chandra's wife, she was invited to sing. She chose a song about the inevitability of dying and of being thrown into the ditch of death. Everyone was taken aback. Babuji muttered how inappropriate the song was and abruptly left the room. Years later, she realized that as a newlywed she should not have sung at all. But on that day she had felt perfectly happy, and Chandra had expressed delight in her singing. He saw an affinity between the song and his being thrown in a ditch by Eddington. He needed to pull himself out and be reborn. That song came to have a special relevance for them; they called it "our song."[25]

Lalitha speaks about Chandra's "generous nature to forgive, to 'overlook,' and continue his friendship with Eddington." Chandra himself said that January 11 "did not affect our personal relations."[26] Eddington continued to take Chandra to cricket matches and to the tennis matches at Wimbledon. They sometimes went cycling together. Eddington confided to Chandra the private code he used to drive himself to cycle farther and farther. The letter n referred to the number of miles cycled that day. The idea was to cycle n miles n times and to raise the figure regularly. He assiduously maintained the discipline. In 1938 he wrote to Chandra, "My cycling is still 75. I was rather unlucky this Easter as I did two rides of 74¾ miles which do not count."[27] In 1943 he noted that his "n is now 77. I think it was 75 when you were here."[28]

Chandra often told Lalitha that "it is because of Eddington that I became the sort of scientist changing my field periodically from one to another. I had to change my field after the controversy."[29] It was not until the 1960s that he returned to the fate of stars. He eased himself back in by attacking the same problem that Eddington had chosen when he began work on the subject, the stability of Cepheid variables.[30] Every five years or so he changed his area of study. Whenever he did, he would ask Lalitha to sing "our song."[31] He considered the transition periods to be fallow times. To guard against depression, he would begin work in the new area of study while he was still completing the previous one.

Recognition of his discovery was slow in coming. He was dismayed to discover that physicists of some repute referred to the upper limit for the mass of white dwarfs as the Landau limit, not the Chandrasekhar limit. He recalled that one colleague who gave lectures on astronomy to undergraduates in the mid-sixties was unaware of his work on the upper limit. Even Otto Struve used to confuse Chandra's limiting mass with the Chandrasekhar-Schönberg limit, the maximum amount of helium a star can produce by burning hydrogen before it begins to bloat and become a red giant.

Eddington was famous for being brutally abrasive when he spoke at the Royal Astronomical Society. Chandra was the same. He chimed in on a discussion whenever he felt like it. People had to be aware beforehand that they might be in for rough treatment from him. One speaker remembers that after his lecture, Chandra made a single brusque comment at the beginning of the discussion session and then strode out, straight across the middle of the room. His colleagues reassured the speaker that this was actually a compliment. If Chandra did not like a lecture, he would walk out well before the end.[32]

Intimidating though he was in public, Chandra was, like Eddington, attentive and supportive in one-to-one sessions with his Ph.D. students. While Eddington had few research students, postdoctoral students, or collaborators, Chandra had many, around forty-five Ph.D. students alone. Among his postdoctoral students was Mario Schönberg, and his collaborators included von Neumann and Fermi. After 1954 Chandra became even more deeply committed to the physics department, and all his students came from that department. In 1964, when he moved to the University of Chicago campus, so did his students, who had previously commuted with him from Yerkes.[33] For twenty years, directors at

Yerkes kept Chandra's office there vacant, hoping he would return. He never did.

Chandra chose his students carefully. A definite prerequisite was mathematical proficiency. The students' research always reflected whatever areas Chandra was currently interested in, and no one ever failed to complete a thesis. Chandra never gave a student a problem that he could not conceive of solving himself. Former students remember him as approachable, attentive, and concerned with their future. He arranged positions for them and followed their progress. He could be a harsh taskmaster if they did not measure up to his standards of research. But no one complained in later years; all agreed that his toughness had been for their own good. Chandra always went out of his way to be cordial to the families of colleagues and students and was especially kind to their children.

Those who worked with him recall Chandra's sense of humor as rich, complex, and sometimes bordering on bitter. He could take a joke, though, and his private persona often differed from the one he presented in public. On one occasion, late in the 1950s, John Sykes, a mathematician, cooked up a parody of Chandra's work with the help of some of Chandra's co-workers at Yerkes. Modeled closely on one of Chandra's papers, "The Instability of a Layer of Fluid Heated Below and Subject to the Simultaneous Action of a Magnetic Field Rotation: II," the parody is titled "On the Imperturbability of Elevator Operators: LVII," by S. Candlestickmaker of the Institute for Studied Advances, Old Cardigan, Whales. It is a hilarious takeoff of Chandra's mode of writing, with intricately complex detail backed up by nested footnotes to many of his previous publications on the topic. Sykes actually sent the paper to the Astrophysical Journal, supposedly for the issue of October 19, 1910, the date of Chandra's birth. The secretary, guessing it was a joke, showed it to Chandra, assuming he would not be amused. In fact he was delighted and took great pleasure in showing it to others. He also agreed to have it printed in the style of the Astrophysical Journal in a reprint format. The "authors" took up a collection to pay for the printing.[34] Sykes had caught Chandra's style so well that Chandra seriously recommended it as an example of how to write a scientific paper.

Jesse Greenstein, the astronomer at the California Institute of Technology who did important work on quasars, left Yerkes for the West Coast in 1948. He sent Chandra a photograph of himself and his family enjoying a beautiful, sunny southern California day. In return Chandra sent him a photograph of a typically gray day at Yerkes. He added a note: "Dear Jesse, This shows how Williams Bay looked on that day. I worked the entire day, and I am now one day ahead of you."[35]

Eddington made it all look easy, moving effortlessly among astrophysics, astronomy, the general theory of relativity, cosmology, quantum mechanics, abstract mathematics, philosophy, and religion, combining his work with trekking, cycling, and voracious reading. Chandra, on the other hand, struggled with a small subset of these endeavors. Yet what really separated him from Eddington was his ability to pour out his heart to Babuji, Balakrishnan, and friends such as Rosenfeld and Cronin. He went to great lengths to spare Lalitha the torment he felt, his private suffering, his yearning to be a theoretical physicist instead of the reluctant astrophysicist he was for so much of his life.

Always looking inward, Chandra examined every moment of his life and his own feelings, providing us with glimpses of the complex human stories that lie behind the greatest scientific discoveries. For even though science deals with abstractions and grand cosmic issues that dwarf our small human lives, those who carry out scientific research and spend their lives absorbed in calculations and theories are human beings. They may be driven by irrational impulses, as Eddington was when he refused to believe a result that did not square with his view of the universe, and, as they go about their intensely solitary work, may be haunted by passions, jealousies, fears, ambitions, and disappointments. Chandra's lifelong quest for personal peace could never be fulfilled.

Appendix A: The Ongoing Tale of Sirius B

TO ASTROPHYSICISTS in 1939, Sirius B, the white dwarf companion of Sirius A, presented some particularly knotty problems. By the time of the Paris conference, new measurements had shown that it was actually brighter than previously thought. Its temperature was now believed to be 1,000 to 2,000 degrees kelvin higher than the figure Eddington had used in 1926, and it had been reclassified as an A-type star in the H-R diagram. The result was a slight change in the calculated radius of the star; combined with the figure for the mass, which was still assumed to be correct, the new radius calculation resulted in a gravitational redshift of about 30 kilometers per second, 10 kilometers per second larger than that measured by Walter Adams in 1925. Gerard Kuiper described all this at the conference, saying that the discrepancy was somewhat irksome but could probably be resolved once it became possible to make more precise measurements.[1]

There was a larger problem. Although Chandra's theory was supported by astronomical observations of Sirius B, it clashed with some exciting new developments on the constitution of stars that had begun to emerge from nuclear physics. Chandra's theory predicted a relationship between the mass and the radius of a white dwarf, from which astrophysicists determined its mean molecular weight. On the basis of the most recent thoughts on the mass, radius, and surface temperature of Sirius B, Kuiper found that according to Chandra's theory the star was 40 percent hydrogen.[2] But only the previous year, Hans Bethe had formulated a theory that the basic force creating energy in stars was the burning of hydrogen by fusion. If the temperature inside white dwarfs was high enough to spark the fusion of hydrogen, why was Sirius B not much, much brighter? Perhaps the mass was actually less than was currently assumed, in which case Chandra's theory would predict a smaller amount of hydrogen inside Sirius B. That might be the case if there was a third star in the orbit of Sirius A and B, which was unlikely.

Kuiper proposed another way out. Suppose there was no hydrogen at all in the interior of white dwarfs? They were, after all, believed to be at the very end of their lives. In that case, Sirius B would have a smaller radius, a surface temperature of 25,000 degrees kelvin, and a gravitational redshift of 80 kilometers per second, thus moving the star into the upper part of spectral class B in the H-R diagram. Kuiper rejected all such suggestions as "equally impossible."[3]

In the long run this extraordinary scenario turned out to be close to the truth. With improved methods for measuring the radius and brightness of Sirius B, it became possible to eliminate the overwhelming brightness of Sirius A. This revealed that the spectrum of Sirius B found by Adams and others was greatly contaminated with lines from Sirius A, showing iron, chromium, and magnesium, which are not present in Sirius B at all. F-type stars are made up of large amounts of these elements, so Eddington had mistakenly classified Sirius B as an F. That in turn had led him to deduce the very low temperature of 8,000 degrees kelvin, from which he worked out the gravitational redshift of Sirius B, having calculated its mass as 0.85 times the mass of the Sun and its radius as 18,800 kilometers, about three times the radius of Earth.

In 1971, however, the astronomers Jesse Greenstein, John Beverley Oke, and Harry Shipman took precise measurements of Sirius B using modern instrumentation and methods. They showed that it had a surface temperature of 32,000 degrees kelvin, placing it in spectral class O; a radius 0.85 times that of Earth (far smaller than Eddington had assumed); and a mass about the same as the Sun's, giving a gravitational redshift of some 89 kilometers per second. Their results have since been verified and refined and remain largely unchallenged.[4] Current measurements suggest that the surface temperature of Sirius B is 27,000 degrees kelvin (making it a B-type star); its radius is about 3,000 miles, a little smaller than Earth's; and its average density is 1 million grams per cubic centimeter, one hundred times greater than the figure Eddington used.

George Gamow took Bethe's results more seriously than the astrophysicists did. He analyzed it thus: since a white dwarf is a burned-out star, it has no hydrogen left. Its average molecular weight is therefore 2, which makes the Chandrasekhar maximum mass 1.4 times the mass of the Sun.[5] This is the figure usually quoted in texts. In 1940 Robert Marshak took the first step toward resolving in detail the problem of hydrogen in Sirius B in his Ph.D. thesis at Cornell University, supervised by Bethe. Using the most up-to-date results from nuclear physics, he calculated that the temperature at the core of Sirius B was at least 10 million degrees kelvin and the density 1 million grams per cubic centimeter. Under such conditions, hydrogen would fuse very quickly, making Sirius B much brighter than shown by astronomers' observations. To explain the discrepancy, he suggested that hydrogen showed up so strongly in the spectral lines because there was a thin layer of hydrogen — at most 1 percent of its total constitution — floating on the surface of the star, held in place by gravity.

If we assume that there is no hydrogen in the interior of Sirius B, Chandra's theory gives a much smaller radius than it did under Kuiper's assumption that the star was 40 percent hydrogen. This results in a sharply increased gravitational redshift, four times that measured by Adams. Ten years later, T. D. Lee substantiated and elaborated on Marshak's results in conversations with Chandra at Yerkes. Lee proved that white dwarfs whose interiors contained any more than .1 percent hydrogen were bound to explode and that energy moved though the degenerate core of the star primarily through movement of the electrons, because nuclear reactions were no longer possible there. These findings were consistent with what was expected to happen at the end of a star's lifetime.[6]

Greenstein recalled the intensity of the debates that raged from the 1930s on concerning "the 'mysterious case of the gravitational redshift of Sirius B.' . . . Few participants were right, except for Chandrasekhar, who remained aloof." Kuiper withdrew early on. Frustrated at not having the proper instrumentation, he abandoned white dwarfs and went into planetary astronomy: he "dropped out in favour of planets."[7]

Appendix B: Updating the Supernova Story

HE COLGATE and White supernova scenario, with all its complexities, was very convincing — the best anyone could have come up with at the time — but eventually it turned out to be a glorious failure. Many astrophysicists wrote responses challenging its assertions. As White emphasizes, "It was the earliest application of digital computers to the study of stellar hydrodynamics. No one previously had treated the dramatic gravitational collapse that leads to a supernova."[1]

All the computer power at Colgate and White's disposal was just not enough. They were limited to modeling a star as if it were a circle instead of a sphere and studying its properties only along its radii — in one dimension. They also had to ignore the way in which elements mix as a result of currents caused by temperature differences in a star, which must be studied in at least two dimensions. The stellar models they were creating were unrealistic.

"Colgate and White started an industry," recalls Stan Woosley, a leading astrophysicist at the University of California, Santa Cruz, who specializes in models of supernovae. As Woosley explains, Colgate and White's deep insight was to take the "gravitational energy of the neutron star, convert it into neutrino energy, then take part of that neutrino energy to blow up the star in a neutrino-powered explosion."[2] As to why the cataclysmic core collapse takes place, Woosley points out that "massive stars are gravitationally confined thermonuclear reactors — they are like gravity bombs."[3]

One problem with Colgate and White's model was their suggestion that a supernova explosion could be caused by neutrinos heating up the mass of nuclear ash falling into the star's center. Fowler and Hoyle, among others, discussed other possibilities. One was that the star's rotation and its magnetic field might provide the push needed to set in motion the shock wave that had been stalled by the falling ash. Computers in the 1960s were simply not fast enough for the calculations necessary to work out modern theories of supernova explosions.

But Colgate and White's scenario had a more fundamental flaw. This did not emerge until the electroweak theory was formulated in the late 1960s by Sheldon Glashow at Harvard, Steven Weinberg at the Massachusetts Institute of Technology, and Abdus Salam, a Pakistani physicist at the International Center for Theoretical Physics in Trieste, Italy, who all shared the 1979 Nobel Prize in physics for discovering it. The electroweak force is a union of the electromagnetic force with the weak nuclear force that controls beta decay, the process of radioactive decay in which a neutron turns into a proton, an electron, and a neutrino. Among other things, it predicts ways that neutrinos can interact.

Neutrinos are loners. They whiz around through space, through Earth, and through us, each on its lonely course, as if we were not there. They interact so weakly that they can fly through space for 3 trillion miles unhampered. According to the electroweak theory, the collision of a neutrino and an electron is like two billiard balls rebounding off each other. Neutrinos collide in a similar way with neutrons and protons. These reactions are relatively rare under most conditions, but in the cramped, hot innards of stars they are common. Even though neutrinos interact weakly, when they become trapped inside the incredibly dense collapsing core of a star they have to fight their way out, and they bang against electrons, nuclei, protons, and neutrons. Just as light struggles to emerge from a star, so do neutrinos. The crowd of barging neutrinos creates what is called neutrino opacity. Eventually the star explodes and the neutrinos burst out of their prison, generating neutrino luminosity or "shine."[4]

There was still no observational evidence for this neutrino scenario. Then, on February 24, 1987, Ian Shelton, a Canadian astronomer working at Las Campanas Observatory in Chile, noticed a spot on his photograph of the Large Magellanic Cloud, a satellite galaxy of the Milky Way. It was startlingly bright, so bright that it had to be a supernova. As it happened, just the previous day Shelton had photographed a star catalogued as Sanduleak-69° 202, twenty times the mass of the Sun and some 100,000 trillion miles away. Shelton went outside and looked up.

The star was now a thousand times brighter: Supernova 1987A (SN 1987A), the death throes of Sanduleak-69° 202, had just exploded. It was the first supernova to be seen with the naked eye since 1604.

A few years earlier, two neutrino detectors — Kamiokande II, near the city of Gifu, right in the middle of Japan, and IMB, near Fairport, Ohio — had been built deep underground and had gone into operation with the aim of testing predictions of certain unified theories of elementary particles. They were on twenty-four hours a day, so scientists were able to go back over their records. What they found was that three hours before SN 1987A lit up the sky, the Japanese detector had spotted and recorded twelve neutrinos and the American one had recorded eight, practically simultaneously. These could only have come from the supernova.[5] It was even possible to trace some of the Kamiokande neutrinos to the region where SN 1987A had erupted. Finally, here was conclusive proof that supernovae emitted neutrinos. And knowing that the neutrinos arrived three hours before the supernova lit up gave vital clues to the way the explosion worked.

Four years after SN 1987A appeared, NASA sent up the orbiting Hubble Space Telescope. One of its tasks was to collect more precise data on the supernova. Studying the characteristics of the light it emitted, astrophysicists discovered that there had been a great stirring up of elements. Elements had moved up and down as well as sideways, which meant that in order to study them, scientists would need models of at least two dimensions. By now there were powerful desktop computers that could handle this sort of analysis. Everyone got to work. One much-debated scenario of the explosion is as follows.

The shock front hurtling outward at 6,000 miles per second smashes into burned-out nuclear ash collapsing inward at 36,000 miles per second and stalls. But instead of being beaten back, the shock front is held in suspension by the combined degeneracy pressures of electrons (which still exist in some abundance in the star's outer regions, where temperatures are not high enough for them to be badly depleted through combination with protons) and neutrinos (from reactions in which nuclei in the core are smashed by high-energy radiation). Interacting with other particles in the core, the neutrinos are trapped and compressed so much that they develop an outward degeneracy pressure, much as electrons do. This is called the delayed shock process, to distinguish it from the "prompt shock" in earlier theories having to do with core bounce. As the falling nuclear ash passes

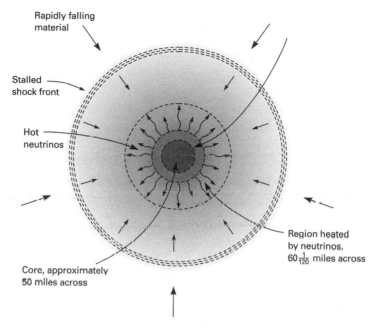

Rapidly falling material

Stalled shock front

Hot neutrinos

Region heated by neutrinos, $60\frac{1}{120}$ miles across

Core, approximately 50 miles across

The iron core of the star, which is at first 1,000 miles in diameter, becomes extremely compressed by matter falling into it. The core "bounces," or springs back, generating a shock front that is then stalled by infalling matter while leaving behind a very hot, dense core about 50 miles across. Neutrinos ooze out of the core and heat up the layer of matter around it. The heat percolates outward and invigorates the stalled shock front, causing an explosion.

through it, the suspended shock front, 360 miles across, accumulates matter from within and without.

At this point the star's iron core is a mere 50 miles across and is made up of electrons, trapped neutrinos trying desperately to escape, neutrons, and heavier nuclear matter. Between the spherical shock front and the highly dense core at its center is a layer of matter (like the ham in a sandwich) that is blocked by the stalled shock front and is waiting to blast its way out. But some neutrinos manage to ooze out of the core, gradually heating up the material in the sandwiched layer right out as far as the stalled shock front. As more and more neutrinos ooze out, the core loses its rock hardness, and in a fraction of a second it shrinks to a dense neutron star just 18 miles across with a mass 1.4 times that of the Sun. The matter heated by the neutrinos energizes the shock front so much that it starts to move again, pushing against the falling nuclear ash like a giant plunger. The remaining neutrinos burst out, driving the shock front and making it move faster.

As the shock front reaches more diffuse regions of the star, it becomes supersonic, plowing into the star's outer layers. Speeding through the burned-out nuclear ash, the shock front compresses the ash so much that it sets off a new round of thermonuclear reactions in a last gigantic ignition, creating all the elements heavier than iron, among them zinc, silver, tin, gold, lead, mercury, and uranium. These reactions require so much energy that they cannot take place until a star goes supernova. The fact that all of these elements are found on Earth shows that everything that makes up the world we live in, including ourselves, is a product of massive supernova explosions. We are indeed children of the stars.

As the ash blasts into space, there is a dazzling flash of light. The energy released is huge — hundreds of times greater than all the energy the Sun has produced over Earth's whole lifetime, 4.6 billion years. The star has become a supernova.

This theory of a delayed shock finally seemed to be a viable scenario. It created a model of a supernova that fitted with what had been observed, suggesting that the material around the stalled shock front was suspended for long enough to be "cooked" by neutrinos. Two-dimensional simulations to analyze the light emitted by SN 1987A showed that when Sanduleak-69° 202 exploded, its onionlike structure was not intact. There was a great deal of swirling and turbulence in the star's interior, as when smoothly flowing water breaks into eddies or smoke rings become fuzzy. This turbulence produced the percolation of heat that reinvigorated the shock front so that it could blast through. The resistance to neutrinos passing through a star (neutrino opacity) is less than the resistance to light, and it decreases as the neutrinos arrive near the star's surface; that is why neutrinos from SN 1987A arrived three hours earlier than its light. SN 1987A actually exploded 160,000 years ago.

Neutrinos reveal a great deal about the way stars die. A straightforward calculation shows that in just 10 seconds, SN 1987A emitted trillions upon trillions upon trillions of neutrinos, a hundred times more energy than the Sun will emit in its entire lifetime.[6] Unlike sunshine, however, this "neutrino shine" is not visible to the human eye. But for a few seconds the luminosity of the neutrinos from SN 1987A was greater than the light from every star in the observable universe.

Every researcher into supernovae emphasizes that much of what we think we know about these exploding stars is still controversial, and much remains to be discovered. One unsolved problem is that the mass of a neutron star as predicted by current models is lower than the masses measured in binary systems in which one of the two stars is a neutron star. The figure depends critically on the equation of state describing the neutron star's interior. No one is sure what constitutes the core of a neutron star, where densities are more than ten times that of nuclear matter. Perhaps neutrons are stripped down to their constituent quarks or even beyond, to unknown forms of matter. The core is so dense that a piece the size of a basketball would weigh a trillion tons on Earth. A neutron star's gravitational pull is 190 billion times that of Earth; an object dropped from three feet above the surface of one would smash into the ground at 70,000 miles per second; on Earth it lands at a gentle .002 mile per second.

As Colgate put it, in words that could have been taken from the pages of *The Observatory* during the debates among Eddington, Jeans, and Milne, "Intellectual honesty demands that we remove the approximations in our models, so that we become confident that our simulations duplicate reality" and are not merely exercises in differential equations.[7] Astrophysicists will need to make more precise simulations to model something as complex as the convection processes that trigger a supernova explosion. At the moment the theory of convection in supernovae is rudimentary and does not sufficiently take into account magnetic fields and rotation. It is also possible that some supernovae may be powered not by neutrinos but totally or partially by bursts of gamma rays.

Stars that go supernova as a result of gravitational collapse in the way we have been considering are what astronomers call Type II supernovae or "gravity bombs," to use Woosley's term. The white dwarfs in binary systems such as Sirius A and Sirius B meet their end in a different way. In these Type I supernovae, the white dwarf accretes matter from its companion, grows until it exceeds the Chandrasekhar limit, then blows itself to pieces.[8] Under certain conditions it can eject the excess mass in a mild supernova explosion and become a white dwarf again. Astrophysicists are finally groping toward an answer to the question that Milne often put to Chandra: what happens if a speck of dust is added to a white dwarf that has reached the Chandrasekhar limit? The answer seems to be that it will go supernova.[9]

Astrophysicists believe they have most of the physics right about white dwarfs, but details are still missing. It is impossible to simulate the conditions in white dwarfs in terrestrial laboratories. Instead scientists have to make theoretical calculations of the properties of the various elements that make them up, predominantly carbon and oxygen. Just as neutron stars have magnetic fields, so do most white dwarfs. Both are the result of a much larger star being greatly compressed. White dwarfs can have a magnetic field over ten thousand times greater than that of a normal star. While neutron stars rotate very fast, white dwarfs rotate much more slowly, taking hours or even days. White dwarfs can also pulsate. All these properties have to be taken into account to understand these stars properly. They also affect Chandrasekhar's maximum mass. Chandra calculated his upper limit for an idealized white dwarf, a sphere of highly compressed electrons that do not interact with one another. In his day it was a stunning leap of the imagination.

Notes

Prologue

1. Eddington 1936b, p. 301.

1. Fatal Collision

1. Details of the Jan. 11, 1935, meeting are in *The Observatory* 58: 37–39. Chandra's lecture was based on Chandrasekhar 1934b, and a summary is in Chandrasekhar 1935a.
2. Chandrasekhar 1977, p. 33.
3. Quoted in Wali 1991, p. 124.
4. Chandrasekhar 1980a, p. 3.
5. The lecture room dates from 1874, when the RAS moved to Burlington House. In 1969 it was partitioned and reconfigured to make two offices.
6. Quoted in Kanigel 1991, p. 277.
7. Chadwick 1961.
8. Letter from Chandra to his father, 8 June 1933.
9. Chandrasekhar 1979b, p. 3.
10. Chandrasekhar 1979a, p. 5.
11. Chandrasekhar 1977, p. 14.
12. Einstein 1923, p. 482.
13. Chandrasekhar n.d.2, p. 5. This passage is based on Chandrasekhar 1934b, p. 377.
14. Eddington 1935a, p. 38.
15. Chandrasekhar 1977, p. 33.
16. Eddington 1935a, p. 38.
17. Chandrasekhar 1977, p. 31.
18. Letter from McCrea to Chandra, 16 Jan. 1935.
19. Chandrasekhar 1980c, p. 35.
20. Chandrasekhar 1977, p. 34.
21. Ibid., p. 29.

22. Ibid., pp. 27–28.
23. Ibid., p. 34.
24. Ibid., p. 34.
25. Milne 1935a.
26. Chandrasekhar n.d.4.
27. Chandrasekhar 1980a, p. 2.
28. McVittie 1978, p. 18.
29. Quoted in Wali 1991, p. 128.
30. Chandrasekhar 1979b, p. 8.

2. A Journey Between Two Worlds

1. Chandrasekhar 1977, p. 4.
2. Chandrasekhar n.d.1, p. 16.
3. Ibid., p. 17.
4. Chandra's six sisters, in order of age, were Rajalakshmi (Raham), the oldest, Balaparvathi (Bala), Sarada, Vidya, Savitri, and Sundari. His three brothers were Visvanathan, Balakrishnan, and Ramanathan (Ramanath).
5. Venkataraman et al. 1983, p. 13.
6. Unless indicated otherwise, quotations from Lalitha Chandrasekhar in this chapter are from interviews with the author in Chicago, 1 Nov. 2002.
7. Gandhi 1983, p. 40.
8. Quoted in Wolpert 1997, p. 295.
9. Balakrishnan 1983a, p. 18.
10. I am indebted to Professor Takeshi Oka of the University of Chicago for allowing me to quote from a fax message sent to him by Ramaseshan on 11 June 1998.
11. Balakrishnan 1983b, p. 2.
12. Ayyar n.d., p. 7.
13. Balakrishnan 1983b, p. 3.
14. Quoted in Venkataraman et al. 1983, p. 16.
15. Chandrasekhar 1977, p. 3.
16. Chandra's recollection quoted in Wali 1991, pp. 56–57.
17. Letter from Chandra to Balakrishnan, 5 June 1931.
18. Letter from Chandra to Balakrishnan, 12 May 1931.
19. Chandrasekhar 1977, p. 7.
20. Wolpert 1997, p. 313.
21. Eddington 1926, p. 393.
22. Ibid., p. 1.
23. Eddington 1936b, p. 388. Professor Harlow Shapley, who was in the audience when Eddington delivered these vivid remarks, was a famous astronomer of the time and director of the Harvard College Observatory, which operated the Oak Ridge reflector telescope located in Harvard, Massachusetts. He will appear in later chapters.
24. Russell 1945, p. 134.
25. Einstein suggested that light could behave as particles in 1905. In the most astounding father-and-son act in the history of physics, J. J. Thomson discovered the electron as a particle, for which he received the Nobel Prize for

Physics in 1907; his son George discovered the wave nature of the electron and won a Nobel in 1937.

26. Sommerfeld 1928a, b.
27. Chandrasekhar 1928.
28. Quoted in Wali 1991, p. 63.
29. Chandrasekhar 1977, p. 10.
30. Letter from Chandra to his father, 4 June 1929. Chandra refers to "A Generalized Form of the New Statistics," 1929a; to "The Equipartition Principle and the New Statistics," which never appeared; and to a note to *Nature*, "The Einstein Method of Deriving Planck's Formula and the New Statistics," also never published.
31. Later published as Chandrasekhar 1929b.
32. Ibid.
33. Chandrasekhar 1930. Despite its title, the *Philosophical Magazine* is a physics journal.
34. Letter from Chandra to his father, 4 June 1929.
35. Rosenfeld 1967, pp. 118–19.
36. Heisenberg 1962, p. 6.
37. Balakrishnan 1983b, p. 6.
38. Letter from Chandra to his father, 16 Oct. 1929.
39. Balakrishnan 1945, p. 74.
40. Letter from Chandra to his father, 4 Jan. 1930.
41. Quoted in Wali 1991, p. 65.
42. Quoted in Wali 1991, p. 69.
43. Wali 1991, pp. 69–71.
44. Venkataraman et al. 1983, p. 17.
45. Wali 1991, pp. 172–77.
46. Lalitha's two older sisters became doctors, and her younger sister did a master's degree in Sanskrit. Lalitha's was in physics.
47. Chandrasekhar 1977, p. 10.
48. Letter from Chandra to Balakrishnan, 31 July 1930.
49. Letter from Raman to Chandra, 28 July 1930.

3. Rival Giants of Astrophysics

1. Eddington 1929b, p. 33.
2. Chandrasekhar 1977, p. 27.
3. Interview with David Dewhirst, 16 Sept. 2002.
4. Chandrasekhar n.d.5.
5. Letter from Williams to Otto Struve, 13 June 1958.
6. Quoted in Douglas 1956, p. 121.
7. Chandrasekhar 1977, p. 36.
8. Cowling 1966, p. 128.
9. Douglas 1956, p. 2.
10. Chandrasekhar 1987a, p. 95.
11. See Warwick 2003 for a discussion of Eddington's training at Cambridge.
12. Bollobás, in Littlewood 1953, p. 12.
13. Bollobás, ibid., p. 18.

NOTES TO PAGES 37–41 ✦ 278

14. Kanigel 1991, p. 143. Kanigel offers an insightful discussion on the personal life of Oxbridge dons during the 1920s and 1930s.
15. See, for example, Littlewood 1953, p. 120.
16. Quoted in Kanigel 1991, p. 139.
17. Quoted ibid.
18. Quoted ibid.
19. Payne-Gaposchkin 1968, pp. 2–3.
20. Chandrasekhar 1980a, p. 16.
21. Douglas 1956, p. 30.
22. Ibid., p. 7.
23. Quoted in *The Blue* (journal of Christ's Hospital school) 75, Sept. 1958, p. 179.
24. From Eddington's "Cycling Diary" at Trinity College, Cambridge.
25. Douglas 1956, p. 34.
26. Kanigel 1991, p. 144.
27. Quoted in Douglas 1956, p. 110.
28. At the time, astrophysicists assumed that stars, including the Sun, had much the same chemical makeup as Earth and thus consisted primarily of oxygen, iron, potassium, calcium, sodium, aluminium, magnesium, and silicon. They did not include hydrogen, of which Earth has little. It would turn out that healthy stars are actually made up primarily of hydrogen.
29. Even before Heisenberg's discovery of the new atomic physics — quantum mechanics — in 1925, physicists were questioning the accuracy of Bohr's theory. They realized that atoms do not behave like minuscule solar systems and that such visualizations were actually holding back progress. One of the discoveries of quantum theory was that because electrons can be waves, they are spread out rather than taking a well-defined orbit around a nucleus. Nevertheless, Bohr's theory was invaluable as a first step and is still useful today as a way of understanding the chemical properties of atoms.
30. Electrons are so minuscule that astrophysicists assumed they could safely ignore their dimensions and the force with which they repel one another, an acceptable simplification as the positive charge of the nucleus had a neutralizing effect.
31. Instead of expressing the perfect gas law as the product of the pressure and volume of a gas in proportion to its temperature, astrophysicists usually express the pressure of a gas in terms of its density, temperature, and chemical composition. This is the equation of state of the perfect gas.

    Scientists also studied imperfect gases, extending the theory of the perfect gas to include the size of the particles that make up a gas and the forces between the particles.
32. How bright a star appears to us on Earth depends on how far away it is. Stars emit a certain amount of light energy per second — their luminosity. The brightness of a star as observed from Earth, referred to as its apparent brightness, is calculated as its luminosity divided by its distance squared. To measure the distance, astronomers observe how the star's position changes in relation to more distant stars over a period of six months, the time Earth takes to complete half an orbit around the Sun. This tiny change in position is called stellar parallax. If the star is so far away that its stellar parallax

is too small to measure, its distance can be found by measuring its brightness, spectral lines, and temperature. Most of this information is shown in an H-R diagram (see pages 44, 181).

33. See Eddington 1919. Modeling a Cepheid as if it were a perfect gas enabled Eddington to calculate that its period times the square root of its density was a constant. This fitted the observation that the star's brightness increased most rapidly after it had passed through its minimum radius. The reason was that its density and temperature were greatest at that point, because its gas particles were squeezed together most tightly.

34. The term "light" is used for the tiny part of the electromagnetic spectrum that we can see. "Radiation" is the more general name for the entire spectrum. Wavelength is the distance between the peak of one wave and that of the next, while frequency is the number of peaks that pass a given point per unit of time; thus the longer the wavelength, the shorter the frequency. In other words, the wavelength of a light wave times its frequency equals the speed of light.

35. To measure a star's surface temperature, astronomers first assume that it radiates as if it were a "blackbody," a hypothetical object that is both a perfect emitter and a perfect absorber of radiation. The radiation spectrum — or wavelengths — for a blackbody at different temperatures can be drawn on a graph that compares brightness, or intensity, with wavelength. The curve rises smoothly from zero, at zero wavelength and zero intensity, to a peak and then tails off to zero at higher wavelengths. The peak is usually not in the middle of the curve but skewed toward low or high wavelengths, depending on the blackbody's temperature; different temperatures give differently shaped curves. Measurements of light intensity from various stars indicate that they do indeed emit light much as blackbodies would. This permits astronomers to use the mathematics of blackbody radiation, developed by the German physicist Max Planck in 1900.

36. Russell 1914, p. 282.

37. Ibid.

38. Quoted in DeVorkin 2000, p. 126.

39. The o is the Greek letter omicron. Some years later this binary system was renamed 40 Eridani A and 40 Eridani B.

40. Russell 1941, p. 6.

41. Adams 1914, 1915.

42. Measurement of orbital parameters, such as the length of time two stars take to make a complete revolution around a common center (the orbital period), the distance of the binary system from Earth (found by measurement of stellar parallax), and the degree to which the shape of the orbit differs from a circle (the orbital eccentricity), gives enough information to calculate the mass of each star. Each star's radius can be calculated from its surface temperature and luminosity. Average density can be calculated from the ratio of the star's mass to its volume; taking the star to be a sphere, its volume is four times pi times its radius cubed, divided by three.

43. Astronomers measure mass in grams, distance in centimeters, and time in seconds. They use the mass, not the weight, of an object because mass remains the same wherever the object happens to be in the universe; its

weight, however, varies depending on gravity. On the Moon, for example, an object weighs a sixth of what it weighs on Earth, because the Moon's gravity is so much weaker.

44. Eddington 1926, p. 171.
45. The calculation is essentially as follows: the amount of energy generated by a collection of gas particles that has been reduced from a ball with an infinitely large radius to the present radius of the Sun is divided by the average amount of energy emitted by the Sun per second.
46. Eddington 1917b. At this time the only known "elementary particles" were electrons and protons. No one speculated seriously on their origin.
47. Until the discovery of the neutron in 1932, physicists assumed that the nucleus was made up of protons and electrons. They assumed that the helium nucleus was made up of four protons and two electrons, which would give a net positive charge to balance the negative charge of the two orbiting electrons. Since the electron's mass is minuscule compared with the proton's, it was ignored in these calculations.
48. Aston 1938, p. 106.
49. Eddington 1920, p. 19.
50. Eddington 1926, p. 301.
51. Ibid., p. 17.
52. Ibid., p. 19.
53. At that point, astrophysicists did not speculate seriously about the origins of the heavier elements, but assumed they were part of the interstellar dust from which stars were formed.
54. Eddington used data on the brighter of the two stars making up the binary system Capella — Capella A, more than 10^{14} miles away in the constellation Auriga. According to Eddington, the observational data for Capella were "unusually complete," resulting in extremely precise estimates on its mass, surface temperature, radius, and brightness.
55. Eddington 1924a, p. 786.
56. Eddington 1926, pp. 146, 243; Chandrasekhar 1939, p. 278.
57. Eddington 1926, p. 244.
58. Letter from Russell to Payne, 14 Jan. 1925, quoted in DeVorkin 2000, p. 240.
59. Payne 1925, p. 248.
60. Payne-Gaposchkin 1968, p. 17. After four years at Cambridge, Payne was given a certificate which essentially said that if she were a man, she would have been awarded a B.A. Impressed with a lecture given by Shapley at the 1923 meeting of the British Astronomical Association, Payne approached him and asked if she could continue her research at Harvard. There she was awarded a Ph.D. from Radcliffe (the women's college at Harvard). Payne was the first person to receive a Ph.D. for research conducted at the Harvard Observatory and the first woman to become a full professor at Harvard. Otto Struve, the powerful director of the Yerkes Observatory, declared her Ph.D. "the most brilliant thesis ever written in astronomy." Struve and Zebergs 1962, pp. 220–21.
61. For details of this controversy see DeVorkin 2000, pp. 205–20.
62. Strömgren 1932, pp. 118–22. See also Strömgren 1976, pp. 20–21.

63. Eddington 1932.
64. Quoted in Douglas 1956, p. 40.
65. Quoted ibid. General relativity predicts that light near the Sun will be deflected by 1.75 seconds, while Newton's theory of gravity predicts half that amount. Difficulties in making the measurements, in particular the very brief time available during an eclipse, led to errors in Eddington's data; his best result was 1.98 seconds plus or minus 0.16 second. Nevertheless, his results clearly supported Einstein's theory over Newton's. Measurements that astronomers have made since 1919 on more than 380 stars support Einstein's prediction even more strongly. See Weinberg 1972, pp. 188–94, and Misner et al. 1970, pp. 1104–5.
66. *New York Times*, 9 Nov. 1919.
67. Quoted in Douglas 1956, p. 43.
68. Chandrasekhar n.d.5.
69. Eddington 1936b, p. 390.
70. Assuming that the gas of electrons inside a white dwarf is a perfect gas, Eddington estimated the internal temperature at about 1 billion degrees kelvin. See Eddington 1926, p. 174.
71. Letter from Eddington to Adams, 13 Jun. 1924.
72. Letters from Eddington to Adams, 13 Jan. 1924 and 22 Mar. 1924.
73. Eddington 1926, p. 6.
74. Plummer 1948, p. 117.
75. Eddington 1926, p. 172.
76. Ibid., p. 173.

4. Stellar Buffoonery

1. Milne 1945a, p. 62.
2. Milne 1923 and 1945b, p. 87.
3. Letter from Chandra to his father, 10 Oct. 1930.
4. Chandrasekhar 1980b, p. 4.
5. See Weston Smith 1990 for details of the development that follows. Meg Weston Smith is Milne's daughter.
6. Quoted ibid., p. 242.
7. When the English scientist H. G. J. Moseley was killed at Gallipoli in 1915 at the age of twenty-eight, the British scientific community raised an uproar. The government realized that in an era when science was becoming more and more relevant to warfare, it should make better use of its scientists. Moseley almost certainly would have won a Nobel Prize for his work in atomic physics.
8. Bohr came up with this model by combining the Newtonian physics for planetary motions with results from quantum physics in a way that guaranteed the stability of atoms. In 1913 other scientists were trying to explain why most atoms were stable. Bohr realized the futility of this approach and instead adopted the daring strategy of taking atomic stability as given. From this he concluded that electrons can occupy only certain orbits, in contrast to Earth, which can wobble in its journey around the Sun.
9. Letter to Bohr from George Hevesy, 23 Sept. 1913, quoted in Pais 1986, p. 208.

10. In numerical terms, electrons can have half a unit of spin.
11. Fowler's theory of white dwarf stars is based on an equation that relates the pressure of a gas of electrons to its density. The perfect gas equation, in contrast, relates the gas's pressure to its temperature as well as its density. The absence of temperature in Fowler's theory is an unexpected result of quantum physics. When a white dwarf has cooled completely, its temperature is effectively zero. If it were a perfect gas, its gas pressure would also be zero and it would collapse to nothing, as Eddington had predicted. But the Pauli exclusion principle averts this undesirable result by bringing in an outward pressure — degeneracy pressure — which operates even when the star's temperature is zero and which prevents it from being crushed as it cools off.

    In other words, even though astronomers measured white dwarfs as being very hot and very dim, physicists could analyze them as if they were cold and dark. When a white dwarf becomes cold and dark, of course, we can no longer see it.
12. Fowler 1926, p. 115.
13. Letter from P. G. Gostline to McCrea, undated, written shortly after Milne's death on 21 Sept. 1950.
14. McCrea 1951, pp. 162–63.
15. Chandrasekhar 1979b, p. 3, and McCrea 1951, pp. 163–64.
16. Eddington 1930a, p. 809.
17. McCrea 1951, p. 164.
18. Interview with Meg Weston Smith, 18 Sept. 2002.
19. Ibid.
20. McCrea 1951, p. 165.
21. Eddington 1929a.
22. Letter from Milne to Geoffrey Milne, 23 Mar. 1930.
23. See, for example, Milne 1930a, pp. 332–34.
24. Eddington 1930b.
25. Letter from Milne to Herbert Dingle, 1 Aug. 1939.
26. Eddington 1930c, p. 210, letter to the editor dated 19 June 1930.
27. Milne 1930b, p. 4.
28. Jeans 1931, p. 37.
29. Quoted in Douglas 1956, p. 88.
30. Letter from Heisenberg to Pauli, 31 July 1928, in Pauli 1979, p. 467. See also Miller 1994, pp. 30–31. In 1931 Dirac brilliantly cut through this apparent inconsistency by showing that the weird particles he had proposed were actually antimatter. They were positrons, antiparticles of electrons, with the same mass as electrons and identical in every respect except that electrons have a negative charge while positrons have a positive charge. Although positrons and protons have the same charge, they are not the same. Among their many differences is that the mass of a proton is 10,000 times that of a positron. When an electron and a positron touch they annihilate each other, producing pure energy in the form of light. That same year Carl Anderson, working at the California Institute of Technology (Caltech), discovered positrons in cosmic rays. Eddington had to rephrase his tentative suggestion that stars shine as a result of the annihilation of

electrons and protons. In fact, stars shine as a result of the annihilation of electrons and positrons.

31. Eddington 1958a, pp. xi–xiii.

32. Eddington never specified what the E in E-numbers stood for. Perhaps it stood for Eddington!

33. Eddington 1959, pp. 230–31.

34. Sommerfeld 1923, p. vii, an extract from the preface to the first German edition of 1919.

35. Eddington 1959, p. 237.

36. Russell 1945, p. 135.

37. Quoted in Kilmister 1994, p. 117.

38. Eddington calculated the rate of recession of galaxies as 467 miles per second for each 19 million trillion miles (1 megaparsec) a galaxy is distant from Earth. Astronomers had measured 330 miles per second per megaparsec. The issue of galaxy recession is taken up in Chapter 13.

39. Eddington assumed that the universe is a degenerate gas, like that inside a metal. He used quantum statistics to calculate the total number of electrons if all levels were filled up to the highest possible level of energy. This he assumed to be the energy of an electron as deduced from the mass-energy relation $E = mc^2$, in which m is the electron's mass. However, he fudged the final result in order to obtain the number of electrons he wanted.

40. Quoted in Chandrasekhar 1987a, p. 136. An apt page number!

41. For an example of Pauli's mysticism, see Pauli 1955.

42. Conversation with the late Victor F. Weisskopf, 10 Apr. 1986. Other versions of this story are in Pais 2000, pp. 250–51, and Enz 2002, pp. 533–34.

43. Letter from Pauli's colleague Oskar Klein to Pauli, 18 Feb. 1929, in Pauli 1979, p. 491.

5. INTO THE CRUCIBLES OF NATURE

1. Quoted in Ramaseshan 1996, p. 104.

2. Letter from Raman to Chandra, 28 July 1930.

3. Chandrasekhar n.d.2, p. 2.

4. Chandrasekhar 1977, p. 13.

5. Raman 1930.

6. Chandrasekhar n.d.2, p. 2.

7. Chandrasekhar 1931c, p. 81.

8. Chandra calculated the limit to be 0.91 times the mass of the Sun. By 1934 he had made a more precise calculation and discovered that the maximum mass depended on the white dwarf's chemical constitution. The exact figure, however, was less important than the fact that there was a maximum mass beyond which a white dwarf would shrink to nothing. The figure currently used as Chandra's maximum mass — 1.4 times the mass of the Sun — first appeared in 1939 in a paper by the Russian physicist George Gamow — see Gamow 1939. By then the physicist Hans Bethe had discovered the nuclear processes that generate stellar energy, greatly enhancing our knowledge of the constitution of a star.

Strictly speaking, degeneracy pressure depends on a star's temperature.

Fowler and Chandra both restricted themselves to the case of a white dwarf effectively at zero kelvin, when the star has cooled off and become rock-hard, and thus is at maximum degeneracy. This was a reasonable approximation and it kept their equations simple. If all white dwarfs were actually at zero kelvin, they would be invisible. Nevertheless, for certain temperatures and densities astrophysicists can model white dwarfs as if they were a gas of electrons at zero kelvin.

9. Chandrasekhar 1980b, p. 3.
10. Balakrishnan 1983b, p. 3.
11. Chandrasekhar 1977, p. 9.
12. Letter from Chandra to his father, 4 Sept. 1930.
13. Chandrasekhar n.d.2, p. 3.
14. In 1929, Wilhelm Anderson, a thirty-nine-year-old Polish-born physicist at the University of Tartu in Estonia, informed Stoner that he had omitted to take special relativity into account in his earlier papers — see Anderson 1929. But Stoner quickly found that his inclusion of special relativity was incorrect. Stoner's subsequent correct result for relativistic degeneracy (Stoner 1930) — the same as Chandra's — is sometimes referred to as the Stoner-Anderson equation.
15. Stoner 1930, p. 953. Eddington had been in correspondence with Stoner regarding new data on the mass of white dwarfs. He recommended for publication two of Stoner's subsequent papers on relativistic degeneracy: Stoner 1932a, b.
16. Chandrasekhar 1977, p. 16.
17. Letter from Chandra to his father, 22 Nov. 1930.
18. Chandra often gleefully told the story (Chandrasekhar n.d.2, p. 3; 1977, pp. 16–18). The referee thought that Chandra's equation of state for a relativistically degenerate gas was incorrect. In the paper as originally submitted, Chandra did not derive this result, but he did so in his reply to the referee, who then withdrew the criticism. Chandra felt that he had to refer to the journal's editor on a point of principle: namely, that although his paper might not be worth publishing because the referee felt it was not interesting, the issue the referee raised was the "supposed invalidity" of an equation, which was not the case. Chandra never forgot this episode. When he became editor of the *Astrophysical Journal* in 1952, he looked up the correspondence files to find out the name of the referee, who, it turned out, had been the physicist Carl Eckart, then at the University of Chicago.
19. Chandrasekhar 1977, pp. 14, 15.
20. Letter from Chandra to his father, 16 Sept. 1930.
21. Letter from Chandra to his father, 10 Oct. 1930.
22. Letter from Chandra to his father, 19 Nov. 1930.
23. Letter from Chandra to Balakrishnan, 16 Oct. 1930.
24. Letter from Chandra to his father, 30 Sept. 1931.
25. Letter from Chandra to his father, 19 Nov. 1930.
26. Letter from Chandra to his father, 26 Nov. 1930.
27. The Dirac equation stunningly represented the electron in terms of both relativity theory, which deals with large-scale phenomena, and quantum theory, which deals with atoms.

28. Letter from Chandra to his father, 16 Jan. 1931.
29. Letter from Chandra to Balakrishnan, 21 Jan. 1931.
30. Chandrasekhar 1979b.
31. Letter from Milne to Chandra, 2 Nov. 1930. These papers are Chandrasekhar 1931a, b.
32. Chandrasekhar 1977, p. 49.
33. Milne 1931, p. 34.
34. Eddington 1931a, pp. 34, 35, 36.
35. Jeans 1931, pp. 36, 37, 38. A differential equation is a mathematical way of representing how one quantity behaves when other quantities change. For example, Newton's physics contains a differential equation that shows how an object's position in space varies over time as the object is falling to the ground. In stellar modeling the differential equations are much more complicated. A star's internal pressure depends on its volume and on the mix and distribution of chemical elements, among other variables. Several different linked equations have to be solved, and in each of them the quantities are often represented in complicated ways.
36. Smart 1931, p. 38.
37. Hardy 1931, p. 40.
38. Cowling 1978, p. 20.
39. Letter from Chandra to George Kenat, 28 June 1983.
40. Merrill 1938, p. 84.
41. Letter from Chandra to his father, 10 Feb. 1931. Chandra made a point of misspelling "occasionally" to irritate his punctilious father. As late as Mar. 1937, when Chandra was at Yerkes Observatory and Otto Struve was director there, Babuji wrote in exasperation, "I am quite pleased over the life you are leading, both Social and University at William's Bay. Well, you ought to spell the word 'occasion' properly I must say, when you are talking of entertaining Dr. and Mrs. Struve and family." Letter from Chandra's father, Mar. 1937.
42. Letter from Milne to Chandra, 16 Jan. 1931.
43. Ibid.
44. Letter from Milne to Chandra, 16 Feb. 1931.
45. Eddington 1931c.
46. Eddington 1931c, p. 446.
47. Chandra would later prove that such temperatures and pressures were indeed possible inside real stars.
48. Eddington 1931b, p. 98.
49. Letter from Eddington to Stoner, 28 Feb. 1932.
50. Letter from Chandra to his father, 30 Sept. 1931.
51. Letter from Chandra to his father, 22 May 1931.
52. Letter from Chandra to Balakrishnan, 12 May 1931.
53. Letter from Chandra to Balakrishnan, 6 June 1931.
54. Letter from Chandra to Balakrishnan, 23 Mar. 1936.
55. Letter from Milne to Chandra, 17 June 1931.
56. Letter from Milne to Dingle, 5 June 1931.
57. Letter from Milne to Dingle, 8 Sept. 1931.
58. Letter from Milne to Dingle, 15 Sept. 1931.

59. Letter from Milne to Chandra, 26 June 1931.
60. Letter from Chandra to his father, 25 June 1931. After returning to Government College in Lahore, Chowla produced a remarkably rich body of research, with several theorems named after him. When India was partitioned, at midnight on 14–15 Aug. 1947, Chowla and his family, who were Hindus, fled to Delhi. From there they moved to the United States, where Chowla continued his distinguished career. He died on 10 Dec. 1995.
61. Letter from Einstein to Born, 4 Dec. 1926. The wave function is also connected to Heisenberg's uncertainty principle, which asserts that an electron's position and velocity cannot both be measured at the same time with absolute accuracy. In Newtonian physics, however, accurate knowledge of both those quantities is essential to predicting an electron's position precisely as it moves through space and time.

   Born always thought he was denied the kudos owed him for the discovery of quantum mechanics. He also believed that the long delay before he received a Nobel Prize (1954) was the result of byzantine plots by major scientists, such as Einstein, who disputed his discovery of the deep meaning of the wave function, combined with the deterioration in relations with his former assistants, who had bypassed him in fame and professional standing. Being Jewish, Born was forced to leave Germany in 1933 and had some problems obtaining a position. He spent two years at Cambridge as Stokes Lecturer (Chandra had little contact with him there), then six months in Bangalore at the Indian Institute of Science, working with Raman. In 1935 Edmund T. Whittaker obtained for Born the Tait Professorship in Edinburgh, from which he retired in 1953. Whittaker also obtained a full pension for Born, who, like others who fled Germany, had lost his possessions and money. Born's corecipient of the Nobel Prize was Walther Bothe. This was not at all to Born's liking because Bothe had been an active worker on the German atomic bomb project. Born died in 1970. See, for example, Miller 2002, discussion and references.
62. Letter from Chandra to his father, 8 July 1931.
63. Letter from Chandra to his father, 22 Jan. 1936.
64. Letter from Chandra to his father, 22 Sept. 1931.
65. Letter from Chandra to his father, 14 Oct. 1931.
66. Letter from Chandra's father to Chandra, 17 Oct. 1931.
67. Singh 1963, p. 41.
68. Letter from Chandra to his father, 28 Oct. 1931.
69. Letter from Milne to Chandra, 16 Oct. 1931.
70. Letter from Chandra to his father, 12 Nov. 1931.
71. Letter from Chandra to Ralph Kenat, 28 June 1983.
72. Letter from Chandra to his father, 12 Jan. 1932.
73. Letter from Chandra to his father, 22 Jan. 1932.
74. Letter from Chandra to his father, 26 Jan. 1932.
75. Letter from Chandra to his father, 27 Mar. 1932.
76. Heisenberg had used group theory for just this purpose in 1926; see Heisenberg 1926. According to quantum mechanics, there are many ways for two or more electrons to combine, depending on their spin and orientation in space. Group theory throws up many combinations, and Pauli's

exclusion principle picks out the correct one to use to explore properties such as the way in which electrons in an atom create spectral lines.

77. Letter from Chandra to his father, 20 Apr. 1932.
78. Letter from Chandra to his father, 16 Aug. 1932.
79. Quoted in Pais 1986, pp. 416–17.
80. Conversation with Victor Weisskopf, 10 Apr. 1986. Sommerfeld's institute was in Munich, and Born's in Göttingen. Those cities, together with Copenhagen, were considered the three major centers of atomic physics research.
81. Chandrasekhar 1977, p. 39.
82. Ibid., p. 20
83. Quoted in Venkataraman et al. 1983, p. 16.
84. Letter from Chandra to Kenat, 28 June 1983. The published paper is Chandrasekhar 1932.
85. Chandrasekhar 1980a, p. 3.
86. Chandrasekhar 1932, p. 327.
87. Ibid., p. 326.
88. Letter from Chandra to Kenat, 28 June 1983.
89. Letter from Chandra to his father, 14 Dec. 1932.
90. Letter from Dirac to Bohr, 6 May 1932.
91. Letter from Chandra to his father, 21 Oct. 1932.
92. Letter from Bohr to Dirac, 14 Nov. 1932.
93. Letter from Chandra to his father, 11 Nov. 1932. Chandra published only one paper on a generalized quantum statistics, in 1929, when he had convinced himself that he was on the trail of something new and exciting in theoretical physics. Now his dream had been destroyed. Perhaps the unpleasant memories of that episode were the reason he did not include the 1929 *Physical Review* paper in a supposedly complete list of published papers in volume 6 of his *Selected Papers*.
94. Letter from Chandra to Balakrishnan, 2 Feb. 1933.
95. Letter from Chowla to Chandra, 8 Sept. 1935.
96. Letter from Chandra to his father, 17 Mar. 1933.
97. Letter from Chandra's father to Chandra, 4 June 1933.
98. Letter from Chandra to his father, 8 June 1933
99. Chandrasekhar 1980b, p. 5. Chandra published his Ph.D. research as a series of four papers packed with detailed calculations — Chandrasekhar 1933a.
100. Chandrasekhar 1980a, p. 5.
101. Chandra quoted Raman's words. Letter from Chandra to his father, 19 Sept. 1933.
102. Letter from Chandra to his father, 19 Sept. 1933. By "over looking" the younger men, Chandra meant that by looking them over, or checking them out, Raman was wasting his time.
103. Chandrasekhar 1933b, pp. 217–18.
104. Chandrasekhar 1977, p. 21.
105. Letter from Chandra to his father, 12 Oct. 1933.
106. Letter from Milne to Chandra, 20 Dec. 1933.
107. Letter from Chandra's father to Chandra, 23 Oct. 1933.
108. Letter from Chandra to his father, 5 Jan. 1934.

109. Letter from Chandra to his father, 24 June 1934.
110. Chandrasekhar 1934a, p. 98.
111. Wali 1991, pp. 116–17.
112. Letter from Chandra to Balakrishnan, 31 Aug. 1934.
113. This method is called polytrope theory, and its most developed form dates from the early twentieth century. The theory provides methods for calculating a star's interior temperature and pressure from what is known about these quantities on the surface, assuming that the star consists of a particular sort of gas, such as a perfect gas. Eddington made extensive use of polytrope theory in his research on the Cepheids, in the course of which he systematized the theory to facilitate its use in stellar modeling. The results are presented in *The Internal Constitution of the Stars*.
114. Letter from Chandra to his father, 14 Sept. 1934.
115. Letter from Chandra's father to Chandra, 24 Sept. 1934.
116. Letter from Chandra to his father, 5 Oct. 1934. There is no known extant correspondence between Chandra and Lalitha for this period.
117. Letter from Chandra's father to Chandra, 19 Oct. 1934.
118. Chandrasekhar 1934b. The three major 1935 papers are Chandrasekhar 1935b, c, and d.
119. For example, Chandrasekhar 1934a.
120. Chandrasekhar 1934b, pp. 374–75.
121. Ibid., p. 377.
122. Chandra deduced that the maximum mass of a white dwarf star would be 5.728 times the mass of the Sun divided by the mean molecular weight squared.

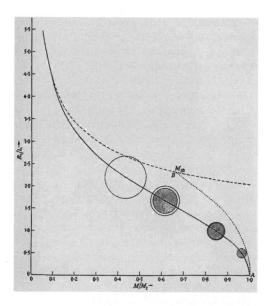

This graph (from Chandrasekhar 1935b, p. 219, courtesy of the Royal Astronomical Society), the one Chandra showed at the RAS in January 1935, is generated from the mathematics of his theory and is the definitive demonstration of how a white dwarf star can expire and what might happen if it exceeds Chandra's maximum mass. The vertical axis shows the radius of a white dwarf (R_1) in units of the number l_1, equal to 7.7×10^8 centimeters divided by the average molecular weight of the gas of electrons that makes up the star. The horizontal axis shows the mass (M) of the white dwarf in units of the maximum mass for a stable white dwarf (M_3), that is, 5.728 divided by the average molecular weight squared. Chandra labeled the vertical and horizontal axes in this very esoteric way for two reasons: to calibrate them using numbers with no units and to show that these ratios emerged naturally from the mathematics of his theory. To draw his curve, he set the average molecular weight at 1. The solid curve represents Chandra's theory and shows very clearly that as the mass of a white dwarf (M) approaches the maximum mass (M_3), the radius (R) becomes zero and the star shrinks to nothing. The broken line (– – –) shows the application of Chandra's theory when the electrons in a star are moving at speeds considerably slower than the speed of light; that is Fowler's theory. We can see that Fowler's inexact theory and Chandra's exact one diverge at a very early point. Electrons move at speeds appreciably less than that of light only in white dwarf stars of very low mass. The circles show stars at different stages in the diagram. The shaded regions show how quickly the incompressible core in Chandra's theory develops. The broken line continues forever, thus Fowler's theory does not predict a maximum mass. The dotted line (.....) shows the transition from the core in Fowler's theory to the one in Chandra's. The transition from B to A was required in Milne's theory, which postulated a nest of core within core in order to prevent complete implosion.

123. Letter from Milne to Chandra, 27 Sept. 1934.
124. Letter from Chandra to Kenat, 28 June 1983.
125. Letter from Chandra to his father, 16 Nov. 1934.
126. Letter from Chandra to Balakrishnan, 4 Jan. 1935.

6. Eddington's Discontents

1. Eddington 1935a, p. 38.
2. Ibid., pp. 38–39.
3. Ibid., p. 98.
4. Chandrasekhar 1977, p. 31.
5. Letter from Chandra to Rosenfeld, 12 Jan. 1935.
6. Letter from Rosenfeld to Chandra, 14 Jan. 1935.
7. Letter from Rosenfeld to Chandra, 19 Jan. 1935.
8. Eddington 1935b and Eddington 1935c, d.
9. Eddington 1935a, p. 38. Eddington also used an equation for the electron's total energy, made up of a nonrelativistic kinetic energy and a relativistic mass-energy part.
10. Eddington 1935b, p. 194.
11. Chandrasekhar 1980b, p. 6.
12. Letter from Chandra to Rosenfeld, 3 Feb. 1935.
13. Letter from Rosenfeld to Chandra, 6 Feb. 1935.

14. Fowler had also dealt with an idealized quantum gas of electrons. Eddington went on to derive an alternative equation for Fowler's theory, writing the gas pressure in terms of the number of electrons in the star and including interactions among all particles in the gas. He claimed that this equation showed that Fowler's theory was in agreement with the "requirements of relativity theory," while Chandra's could not produce a result of such "simplicity." Eddington 1935b, p. 206.
15. Letter from Rosenfeld to Chandra, 6 Feb. 1935.
16. Letter from Chandra to Rosenfeld, 3 Feb. 1935; Chandrasekhar and Möller 1935d, p. 673.
17. Letter from Rosenfeld to Chandra, 29 Jan. 1935.
18. Letter from McCrea to Chandra, 20 Feb. 1935.
19. Interview with Peter Vandervoort, 5 Nov. 2002.
20. See Chandrasekhar 1980b, p. 6; Wali 1991, p. 145; and Fowler 1936, p. 652.
21. Milne 1935a, p. 52.
22. Letter from Chandra to his father, 9 Feb. 1935.
23. Letter from Chandra's father to Chandra, 18 Feb. 1935.
24. Letter from Chandra to his father, 15 Feb. 1935.
25. Letter from Milne to Chandra, 26 Feb. 1935.
26. See, for example, Chandra 1979b — a very personal memoir.
27. Letter from Chandra to his father, 14 Mar. 1935.
28. Letter from Chandra to his father, 20 Mar. 1935.
29. Letter from Chandra's father to Chandra, 9 Mar. 1935.
30. Letter from Chandra to his father, 22 Mar. 1935.
31. Letter from Chandra's father to Chandra, 6 May 1935.
32. Letter from Chandra to Rosenfeld, 2 July 1935.
33. Milne 1935c, p. 174.
34. Eddington 1935e, p. 176.
35. Letter from Chandra to his father, 31 May 1935.
36. Letter from Chandra to his father, 7 June 1935.
37. Quoted in Frisch 1979, p. 174.
38. Chandrasekhar 1977, p. 92.
39. Letter from Milne to Chandra, 1 June 1935.
40. Chandrasekhar 1935d, p. 676.
41. Letter from Chandra to Rosenfeld, 24 Sept. 1935.
42. Letter from Chandra to his father, 21 June 1935.
43. Letter from Chandra to his father, 5 July 1935.
44. Chandrasekhar and Möller 1935.
45. Ibid., p. 676.
46. Eddington 1935c, p. 20.
47. Derivation of the relativistic degeneracy formula rests on a fundamental result of quantum mechanics in which one assumes the electrons to be in a particular space. In this situation their momenta take on certain definite values; in other words, they can be quantized. Eddington wrongly assumed that this violated the uncertainty principle, according to which position and momentum cannot both be taken as arbitrarily precise.

Eddington's assertion also had to do with the basic assumption, in order to make calculation simpler, that the shape of the space in which electrons are confined is arbitrary. The space can be a sphere or a cube, whichever is more convenient; in the end all references to the exact shape disappear. But Eddington disagreed, as he implied in his Nov. paper (Eddington 1935d, p. 258).

Chandra then persuaded the German physicist Rudolf Peierls to investigate "whether the pressure-density relation of such a gas enclosed in a certain volume would be independent of the shape of the volume" (Peierls 1936, p. 780). Using relativistic quantum mechanics, Peierls proved this in great detail. He even covered the situation in which, according to relativistic quantum mechanics, forcing an electron to remain enclosed introduces the possibility of creating electron-positron pairs, with the positron escaping and the electron remaining inside. Peierls showed that this deviation from independence of shape was completely negligible.

48. Letter from Chandra to Rosenfeld, 2 July 1935.
49. Letter from Rosenfeld to Chandra, 5 July 1935.
50. According to Einstein's general theory of relativity, the laws of physics are the same for all observers, wherever they are in the universe and however they are moving. Milne contended that that was not true because the distribution of matter is not the same everywhere in the universe. It is concentrated in galaxies, so only observers at the nuclei of galaxies would have the same law of gravitation. Milne's aim was to deduce a new law of gravity which he claimed explained the shapes of galaxies. He did not succeed.
51. Letter from Milne to Geoffrey Milne, 10 Aug. 1932.
52. Einstein 1949, p. 684.
53. Chandrasekhar 1979b, p. 12.
54. Letter from Chandra to Rosenfeld, 26 Mar. 1935.
55. Letter from Chandra to his father, 15 July 1935.
56. McCrea 1935, p. 259.
57. Ibid., p. 259.
58. Chandrasekhar 1977, p. 29.

7. American Adventure

1. Letter from Chandra to his father, 29 Aug. 1935.
2. Letter from Chandra to his father, 11 Aug. 1935.
3. Letter from Chandra to his father, 12 Dec. 1935.
4. Letter from Chandra to his father, 18 Dec. 1935.
5. Ibid.
6. Letter from Chandra to Balakrishnan, 20 Dec. 1935.
7. Letter from Chandra to his father, 31 Dec. 1935.
8. Letter from Chandra to Balakrishnan, 7 Jan. 1936.
9. Letter from Chandra to his father, 22 Jan. 1936.
10. Struve's letter quoted in a letter from Chandra to his father, 27 Feb. 1936.
11. Letter from Chandra to his father, 27 Feb. 1936.
12. Letter from Chandra's father to Chandra, 20 Mar. 1936.
13. Letter from Chandra to his father, 1 Mar. 1936.
14. Letter from Chandra to his father, 23 Mar. 1936. No one by that name is listed in the Oxford University archives as being on their faculty at that

time. Taking into account Chandra's poor English spelling and the different ways in which Indian names were spelled in the West, he was probably referring to Sarvepalli Radhakrishnan (1888–1975), Spalding Professor of Eastern Religions and Ethics at Oxford from 1936 to 1939. Like Chandra, Radhakrishnan came from Madras. He was a brilliant philosopher-statesman and was president of India from 1962 to 1967.

15. Letter from Chandra to his father, 5 June 1936.
16. Letter from Chandra's father to Chandra, 14 Oct. 1935.
17. Wali 1991, p. 168.
18. Hardy gave a series of lectures at Harvard on Ramanujan's life and work. Earlier that year he had contacted Chandra to discuss his thoughts. He also asked whether, on Chandra's next trip to India, he would look for a good photograph of Ramanujan for the published version of his lectures (Hardy 1940). While he was in India, Chandra sought out Ramanujan's wife and found her living in a small apartment in a narrow back alley in Triplicane. She could not find any photographs of her husband other than the small one in his passport, which in the end Hardy used as the frontispiece for his book. See Wali 1991, pp. 260–65.
19. Quoted in Douglas 1956, p. 105.
20. Eddington 1936b, p. 390.
21. Ibid., p. 390–91.
22. Ibid., p. 391.
23. Kanigel 1991, p. 373.
24. Eddington 1936a, p. 235.
25. Letter from Rosenfeld to Chandra, 11 Nov. 1936.
26. Conversation with Donald E. Osterbrock, 17 June 2004.
27. Letter from Chandra to his father, 12 Sept. 1937.
28. Letter from Raman to Chandra, 28 Feb. 1937.
29. Letter from Trimble to Chandra, 26 May 1953.
30. Eddington 1943.
31. Eddington 1939a, p. 595. In this paper, presented at the meeting of the RAS on 9 June, Eddington used Fowler's result to conclude that Sirius B has a high hydrogen content. He admitted that this ran counter to the usual notion of a white dwarf as a burned-out star. His argument was that a star goes through a quasi–white dwarf stage at the beginning of its life, and in this state it is almost all hydrogen. When the star contracts, hydrogen begins to burn, eventually forming carbon and nitrogen. At this point the star expands, enters the main sequence, and evolves in the usual way. Thus stars are white dwarfs twice in their lifespan — a very unlikely scenario.

Eddington concluded by stating forcefully that the Stoner-Anderson formula "generally used in the theory of white dwarfs is fallacious, and that Fowler's original formula was probably correct" (1939a, p. 605). On the anomalously high hydrogen content of Sirius B, he speculated that this would seem reasonable if Sirius B had split off from the outer layers of Sirius A.

32. Letter from Chandra to Trimble, 24 June 1953.
33. Letter from Amos Shaler to Chandra, 12 Dec. 1938.
34. Letter from Eddington to Chandra, 22 Jan. 1939.

35. Chandrasekhar 1935d, p. 690. Today astronomers still do not completely understand the evolution of Wolf-Rayet stars. The current thinking is that they gradually lose their envelopes of atmospheric gases, leaving exposed an interior of nuclear-processed material.
36. Letter from Eddington to Chandra, 22 Jan. 1939.
37. Ibid.
38. Letter from Shaler to Chandra, 4 Feb. 1939.
39. Letter from Shaler to Chandra, 11 Feb. 1939.
40. Letter from Chandra to Balakrishnan, 17 June 1939.
41. Chandrasekhar 1980a, pp. 10–11.
42. Chandrasekhar 1977, p. 35.
43. Kuiper 1941, p. 5. Discovered in 1917 by the Dutch astronomer Adriaan van Maanen (van Maanen 1917).
44. For white dwarfs that are not part of a binary system, such as van Maanen's star, astrophysicists calculate the mass from the measured gravitational redshift and the radius from the stellar parallax, apparent brightness, and temperature.
45. Kuiper 1941, p. 37. According to Fowler's theory, a white dwarf's mass times its radius cubed makes a constant, giving the result that the smaller the radius, the larger the mass. In other words, there can be tiny white dwarfs with a huge mass — much bigger than Chandra's limiting mass.
46. Chandrasekhar 1941, p. 42.
47. Quoted in Wali 1991, pp. 136–37.
48. Eddington 1941, pp. 66–67.
49. See note 14 to Chapter 5.
50. Kuiper 1941, p. 69.
51. Chandrasekhar 1941, p. 64.
52. Eddington 1941, p. 66. Here Eddington showed great prescience. Physicists were already debating the possibility of neutron stars, and some thirty years later, they were actually discovered.
53. Atkinson 1939, p. 284.
54. Chandrasekhar 1977, p. 34.
55. Ibid., p. 35.
56. Letter from Chandra to his father, 30 July 1939.
57. Eddington 1936a, p. 235.
58. Letter from Chandra's father to Chandra, 15 Aug. 1939. Babuji later wrote a longer letter to Chandra on the subject, dated 1 Jan. 1940.
59. Letter from Chandra to his father, 14 Feb. 1940.
60. Peierls, a brilliant young physicist, had already written a paper in response to Eddington's criticism of Chandra's application of quantum mechanics to an enclosed gas of electrons. See Peierls 1936 and note 47 to Chapter 6.
61. Dirac, Peierls, and Pryce 1942, p. 193.
62. Eddington 1942, pp. 208–9.
63. Salpeter 1961. Peierls had begun to address this (Peierls 1936). He was Salpeter's thesis supervisor in Birmingham in 1946–1948, and had interested Salpeter in this problem. Interview with Salpeter, 31 July 2002, and Salpeter 1999, p. 28.
64. Letter from Struve to Raman, 24 May 1941.

65. This correspondence has been collected in Ramaseshan 1996.
66. Ramaseshan 1996, p. 105.
67. Quoted in Wali 1991, p. 253.
68. Ramaseshan 1997, p. 105.
69. Letter from Chandra to Raman, 29 Nov. 1948.
70. Letter from Milne to Chandra, 15 Mar. 1944.

8. An Era Ends

1. Milne 1947, p. 584.
2. Ibid., p. 586.
3. Chandrasekhar 1980a, p. 8.
4. Chandrasekhar 1977, p. 36.
5. Ibid., pp. 30, 36.
6. Chandrasekhar 1979a, p. 7.
7. Letter from Milne to Chandra, 13 Oct. 1939.
8. Letter from Milne to Chandra, 2 Sept. 1944.
9. Ibid.
10. Letter from Milne to Chandra, 14 Nov. 1936.
11. Chandrasekhar 1979b, p. 9.
12. Letter from Milne to Chandra, 24 Feb. 1935.
13. Chandrasekhar 1979b, p. 8.
14. Letter from Chandra to Cowling, 18 Feb. 1977.
15. Letter from Cowling to Chandra, 2 Feb. 1977.
16. Chandrasekhar 1979a, pp. 5–6.
17. Ibid., p. 7.
18. Ibid., pp. 7–8.
19. Plaskett 1951, p. 172.
20. Chandrasekhar 1977, p. 37.
21. Quoted in Douglas 1956, p. 170.
22. Quoted ibid., p. 152. See Eddington 1946 for his fullest development of the fundamental theory.
23. Chandrasekhar 1987a, p. 139.
24. Eddington 1958b, p. 126.
25. Quoted in Douglas 1956, p. 101.
26. Dirac 1937, p. 323. See also Kragh 1990, chap. 11.
27. Bohr's remark, reported by Gamow in a letter of 1 Sept. 1967 to the editor of *Science*, Philip H. Abelson, is quoted in Kragh 1990, p. 233.
28. Chandrasekhar 1977, p. 51.
29. Chandrasekhar 1937, p. 757.
30. Ibid.
31. Chandrasekhar 1977, p. 50.
32. Chandrasekhar 1983a.
33. Chandrasekhar 1988, p. vii.
34. The Lorentz transformation, named after the brilliant Dutch physicist Hendrik Lorentz, is a mathematical means for observers in laboratories in relative motion to compare measurements such as lengths and times. Each observer would, for example, measure the same velocity of light. The Lo-

rentz transformation is at the heart of the special theory of relativity and so too of Chandra's theory of white dwarf stars.

35. Letter from Eddington to Walter Adams, 20 Apr. 1941.
36. Letter from Eddington to Adams, 29 Jan. 1941.
37. Quoted in Chandrasekhar 1987a, p. 140.
38. Douglas 1956, p. 183.
39. Letter from Winifred Eddington to Shapley, 27 Nov. 1944.
40. Russell 1945, p. 133.
41. Chandrasekhar 1977, p. 49.
42. Compton 1904, p. 247; see also Kanigel 1991, pp. 99–102.
43. Slater 1936, p. 47.
44. Chandrasekhar 1977, p. 54.
45. Wali 1991, p. 114.
46. Chandrasekhar 1977, p. 54.
47. Interview with Michael Turner, 30 Oct. 2002.
48. Littlewood 1953, p. 136.
49. Ibid.
50. Letter from Chandra to his father, 24 Apr. 1936.
51. Wali 1991, p. 237. Chandra's uncle Raman suffered similar humiliating experiences during a visit to America in the 1920s. In Boston he was turned away from several hotels; finally a taxi driver took him to one run by a Japanese couple in the suburbs (Wali 1991, pp. 236–37).
52. Quoted in Douglas 1956, p. 155.

9. HOW STARS SHINE

1. Milne 1932, pp. 346–47.
2. Zwicky studied with Peter Debye and Paul Scherrer, two leaders in the field of crystal structure. Debye was awarded the 1936 Nobel Prize in chemistry.
3. Letter from Chandra to his father, 10 Oct. 1930.
4. Osterbrock 1997, pp. 58–61.
5. Osterbrock 1996, p. 310.
6. Greenstein 1982, p. 26.
7. Osterbrock 2001.
8. A supernova was also discovered in 1572 by the Danish astronomer Tycho Brahe, and another in 1604 by his former assistant, Johannes Kepler. All these events were, of course, seen by the naked eye without the aid of a telescope, which had not yet been invented. The first supernova to be observed with a telescope was in the Andromeda Galaxy in Aug. 1885.
9. Since light takes time to travel across space, we observe the universe in the distant past.
10. See, for example, Struve and Zebergs 1962, p. 340.
11. Baade and Zwicky 1934.
12. Bethe 1993, p. 29.
13. Osterbrock 1996, pp. 58–64.
14. Zwicky 1939, p. 726.
15. Chandrasekhar 1935c, p. 258.
16. Chandrasekhar 1941, p. 47.

17. Quoted in Danin 1989, p. 81.
18. Pellam 1989, p. 198.
19. Landau 1932, p. 458.
20. Ibid.
21. Ibid., p. 459.
22. See also Thorne 1994, pp. 184–87, and correspondence in Khalatnikov 1989, pp. 308–19.
23. Landau did go on to solve the puzzle of superconductivity, for which he won a Nobel Prize in 1962. Kapitza received his in 1978.
24. Bethe 1966, p. 49.
25. Letter from Gamow to Chandra, 7 Jan. 1938.
26. Letter from Gamow to Struve, 17 Sept. 1938, quoted in DeVorkin 2000, p. 254.
27. Rosenfeld 1966, p. 483.
28. This was before the discovery of the neutron, so the alpha particle was assumed to be made up of four protons and two electrons, giving it a double positive charge. When the alpha particle is the nucleus of a helium atom, its double positive charge neutralizes the double negative charge of the two atomic electrons. Rutherford obtained alpha particles from a gas of radon atoms, although he did not understand the process that caused this emission of particles.

    That process was not explained until after the discovery of the neutron in the 1930s, when physicists reached a consistent understanding of the nucleus. They reasoned that certain nuclei are unstable or "radioactive," depending on the numbers of protons and neutrons. To become more stable, the nuclei emit alpha particles, electrons, positrons, neutrons, protons, or gamma rays.
29. See also Stuewer 1986.
30. Frisch 1979, p. 72. See also Rhodes 1986, pp. 370–71, and Khriplovich 1992.
31. Khriplovich 1992, p. 30.
32. Bethe and Critchfield 1938.
33. Chandrasekhar 1977, p. 77.
34. Bethe 1966, p. 4.
35. Ibid.
36. Ibid., p. 50.
37. Gamow 1968, p. 68.
38. First Bethe had to prove that the proton-proton reaction could proceed fast enough to produce helium and thus start the process of fusion. That was the main conclusion of the Bethe-Critchfield paper, which established a chain of nuclear reactions beginning with two protons fusing together and momentarily forming a "diproton." One of the protons is transformed into a neutron, a positron, and a neutrino through one sort of beta decay. The positron and neutrino fly around the interior of the star, but the other proton stays with the neutron to form a deuteron, which then fuses with another proton and neutron to form the nucleus of a helium atom. This is how stars with masses comparable to the Sun's produce energy and shine.

Pauli proposed the neutrino in 1930 in response to problems thrown up by the process inside a nucleus in which a neutron is transformed into a proton and an electron — that is, beta decay. When it was first discovered, beta decay seemed to violate the law of conservation of energy in that the energy content of the nucleus beforehand was greater than the sum of the nucleus's energy afterward plus the energy of the escaping electron. (Electrons do not exist in the nucleus; the electron is created and ejected at the instant the neutron decays.) Pauli made the audacious suggestion that the simplest way to balance these two energies was to assume that a hitherto unknown particle with no mass or charge must exist, and that what actually happens in beta decay is that a neutron becomes a proton, an electron, *and* a neutrino. The neutrino was discovered more than two decades later, in 1956.

39. As Bethe described it, he systematically examined "reactions between protons and other nuclei, going up the periodic system" (Bethe 1972, p. 222) that were possible at the high temperatures inside stars much brighter than the Sun. He discovered a self-sustaining chain of reactions in which hydrogen is burned to produce helium over millions of years, a time scale in agreement with the lifetime of main-sequence stars of this type. The chain begins with hydrogen interacting with carbon, which makes up about 1 percent of a young star. Helium is the end product of a sequence of reactions that produce nitrogen and oxygen. The sequence is, in fact, a cycle. Carbon is reproduced and acts as a catalyst — keeping the reaction going and using up protons. This carbon-nitrogen-oxygen (CNO) cycle can continue for a million years. Bethe's theory fit with the lifetime of stars such as Sirius A, around twice the mass of the Sun, and the hot young Y Cygni, in the constellation Cygnus, some 20 times the mass of the Sun and 600 times more luminous (more recent measurements have raised the figure to 10,000 times more luminous).

40. Eddington 1926, p. 306; Chandrasekhar 1977, p. 39.

41. Bethe 1939. His paper was not published until March of that year, though it had been received at the *Physical Review* on 7 Sept. 1938. It had been entered for the A. Cressy Morrison Prize of the New York Academy of Sciences, which required that papers be unpublished. Bethe won.

42. Chandrasekhar, Gamow, and Tuve 1938.

43. Bethe 1939, p. 156.

44. Rhodes 1986, p. 444.

45. Oppenheimer 1963, p. 18.

46. See note 30 to Chapter 4.

47. After Bethe departed, Oppenheimer and his group began looking at the problem of how energy is produced in stars more massive than the Sun. Oppenheimer insisted on checking the various nuclear reactions his group was investigating against experimental data. According to Robert Serber, someone gave Oppenheimer incorrect data, and this led the group away from a carbon cycle. "Bethe didn't have that data" and therefore assumed that this set of reactions would work. See Serber 1992, p. xxvi.

48. Herken 2002, p. 290.

49. Serber 1992, p. xxxii.

50. Oppenheimer and Serber 1938.
51. Oppenheimer and Volkoff 1939.
52. Zwicky 1939, p. 743.
53. Chandrasekhar 1977, p. 92.
54. Ibid., p. 79.
55. Interview with Donald E. Osterbrock, 24 Mar. 2003. He was a student of Chandra's from 1949 to 1952.
56. Letter from Einstein to Schwarzschild, 9 Jan. 1916.
57. Letter from Eddington to his mother, 5 Aug. 1913.
58. Eddington 1926, p. 6.
59. Oppenheimer and Snyder 1939, p. 456.
60. Einstein 1939, p. 936.
61. Alvarez 1987, p. 78.
62. Letter from Bethe to Chandra, 26 Apr. 1944.
63. Chandrasekhar 1977, p. 98.
64. Letter from Bethe to Chandra, 27 Sept. 1944.
65. Chandrasekhar 1977, p. 98.
66. Ibid., p. 99.

10. Supernovae in the Heavens and on Earth

1. Chandrasekhar 1977, p. 34.
2. Ibid., p. 78.
3. Chandrasekhar 1975a, p. 3; Chandrasekhar 1977, pp. 34, 78, 88.
4. Chandrasekhar 1943.
5. Gamow 1939b.
6. Chandrasekhar and Schönberg 1942.
7. See Sandage and Schwarzschild 1952, and Osterbrock 2001.
8. Astrophysicists first predicted that some stars might expire as diamonds in the 1960s, but it was only in Feb. 2004 that astronomers found one. In the constellation Centaurus, 300 trillion miles away, the white dwarf star BPM 37093 has a diamond core of some 10 billion trillion trillion carats. BPM 37093 is estimated to be 1.1 times the mass of the Sun and 90 percent crystallized. It has been nicknamed Lucy, after the Beatles song.
9. Chandra's astrophysical research entailed much numerical computing, carried out on electric computing machines. In his early days at Yerkes, his graduate students assisted him on particularly lengthy calculations. He also hired full-time computers, as the people who used these machines were called, whose salaries were paid out of observatory funds. Donna Elbert, a native of Williams Bay, was Chandra's best-known computer. She worked for him for more than thirty years, starting in 1949 at Yerkes, then at the Chicago campus. See Elbert 1997; and Osterbrock 1999, pp. 212–13.
10. Shock waves are undulations that travel through a medium faster than sound. Their familiar signature in the skies is the sonic boom produced by jet planes breaking the sound barrier. Radiative transfer is the study of how heat and radiation moves through material, like the radiation escaping from a star.

11. Chandrasekhar 1977, pp. 106, 101.
12. Chandrasekhar n.d.3, p. 4.
13. Telegdi 1997, p. 207.
14. Vandervoort n.d., p. 12.
15. Ibid., p. 18.
16. In daily life we distinguish between left and right; our heart is on our left side. But it was always assumed that the laws of physics made no such distinction; this principle is known as "conservation of parity." Violation of it would mean, for example, that an experiment and its mirror image might yield different results. Lee and Yang predicted in 1956 that on the atomic level, in reactions such as beta decay, nature could in fact distinguish between left and right and parity conservation could be violated. Later that year their theory was verified. In two sets of experiments set up as mirror images of each other, nuclei of cobalt underwent beta decay. The meters in the experiments registered different numbers of electrons emitted by the cobalt nuclei, thus providing definite proof of an asymmetry between left and right that quantum physics would have to allow for. Lee and Yang's work won the Nobel Prize in physics the following year.
17. Osterbrock 1999, pp. 205–6, and personal communication.
18. Chandrasekhar 1977, p. 151.
19. Comment by the physicist Valentine Telegdi, a colleague of Chandra's and Fermi's at the University of Chicago, quoted in Wali 1991, p. 19. Their papers were Chandrasekhar and Fermi 1953a, b.
20. As recalled by Hutchins some years later and quoted in Wali 1991, p. 236.
21. Letter from Chandra to Hutchins, 20 Dec. 1950.
22. Letter from Hutchins to Chandra, 28 Dec. 1950.
23. Although Fermi invited Chandra to transfer to the physics department in 1952, Chandra may simply have delayed making the switch final until 1964. Conversation with Donald E. Osterbrock.
24. Urey won the 1934 Nobel Prize in chemistry for his discovery of deuterium, or heavy water, essential for the production of nuclear weapons. Libby became a Nobel laureate in chemistry in 1960 for discovering radiocarbon dating, and Goeppert-Mayer in physics in 1963, for research on nuclear structure.
25. Interview with Roger Hildebrand, 31 Oct. 2002.
26. After a certain amount of time, nuclei of radioactive elements emit elementary particles and radiation and transform themselves into other elements, thus achieving stability. The time that it takes half a sample of radioactive nuclei to decay is known as the half-life of the nucleus. The half-life of any nucleus is a property of the nucleus itself and does not depend on how many nuclei are packed together or on the temperature of the sample. The half-life of uranium is 4 billion years. That is the age of the planet, which means that only half of the original amount of uranium on Earth has decayed so far.
27. Fermi had assumed that he had discovered an element containing one more proton than uranium and with a half-life of 13 minutes. He took this to be element 93, one beyond uranium, which is designated element 92 in

the periodic table because it has 92 protons. In his data analysis Fermi failed to compare what he thought was a new element with all known elements; he stopped at lead, which was assumed to be the final step in the disintegration of a radioactive element. In fact he had split the uranium nucleus into two large fragments, each radioactive and lighter than lead, namely barium and krypton.

28. Any sample of naturally occurring uranium contains 99.3 percent of ^{238}U and 0.7 percent of its isotope ^{235}U. Bohr had discovered in 1939 that ^{235}U can be split more easily. The neutrons required to split it travel more slowly than those required for ^{238}U.

29. Quoted in Rhodes 1986, p. 417.

30. Quoted in Rhodes 1995, p. 303.

31. Letter from Wheeler to the head of the Princeton University physics department, 8 Jan. 1951. Quoted in Galison and Bernstein 1989, p. 320.

32. ^{239}Pu, produced in reactors by colliding fast neutrons with ^{238}U, undergoes fission as efficiently as ^{235}U and, it turned out, was easier to produce. ^{235}U was manufactured mainly by separating it from the heavier isotope ^{238}U by gaseous diffusion techniques implemented on a grand scale at Oak Ridge, Tennessee. ^{239}Pu was produced in nuclear reactors at Hanford and then chemically separated at the University of Chicago's metallurgical laboratory.

33. Fuchs's Communist leanings led him to flee Germany in 1933. In England he impressed several important scientists, including Max Born, who was then in Edinburgh. Eventually Fuchs was recruited into the British atomic bomb project and then the Manhattan Project at Los Alamos. In 1943, as a reward for his research, Fuchs was granted British citizenship. All the while he was passing key information to the Soviets. After the war he supplied them with details of the development of the hydrogen bomb. Fuchs's espionage activities were suspected when the Soviet documents known as the Verona cables were decoded in 1948, and in 1950 he was arrested. At the time he was head of the theoretical physics division of the Atomic Energy Establishment at Harwell in England. Fuchs was released after serving nine years of a fourteen-year sentence and returned to East Germany, where he resumed his physics career. In 1953, in the United States, Julius and Ethel Rosenberg were executed for committing espionage at a level that paled in comparison to Fuchs's.

34. Quoted in Herken 2002, p. 219.

35. Ulam 1976, p. 60.

36. Ibid., p. 61.

37. As Ulam's wife, Françoise, recalled: see Rhodes 1995, p. 463.

38. United States Atomic Energy Commission 1954, p. 251.

39. Quoted in Holloway 1994.

40. Ibid.

41. Sakharov left weapons work and returned to physics research in 1965, publishing several important papers on particle physics and cosmology. At the end of the 1960s he decided to take a bold stand on violations of human rights in the Soviet Union, for which he was awarded the Nobel Peace Prize in 1975. He died in Moscow on 14 Dec. 1989.

42. Sakharov 1990, p. 94.
43. See Khariton and Sakharov 1993, Khariton, Adamskii, and Smirnov 1998. American scientists claimed that it was not a true hydrogen bomb but a conventional atomic bomb boosted by thermonuclear fuel, set off by chemical explosives. The Americans took its low yield as proof that the Russians had not yet discovered the Teller-Ulam design. Soviet scientists who took part in the project have recently claimed the credit for exploding the first "real" hydrogen bomb, because theirs was the first with the potential to be weaponized.
44. Sakharov 1990, p. 102.
45. Wheeler and Ford 1998, p. 208.
46. Khariton, Adamskii, and Smirnov 1998.
47. Russian scientists claim that the Czar bomb's yield remains unsurpassed; it derived 97 percent of its energy from thermonuclear reactions. In contrast, more than 75 percent of Mike's yield originated in the fission of its atomic-bomb trigger, followed by another round of fission reactions as fast neutrons from the combined trigger and fusion explosion in turn fissioned the uranium tamper (made of ^{238}U) encasing the bomb.
48. Interview by the author with Blinnikov, 27 Nov. 2002.
49. Bethe 1993. "Gadget" was the bomb's nickname at Los Alamos.
50. Christy 1994.
51. Wheeler, Harrison, and Wakano 1958, p. 137.
52. The computer simulation being described is for a star with a mass eighteen times that of the Sun — see Herant, Colgate, and Fryer 1997. See Woosley et al. 2002 for the milestones in the life of a star twenty-five times the mass of the Sun. See also Harrison et al. 1965.
53. The mass of the iron core at this point is actually less than the Chandrasekhar limit, and thus it should not begin to collapse yet. This seeming anomaly puzzled astrophysicists for several years. By the late 1980s they realized that the Chandrasekhar limit has to be handled with care. Chandra had deduced it for the "pristine" case of relativistic electrons that are not interacting with one another — as Chandra stated in his early papers, for an "ideal" white dwarf.

    Various corrections had been introduced from the early 1960s. Salpeter, for example, calculated corrections to take account of electron interactions (Salpeter 1961, Hamada and Salpeter 1961). One of Eddington's criticisms had been that Chandra ignored the interactions between electrons.

    Various factors have to be taken into account to calculate a "realistic" Chandrasekhar limit for cores inside stars. Although the surface of a white dwarf is not under pressure, the iron core is. That fact alone produces a Chandrasekhar maximum mass below the expected core mass, which turns out to be about 1.3 times the mass of the Sun. The electrical interactions between the tightly packed iron nuclei, the heat content of the core, and the effect of general relativity also have to be taken into account. All of this alters the "traditional" Chandrasekhar maximum mass from 1.4 times the mass of the Sun for white dwarfs to as low as 1.15 times the mass of the Sun for the degenerate iron core. See Woosley et al. 2002, especially pp. 1045–47.

54. Wheeler, Harrison, and Wakano 1958, p. 137.

11. How the Unthinkable Became Thinkable

1. Quoted in Wheeler, Harrison, and Wakano 1958, p. 147.
2. Quotations from Stirling Colgate in this chapter are from interviews held by the author in Los Alamos, 29–30 Mar. 2003, and e-mail communications.
3. No one knows exactly what would happen if a hydrogen bomb were exploded in space. Almost certainly gamma rays, radioactivity, and many high-energy particles would burst out — from lighter ones, such as electrons and protons, to heavy nuclei created by the fission-fusion-fission process. In this process, the first fission ignites the fusion fuel; the second is caused by additional fast neutrons that blow off the outer layer (tamper) of ^{238}U and reflect neutrons back into the original fission.
4. The neutrino, a neutral particle with no mass, was postulated in 1930 by Pauli and subsequently discovered, to great excitement, in 1956, a year before the B^{2}FH paper was published.
5. Physicists assume that neutrons and protons are made up of quarks, fundamental building blocks that cannot be broken down any further. But quarks have never been observed as free, isolated objects. To account for this, and to create a theory of how elementary particles such as neutrons and protons interact to form a nucleus, physicists postulated that quarks were always "confined" within the particles they constitute. Observing them as free objects would be a great advance in understanding how to formulate a theory of elementary particles. The stars may present that opportunity.
6. E-mail from White to the author, 20 Sept. 2003.
7. Fowler and Hoyle proposed that core bounce occurs after alpha particles (helium nuclei) have been cut to pieces by high-energy radiation, thus increasing the density and number of particles in the core and preventing further compression. But they did not do any computer simulations. They were amazed when Colgate and White found that the star does not stop collapsing at that point. The explosion that Fowler and Hoyle had predicted turned out to be an implosion. See Fowler and Hoyle 1964.
8. E-mail from White to the author, 6 June 2003.
9. Interview by the author with Colgate, 29–30 Mar. 2003.
10. Thorne 1994, p. 239.
11. Colgate and White 1966, p. 643. My thanks to Richard White for pointing out Chandra's added reference. Despite the fact that Colgate and White had adapted bomb codes for their research, they had no problems with the government censor, who studied every paper published by Livermore personnel. By 1965 it was clear to everyone that compressible hydrodynamics, which was taken into account in their equation of state, was an important aspect of nuclear weapons.
12. Like planets, stars spin. The Sun completes a rotation every twenty-five days; white dwarfs, from once every few hours to once every few days.

13. Astrophysicists suspect that the radio signal from a pulsar comes from charged particles near its north and south magnetic poles, accelerated by the rapidly rotating star's intense magnetic field. The line connecting the north and south poles is tilted away from the line around which the star rotates, and as a result the pulsar's beam sweeps across the sky. When Earth is in the path of the beam, the star appears to pulse on and off.

14. In 1974 Hewish received the Nobel Prize in physics for discovering pulsars. There was some controversy over why Bell did not share the prize. Martin Ryle, another Cambridge radio astronomer, was cowinner.

12. THE JAWS OF DARKNESS

1. The Schwarzschild radius is the distance from the center of the black hole to the "event horizon" — the surface where the event occurs, so to speak. A star does not disappear when it falls into a black hole; it continues to exist in the same way that the Moon does when it is over the horizon.

2. Thorne 1994, pp. 248–51.

3. Eddington 1924b. For details see Misner, Thorne, and Wheeler 1970, pp. 828–32.

4. Penrose 1969.

5. It is difficult to measure the distances of the different galaxies from Earth, with the result that there may be a 10 percent margin of error in the Hubble constant. Moreover, because the universe may not always have expanded at the same rate, it is possible that the Hubble constant was different in the past. Having established the Hubble constant, we can very easily work out the age of the universe to be 13 to 16 billion years (taking into account the 10 percent margin of error).

In the 1950s there were two conflicting theories of the origin of the universe. According to the steady state theory, the universe is the same today as it always was. The Big Bang theory attempted to incorporate Hubble's discovery and predicted a cosmic microwave background temperature of about 3 degrees kelvin. (An advocate of the steady state theory, Fred Hoyle, coined the term "Big Bang" in a somewhat derogatory manner on a BBC radio program in 1950, to distinguish it from his own theory.) The most vigorous proponent of the Big Bang was Gamow. See Singh 2004 for an enlightening discussion of the history of the Big Bang theory.

In 1965 Arno Penzias and Robert Wilson discovered radiation that seems to permeate the universe and that is believed to be the "echo" of the Big Bang; it has exactly the temperature predicted by the Big Bang theory, 3 degrees kelvin. Beginning as an incredibly hot, small, dense mass, the universe cooled as it expanded, to the 3 degrees kelvin it is today. Penzias and Wilson's work was the impetus for modern research into the origin of the universe and is considered to be the turning point in modern cosmology. See Weinberg 1977 for an account of the discovery of the 3-degree background radiation and its immediate effect on astrophysics.

In 2003 the Wilkinson Microwave Anisotropy Probe satellite made the most precise measurements to date of the universe's temperature and thus of its age: in line with previous estimates, it turned out to be 13.7 billion years, with a margin of error of a few hundred million years.

6. Radar is a form of radio waves emitted from a radio transmitter-receiver that is like a two-way telescope; if the waves encounter an object, they are reflected back.

7. One of these was the radio wave source first detected by Jansky back in 1932. As in Chandra's case, no one had believed his discovery.

8. The term "optical" is used because the counterparts can be "seen" with a telescope and photographed, even though they cannot be seen with the naked eye.

9. Baade and Minkowski 1954.

10. In the 1980s infrared telescopes showed that what Baade and Minkowski had interpreted as two galaxies in collision is in fact an optical illusion caused by the absorption of light by the dust band running across the face of the galaxy containing Cygnus A.

11. The numbers are the designations in the *Third Cambridge Catalogue* of radio wave sources, assembled by a group working under the radio astronomer Martin Ryle.

12. Conversation with Gary Steigman, 9 July 2003.

13. Schmidt 1963.

14. Ibid.

15. Greenstein and Matthews 1963.

16. Schmidt 1963, p. 1040.

17. Greenstein and Schmidt 1964.

18. Colgate 1967. See also comments in Burbridge and Burbridge 1967, p. 193. Richard White recalls that Colgate's paper was so complicated that the original referee at Livermore asked for his help. White in turn requested advice from experts at Livermore in the various fields on which Colgate drew. The collective recommendation of this "board" of referees was that although parts of the argument needed clarification, the paper was worthy of publication. No one ever managed to read it through completely.

19. Hoyle and Fowler 1963, p. 170. See Burbridge and Burbridge 1967, chap. 17, for other proposals for energy generation.

20. Fowler 1984, p. 141.

21. Fowler 1966.

22. Chandrasekhar 1977, pp. 149–50.

23. Ibid., p. 79.

24. Ibid., p. 154.

25. Chandrasekhar 1964a, b.

26. Chandrasekhar 1977, p. 153.

27. Interview with Richard White, 8 July 2003.

28. Ibid. At this stage Halton C. Arp, a thirty-seven-year-old observational astronomer at the Hale Observatory, put himself forward as the leading advocate of gravitational redshift as part of the explanation for quasars. If that was the correct scenario, the quasar would be extremely dense but would not have to be so massive as to be unstable. The gravitational redshift would be great enough to remove the need to claim that quasars were moving

at very high speeds or were very far away from our galaxy. Greenstein and Schmidt had already advanced strong arguments against this position (Greenstein and Schmidt 1964). But Arp soldiered on and even produced photographs showing quasars apparently in nearby galaxies. Hubble's law, he argued, could not apply to quasars because their redshifts suggested that they were farther away. There had to be some as yet undiscovered law that would place them closer to Earth and prove that they were not as luminous as they appeared. His arguments were so persuasive that even astronomers like the Burbridges were prepared to hedge their bets and declare that they could "see no conclusive answer." Cecilia Payne, then America's premier woman astronomer, wrote in her 1970 astronomy textbook that scientists might have to alter their view of galaxy recession. See Burbridge and Burbridge 1967, p. 220, and Payne-Gaposchkin and Haramundanis 1970, p. 562.

Chandra was less charitable toward Arp's proposal. He remembered that as managing editor of the *Astrophysical Journal,* he "used to reject Arp's papers outright, several of them." He agreed to publish only Arp's data and photographs, but with no theoretical interpretation. Eventually astronomers realized that most of Arp's data could be explained away. In the two-dimensional photographs, quasars that happened to be in the background of low-redshift galaxies could look as if they were part of the galaxy in the foreground. See Chandrasekhar 1977, pp. 133, 157.

29. See Thorne 1994, pp. 256–57, and Wheeler and Ford 1998, pp. 296–97, for this story.
30. Kerr 1963.
31. According to Euclidean geometry, the sum of the three angles inside a triangle is always 180°. Such a triangle can be drawn on any flat surface, no matter how the surface is oriented in relation to Earth. In 1838 the astronomer-mathematician Friedrich Bessel — who predicted that Sirius A had a companion — applied this law of Euclidean geometry to the triangle employed in measuring the stellar parallax of the star 61 Cygni. Two sides of the triangle were the distances from the star to Earth. The sightings were taken six months apart so that Earth was at opposite sides of its orbit around the Sun. The third side was the diameter of Earth's orbit. Bessel found that the sum of the angles was still 180°, with little margin of error. From this he concluded that the physical nature of space was Euclidean. Thus we say that three-dimensional space is Euclidean and flat. If Bessel had obtained a result greater than 180°, he would have inferred that space is curved like the surface of a sphere.

One of the extraordinary things about general relativity is Einstein's discovery that he had to write its equations in a geometry in which the interior angles of a triangle did not necessarily add up to 180° — a "non-Euclidean" geometry. This turned out to be the proper way to represent the fabric of spacetime, and it led to the prediction of physical phenomena caused by the extreme curvature of space around massive bodies such as stars. One such phenomenon was the deflection of starlight as it passed by the Sun, which Eddington observed in 1919. See Miller 1972, 2000.

32. Wheeler came up with phrase in about 1975. See Thorne 1994, p. 275.
33. Chandrasekhar 1975b, p. 54.
34. Electric and magnetic disturbances travel through space in the form of waves at 186,000 miles per second, which we observe as light waves.
35. In 1978 Russell Hulse and Joseph Taylor obtained indirect evidence for gravitational waves by observing the binary pulsar PSR 1913+16, made up of two neutron stars. General relativity has to be used in the analysis of such a system. Hulse and Taylor found that the rate of decrease of the system's orbital period was consistent with the emission of gravitational radiation. For discoveries such as the exploration of this highly interesting binary system, they shared the 1993 Nobel Prize in physics. More direct tests for gravitational waves are discussed in Chapter 14.
36. Chandrasekhar 1983b, Prologue.
37. Chandrasekhar 1972, p. 173.
38. Supergiants such as HDE 226868 and Betelgeuse are rare stars a thousand times larger than the Sun and a hundred thousand times as luminous; if the Sun were replaced by a supergiant, Earth's orbit would be inside the star. HDE 226868 appears bluish.
39. Scott and Leonov 2004, pp. 317–18.
40. NASA 1972, chap. 17, pp. 13–17.
41. Chandrasekhar 1978, esp. pp. 405, 415–17.
42. Chandrasekhar 1977, p. 77.
43. The jets are the means by which energy is extracted from the region around a supermassive black hole and transferred to the radio lobes. How does the black hole produce these jets? One possibility is that the crowding of gas particles in the inner portion of the accretion disk leads to such high pressures that matter has to be expelled to prevent an explosion. Particles seek the path of least resistance, so rather than plowing back through the accretion disk, excess matter shoots off at right angles to it.

13. Shuddering Before the Beautiful

1. Quoted in Wali 1991, p. 300.
2. Quoted ibid., p. 301.
3. Chandrasekhar, 20 Jan. 1976, Box 77 (Addenda), Joseph Regenstein Library.
4. These were the Reissner-Nordström equations, representing a Schwarzschild black hole with electrical charge. Chandra was also working on the Kerr-Newman equations, which are analogues of the Reissner-Nordström equations but for a Kerr black hole. See Chandrasekhar 1977, p. 166.
5. Chandrasekhar, 16 Nov. 1976, Box 77 (Addenda), Joseph Regenstein Library.
6. Chandrasekhar 1977, p. 173.
7. Chandrasekhar, 18, 19 Oct. 1980, Box 77 (Addenda), Joseph Regenstein Library.
8. Chandrasekhar 1977, p. 56.
9. Interview with Lalitha Chandrasekhar, 1 Nov. 2002.

10. Chandrasekhar, 5 June 1982, Box 77 (Addenda), Joseph Regenstein Library.
11. Chandrasekhar, 28 Nov. 1982, Box 77 (Addenda), Joseph Regenstein Library.
12. Chandrasekhar 1983b, p. 529.
13. Quoted in Rees 1997, p. 184.
14. Ibid., p. 185.
15. Chandrasekhar 1977, p. 155.
16. Johansson 1993, p. 136.
17. Chandrasekhar 1983b, p. 140.
18. Wali 1991, pp. 294–98.
19. Interviews with Gary Steigman, 1 Aug. 2002, and with Michael Turner, 30 Oct. 2002.
20. Vandervoort n.d., p. 39.
21. In his acceptance speech at Yerkes, Fowler lamented that Fred Hoyle had not shared the prize, especially since he had made such critical contributions to their earlier work, for example, the B^2FH paper. Most probably Hoyle was excluded for two reasons. He continued to support the Steady State universe in the face of overwhelming evidence for the Big Bang, and he had complained vociferously to the Nobel committee that they had overlooked Jocelyn Bell's part in the discovery of pulsars.
22. Quoted in Wali 1991, p. 295.
23. Interview with Valeria Ferrari, 8 Dec. 2002.
24. Xanthopoulos 1991, p. xvi; Chandrasekhar and Ferrari 1984.
25. Hawking and Israel 1987.
26. Quoted in Dalitz 1997, p. 152.
27. Swerdlow 1997.
28. Chandrasekhar, 4 May 1991, Box 77 (Addenda), Joseph Regenstein Library.
29. Chandrasekhar n.d.3, p. 2.
30. Chandrasekhar 1995, Prologue.
31. Interview with Chandra by the author, 31 Oct. 1993.
32. Westfall 1996, p. 702. See Gleick 2003, an informative biography of Newton.
33. See, for example, Chandrasekhar 1987c, p. ix, and Chandrasekhar 1983a, p. 140.
34. Hardy 1941, p. 91.
35. Letter from Chandra to Balakrishnan, 17 June 1939.
36. Chandrasekhar n.d.4, p. 1.
37. Chandrasekhar 1977, p. 161.
38. Chandrasekhar 1992, p. 135.
39. Chandrasekhar 1979c, p. 69; Chandrasekhar 1992, p. 135. See Francis Bacon, *Essays on Beauty*, Essay 43.
40. According to quantum theory there is no such thing as empty space. Space is seething with matter and antimatter, appearing and disappearing, constantly being created and annihilated. Imagine that somewhere near the event horizon, the black hole's intense gravitational field spawns a particle

and its antiparticle. During their fleeting existence, one falls through the event horizon into the black hole. The other, unable to combine with its partner and be annihilated, wanders off into space. According to Einstein's equation $E = mc^2$, the particle that escapes removes energy from the black hole. The same thing happens when two quanta (particles of light) are created and one escapes. Hawking hypothesizes that black holes will — given time — evaporate by leaking energy as particles escape, leaving their partners behind. The escaping particles constitute Hawking radiation.

The evaporation time — more than 10^{67} years — is so long compared to the age of the universe (13 billion, or just over 10^9 years) that black holes formed from collapsed stars simply do not have time to evaporate in this way. Strangely enough, it turns out that lightweight black holes — those with masses a billion billion (10^{18}) times less than the Sun's — have an evaporation time comparable to the age of the universe. These "mini" black holes could have formed at the time of the Big Bang, when the universe was an ultradense, incredibly hot, seething soup in which quantum fluctuations occurred continually. Some of them should now be on the verge of evaporating in a blast of Hawking radiation, made up of particles and gamma rays, that ought to be picked up by orbiting observatories. So far, however, nothing like this has been observed.

41. Chandrasekhar 1992, p. 135; Chandrasekhar 1979c, p. 70; Heisenberg 1974, chap. 13.
42. Chandrasekhar 1992, p. 135.
43. Chandrasekhar 1977, p. 161.
44. Chandrasekhar 1992, p. 138.
45. Khan and Penrose 1971.
46. Interview with Valeria Ferrari, 8 Dec. 2002.
47. Chandrasekhar 1992, p. 143.
48. $\mathcal{E}$ is in fact a quantity that plays a role in the mathematics of general relativity. It is called the Ernst function.
49. Chandrasekhar 1992, p. 138; Heisenberg 1971, p. 68.
50. Interview with Valeria Ferrari, 16 July 2003.
51. Chandrasekhar 1983b, p. 637.
52. Interview with Lalitha Chandrasekhar, 1 Nov. 2002, and Lalitha Chandrasekhar n.d.

14. Into a Black Hole

1. The orbiting observatory, originally known as the Advanced X-ray Astrophysics Facility, had been in planning since 1977. The new name was chosen in a competition won by Jaltila van der Veen of the Physics Department at the University of California, Santa Barbara, and Tyrel Johns, a high school student from Laclede, Idaho. Their entries paid tribute to Chandra's discovery of the maximum mass for white dwarfs.

The Chandra X-ray Observatory was placed in a highly elliptical orbit: at its closest point it is 6,000 miles above Earth; at its farthest, 86,400 miles, a third of the way to the Moon. It takes 64 hours to circle Earth once. The elongated orbit carries the observatory's sensitive x-ray detectors well beyond interference from Earth's radiation belts, permitting it to make 55 hours of continuous observations in each orbit. The observatory itself is 45 feet long and 64 feet wide with its solar panels deployed, and weighs over 4 tons. It has a minimum lifetime of five years. At its heart is an x-ray telescope made up of exquisitely machined and polished mirrors weighing about a ton. The mirrors focus x-rays into a high-resolution camera with the accuracy you would need to hit a hole-in-one from Los Angeles to San Francisco.

The Hubble Space Telescope and the Chandra X-ray Observatory often complement each other. While Hubble operates in the near infrared, visible, and ultraviolet ranges, Chandra is used for the x-ray range. To obtain a picture of the full spectrum of radiation, data from Hubble and Chandra are combined with observations made by the Very Large Array, twenty-seven linked radio telescopes in the desert near Socorro, New Mexico. Hubble, the first of NASA's so-called Great Observatories, was placed in orbit by the space shuttle *Discovery* in 1990. Next up was the Compton Gamma Ray Observatory, placed in orbit by *Atlantis* in 1991. After nine productive years that mission ended. The Spitzer Space Telescope (formerly the Space Infrared Telescope Facility) was launched on 25 Aug. 2003.

The Chandra X-ray Observatory also often works in conjunction with XMM-Newton, built by the European Space Agency (ESA) and placed in orbit by *Ariane 504* on 10 Dec. 1999. The proximity of the two launch dates was a historical accident. The launch of Chandra was delayed because of the *Challenger* disaster in 1986, initial problems with Hubble's mirrors, and the vagaries of U.S. funding of NASA programs. XMM stands for X-ray Multi-Mirror Mission, referring to its combination of three x-ray telescopes. Aware of NASA's plans, ESA designed XMM-Newton to complement Chandra. XMM-Newton's detecting apparatus was set up to collect the maximum number of x-rays and to make more detailed measurements of faint x-ray sources than Chandra can.

2. Lalitha Chandrasekhar n.d.

3. An active galaxy is one that is emitting huge amounts of radiation and radio waves; the active galactic nucleus is the energy source at its heart, generating more energy than can be explained simply as the product of nuclear reactions. Quasars are one sort of active galaxy. Other galaxies, less luminous than quasars, show signs of intense and violent activity at their centers; they also emit radio waves. These are known as Seyfert galaxies because they were first identified by Carl Keener Seyfert at the Mount Wilson Observatory in 1943.

4. Thorne 1994, chap. 10; Thorne 1998; and conversations with Keith Mason, 23 Feb. 2004.

5. Rees 1998, p. 99.

6. Schödel et al. 2002.

7. ROSAT (Röntgen Satellite) — named after Konrad Röntgen, the discoverer of x-rays in 1895 — was launched on 1 June 1990 and ended its mission on 12 Feb. 1999.

8. Giddings and Thomas 2002; Dimopoulos and Landsberg 2001.

9. For more on Hawking radiation see Chapter 13, note 40. In an extraordi-

nary about-face, Hawking recently retracted his breathtaking claim that the very existence of black holes violates the most fundamental theory of matter, quantum theory. According to quantum theory, information can become inaccessible but can never be lost. Any process must be traceable back to its origins; it should always be possible to "run the film backward." But black holes as predicted by general relativity — Kerr black holes — are featureless, except for their mass and spin. All the information about whatever falls in — be it elephants, cars or stars — is lost because nothing can escape from a black hole. But although this information is inaccessible, it must remain inside in some form or other. The problem arose when Hawking brought quantum theory into the picture and proved that black holes can evaporate in the form of Hawking radiation, which is random and featureless and so offers no clue as to what is inside. He predicted that eventually all black holes will evaporate along with their information. But if they do destroy information in this way, that subverts quantum theory. This was Hawking's conclusion thirty years ago.

Hawking now claims that an event horizon never actually forms, so there are no real black holes. Instead, after emitting a certain amount of Hawking radiation, black holes will open up and release the information that fell in. Details have yet to be sorted out and alternatives eliminated.

10. Current theories on the cutting edge of astrophysics and physics, such as string theory, assume that there are seven dimensions in addition to the four we experience, making eleven in all. The new equations that emerge from formulating physics in more than four dimensions are expected to be those at the basis of electromagnetic theory and the electroweak theory, among others. Theories based on eleven dimensions are spoken of as grand unified theories because they can deal with quantum and gravitational phenomena in a way that avoids singularities. See, for example, Greene 2000.

11. This singularity could arise from the collapse of spacetime due to the bunching up of high-energy x-rays from matter sucked in at speeds approaching that of light.

12. Light takes 26,000 years to reach Sagittarius A* from Earth, but only objects with no mass can travel that fast. A spaceship has to accelerate gently over a very long period of time to reach a speed close to that of light so that astronauts can survive. Astronauts' training to go to the Moon involved whirling them in giant centrifuges fast enough to put them under a crushing gravitational force of twelve to fourteen times their body weight, about the maximum humans can endure. For further discussion of black holes, wormholes, and time travel, see Thorne 1994, chap. 14, "Wormholes and Time Machines"; and Gott 2002. Or just watch the film *Back to the Future!*

13. This slowing down of time has been verified in experiments. Elementary particles moving at speeds close to that of light — as in cosmic rays — decay more slowly than those at rest in the laboratory.

14. For amusing and insightful discussions of the conundrums involved in backward time travel — such as killing your mother before you were born — see Thorne 1994 and Gott 2002. Again, I recommend *Back to the Future.*

15. Chandrasekhar, 9 Sept. 1978, Box 77 (Addenda), Joseph Regenstein Library.

16. Interview with Lalitha Chandrasekhar, 1 Nov. 2002, and Lalitha Chandrasekhar n.d.
17. From Chandrasekhar 1980a, p. 14. Lalitha added pp. 14–15 and note 6 on pp. 17–18. They were driving from Chicago through New York State, toward Binghampton.
18. Chandrasekhar 1980a, p. 17.
19. McCrea 1996, p. 123.
20. Ibid., p. 123.
21. Cronin 1998.
22. McCrea 1998. McCrea had a distinguished career in a series of universities and had the reputation of being a diligent ambassador for science.
23. From interviews with James Cronin, 4 Nov. 2002, and Michael Turner, 30 Oct. 2002.
24. Interview with Cronin, 4 Nov. 2002. The biography of Schrödinger that Chandra had read was Moore 1991.
25. Lalitha Chandrasekhar 1996.
26. Chandrasekhar 1977, p. 29.
27. Letter from Eddington to Chandra, 4 July 1938.
28. Letter from Eddington to Chandra, 2 Sept. 1943.
29. Lalitha Chandrasekhar 1996, p. 276. See also Chandrasekhar n.d.3.
30. Chandrasekhar 1977, p. 156.
31. Lalitha Chandrasekhar 1996, p. 276.
32. Conversation with Gary Steigman, 1 Aug. 2002.
33. See Osterbrock 1999 for a fairly complete list of Chandra's Ph.D. students, postdocs, and collaborators.
34. This is Candlestickmaker 1972. See also Osterbrock 1999, pp. 228–29.
35. Vandervoort n.d., p. 33.

APPENDIX A. THE ONGOING TALE OF SIRIUS B

1. Kuiper 1941, p. 33.
2. For case of presentation, Chandra based the graph for his theory on a mean molecular weight of 1, which happens to represent a white dwarf made up of 100 percent hydrogen. But Kuiper found that the mean molecular weight as calculated from the measured values for the radius and mass of Sirius B was not on the solid-line curve in Chandra's graph (that is, for a mean molecular weight of 1) but on a curve drawn for a mean molecular weight of about 1.43 (giving a hydrogen content of 40 percent). This value of the mean molecular weight gives a maximum mass for Sirius B of 2.86 times the mass of the Sun. See Kuiper 1941, p. 32; and see note 122 to Chapter 5.
3. Kuiper 1941, p. 33.
4. Greenstein, Oke, and Shipman 1971. For further discussion of Adams's problems in taking measurements of Sirius B, see Greenstein 1985.
5. Gamow 1939b, p. 722. Astrophysicists can derive an equation for the mean molecular weight in such a way that it depends only on the star's hydrogen content, by taking an average of the amounts of chemical elements in the star. The result depends on the degree to which atoms are stripped of their

electrons. If they are assumed to be totally stripped — as is pretty much the case in the hot interior of a white dwarf — then the mean molecular weight is 2 divided by the sum of 1 plus the percentage of hydrogen. Gamow's paper was not mentioned at the Paris meeting and was not cited by Kuiper in the proceedings, published two years later.

6. Marshak 1940, Lee 1950.
7. The quotations are from Greenstein 1985, p. 280; Greenstein 1974, p. 35.

APPENDIX B. UPDATING THE SUPERNOVA STORY

1. E-mail from White to the author, 30 May 2003.
2. Interview by the author, 25 Mar. 2003.
3. Woosley, Heger, and Weaver 2002, p. 1015.
4. In the years after the Colgate-White paper, much work was done in refining neutrino mechanisms and other matters, such as the rotation and magnetic field of a star, to see how a supernova could be generated (Bethe 1990, Woosley, Heger, and Weaver 2002). Among the major figures in this research were David Arnett, Hans Bethe, Jim Wilson, T. A. Weaver, and Stan Woosley.
5. The neutrino detectors were in the northern hemisphere, and SN 1987A was visible only from the southern. But this is irrelevant when it comes to detecting neutrinos, which pass through Earth as if it were not there.
6. To be more precise, the number of neutrinos emitted was one followed by fifty-eight zeros, or 10^{58}.
7. "New Supernova Models Take On Third Dimension," interview in News and Public Affairs — News Releases, Los Alamos National Laboratory, 4 June 2002: www.lanl.gov/worldview/news/releases/archive/02-061.shtml.
8. In a degenerate gas of electrons, the pressure and volume do not depend on its temperature. As a result, when a white dwarf's mass exceeds the Chandrasekhar limit, there is no safety valve allowing the pressure to decrease as the temperature increases. The star's core cannot expand and cool off. This is not the case in a perfect gas, where an increase in the gas's temperature brings about changes in its volume and pressure. When a white dwarf accretes enough mass from its companion to exceed the Chandrasekhar limit, the rise in temperature leads to a series of uncontrolled reactions in the carbon and oxygen core, which eventually blows the star to pieces.
9. Type II supernova explosions take place in massively heavy stars. The light that the explosion emits contains spectral lines of hydrogen. White dwarfs have only a little hydrogen on their surface, so these lines are missing from the spectra of Type Ia supernovae. Types Ib and Ic supernovae are the explosions of heavier stars stripped of their hydrogen layer by stellar winds or by having it drawn off by a companion star; the light from these explosions has spectral lines characteristic of helium.

Bibliography

Chandra's letters and archival material are in the Chandrasekhar Archives, Joseph Regenstein Library, University of Chicago. Milne's correspondence and papers are on deposit at the Bodleian Library, Oxford University. Eddington's letters to Walter Adams are in the Huntington Library, San Marino, California. Eddington's cycling diaries and other correspondence are on deposit at Trinity College, Cambridge University. Stoner's correspondence is at the Stoner Archives, Leeds University.

ABBREVIATIONS
AIP: Niels Bohr Library, American Institute of Physics, College Park, Maryland
CA: Chandrasekhar Archives, Joseph Regenstein Library, University of Chicago
OHP: Oral History Project, California Institute of Technology Archives

Adams, Walter Sydney. 1914. "An A-type Star of Very Low Luminosity." *Publications of the Astronomical Society of the Pacific* 26: 198.
———. 1915. "The Spectrum of the Companion of Sirius." *Publications of the Astronomical Society of the Pacific* 27: 236–37.
Alvarez, Luis. 1987. *Alvarez: Adventures of a Physicist.* New York: Basic Books.
Anderson, Wilhelm. 1929. "Über die Grenzdichte der Materie und der Energie." *Zeitschrift für Physik* 56: 851–56.
Aston, Francis William. 1938. "Forty Years of Atomic Theory." In Joseph Needham and Walter Pagel, eds., *Background to Modern Science.* London: Macmillan.
Atkinson, Robert d'Escourt. 1931. "Atomic Synthesis and Stellar Energy I." *Astrophysical Journal* 7: 250–95.
———. 1939. "A Model Symposium." *The Observatory* 62: 281–84.
Ayyar, C. Subramanyan. n.d. "Details of the Biography of C. Subramanya Ayyar, B. A." Chandrasekhar Archives, Box 6, Folder 5.
Baade, Walter, and Fritz Zwicky. 1934. "Supernovae and Cosmic Rays." *Physical Review* 45: 138.
Baade, Walter, and Rudolph Minkowski. 1954. "Identification of the Radio

Sources in Cassiopeia, Cygnus A, and Puppis A." *Astrophysical Journal* 119: 206–14.

Balakrishnan. 1945. "Subrahmanya Chandrasekhar." *Trevini* 17: 73–85.

———. 1983a. "My Brother Chandra." *Aside*, Dec., pp. 18–29.

———. 1983b. "Chandrasekhar Re-reminisced." Typescript manuscript, Chandrasekhar Archives, Box 1, Folder 10.

Bethe, Hans. 1939. "Energy Production in Stars." *Physical Review* 55: 434–56.

———. 1966. Interview by Charles Weiner and Jagdesh Mehra, 27 Oct. 1966, AIP.

———. 1972. Interview by Charles Weiner, 8–9 May 1972, AIP.

———. 1990. "Supernova Mechanisms." *Reviews of Modern Physics* 62: 801–66.

———. 1993. Interview by Judith Goodstein, 28 Feb. 1993, OHP.

Bethe, Hans, and Charles L. Critchfield. 1938. "The Creation of Deuterons by Proton Combination." *Physical Review* 54: 248–54.

Burbridge, Geoffrey, and Margaret Burbridge. 1967. *Quasi-Stellar Objects*. New York: Freeman & Company.

Candlestickmaker, S. 1972. "On the Imperturbability of Elevator Operators: LVII." *Quarterly Journal of the Royal Astronomical Society* 13: 63–66. (The real author was J. Sykes.)

Chadwick, James. 1961. "Frederick John Marrian Stratton." *Biographical Memoirs of Fellows of the Royal Society* 7: 281–93.

Chandrasekhar, Lalitha. 1996. "Our Song." Talk given after symposium banquet, 14 Dec. 1996. Videotape on deposit, Chandrasekhar Archives. Also in Robert M. Wald, ed., *Black Holes and Relativistic Stars*, pp. 273–78. Chicago: University of Chicago Press. Page numbers in the notes refer to that book.

———. n.d. (probably 1999). Unpublished speech and other notes on the occasion of the TRW hospitality reception for winners of the competition to name NASA's Orbiting X-ray Observatory.

Chandrasekhar, Subrahmanyan. 1928. "The Thermodynamics of the Compton Effect with Reference to the Interior of Stars." *Indian Journal of Physics* 3: 241.

———. 1929a. "A Generalized Form of the New Statistics." *Physical Review* 34: 1204–11.

———. 1929b. "The Compton Scattering and the New Statistics." *Proceedings of the Royal Society* A 125: 231–37.

———. 1930. "The Ionization-Formula and the New Statistics." *Philosophical Magazine* 9: 292–99.

———. 1931a. "The Highly Collapsed Configurations of a Stellar Mass." *Monthly Notices of the Royal Astronomical Society* 91: 456–66.

———. 1931b. "The Dissociation Formula According to Relativistic Mechanics." *The Observatory* 91: 446–55.

———. 1931c. "The Maximum Mass of Ideal White Dwarfs." *Astrophysical Journal* 74: 81–82.

———. 1932. "Some Remarks on the State of Matter in the Interior of Stars." *Zeitschrift für Astrophysik* 5: 321–27.

———. 1933a. "The Equilibrium of Distorted Polytropes." *Monthly Notices of the Royal Astronomical Society* 93: 390–405, 449–61, 462–71, 539–74.

———. 1933b. "The Equilibrium of Distorted Polytropes." *The Observatory* 56: 215–19.

————. 1934a. "The Physical State of Matter in the Interior of Stars." *The Observatory* 57: 93–99.

————. 1934b. "Stellar Configurations with Degenerate Cores." *The Observatory* 57: 373–77.

————. 1935a. "Stellar Configurations with Degenerate Cores." *The Observatory* 58: 37.

————. 1935b. "The Highly Collapsed Configurations of a Stellar Mass. Second Paper." *Monthly Notices of the Royal Astronomical Society* 95: 207–25.

————. 1935c. "Stellar Configurations with Degenerate Cores." *Monthly Notices of the Royal Astronomical Society* 95: 226–60.

————. 1935d. "Stellar Configurations with Degenerate Cores. Second Paper." *Monthly Notices of the Royal Astronomical Society* 95: 676–93.

————. 1937. "The Cosmological Constants." *Nature* 139: 757–58.

————. 1939. *An Introduction to the Study of Stellar Structure*. Chicago: University of Chicago Press.

————. 1941. "The White Dwarfs and Their Importance for Theories of Stellar Evolution." In Lundmark et al., pp. 41–48; discussion, pp. 49–50; discussion of Eddington's lecture, pp. 64–69.

————. 1943. *The Principles of Stellar Dynamics*. Chicago: University of Chicago Press.

————. 1950. *Radiative Transfer*. London: Oxford University Press.

————. 1961. *Hydrodynamic and Hydromagnetic Stability*. Oxford: Clarandon Press.

————. 1964a. "Dynamical Instability of Gaseous Masses Approaching the Schwarzschild Limit in General Relativity." *Physical Review Letters* 12: 114–16.

————. 1964b. "The Dynamical Instability of Gaseous Masses Approaching the Schwarzschild Limit in General Relativity." *Astrophysical Journal* 140: 417–33.

————. 1972. "The Increasing Rôle of General Relativity in Astronomy." *The Observatory* 92, 160–74. Chandra's Halley Lecture for 1972, Oxford University, 2 May 1972.

————. 1975a. "How I Came Periodically to Change the Area of My Active Interest After Writing a Book," CA, Addenda, Box 70, Folder 6, dated 8 Jan. 1975.

————. 1975b. "Shakespeare, Newton, and Beethoven, or Patterns of Creativity." Nora and Edward Ryerson Lecture, University of Chicago, 22 Apr. 1975. Reprinted in Chandrasekhar 1987c, pp. 29–58. Page numbers in notes refer to Chandra's book.

————. 1977. Interview by Spencer Weart, 17 May 1977, AIP.

————. 1978. "The Kerr Metric and Stationary Axisymmetric Gravitational Fields." *Proceedings of the Royal Society* A 358: 405–20.

————. 1979a. "I. Edward Arthur Milne." Dictated to Lalitha on 22 Jan. 1979, edited and transcribed by her, CA, Addenda, Box 77, Folder 5: "Recollections of Milne, Eddington and Fowler."

————. 1979b. "Edward Arthur Milne — Recollections and Reflections." Manuscript in AIP and Chandrasekhar Archives. Written at the request of Milne's granddaughter, Miranda Weston-Smith. I have assigned the date 1979 because this is a manuscript version of Chandrasekhar 1987b, presented at Oxford University 6 Dec. 1979.

————. 1979c. "Beauty and the Quest for Beauty in Science." Lecture delivered at the Fermi Accelerator Laboratory, Batavia, Illinois, at symposium in honor of Robert R. Wilson, 27 Apr. 1979. Reprinted in Chandrasekhar 1987c, pp. 59–73. Page numbers in notes refer to that book.

———. 1980a. "II. Sir Arthur Stanley Eddington." Dictated to Lalitha 3 July 1980, edited and transcribed by her. CA, Addenda, Box 77, Folder 5: "Recollections of Milne, Eddington and Fowler."

———. 1980b. "III. Ralph Howard Fowler." Dictated to Lalitha 4 Sept. 1980, edited and transcribed by her. CA, Addenda Box 77, Folder 5: "Recollections of Milne, Eddington and Fowler."

———. 1980c. Interview by D. J. R. Bruckner, 6 Aug. 1980, CA, Box 2, Folder 10.

———. 1983a. "On Stars, Their Evolution and Their Stability." Nobel lecture, 8 Dec. 1983. In Tore Frängsmyr, ed., *Nobel Lectures in Physics: 1981–1990*, pp. 142–63; "Subrahmanyan Chandrasekhar," autobiographical statement, pp. 139–41. Singapore: World Scientific, 1993.

———. 1983b. *The Mathematical Theory of Black Holes*. Oxford: Clarendon Press.

———. 1987a. "Eddington: The Most Distinguished Astrophysicist of His Time." In Chandrasekhar 1987c, pp. 92–143. Centenary Lectures in Memory of Arthur Stanley Eddington, Cambridge University, 19 and 21 Oct. 1982.

———. 1987b. "Edward Arthur Milne: His Part in the Development of Modern Astrophysics." In Chandrasekhar 1987c, pp. 74–91. Chandra's Milne Lecture, Oxford University, 6 Dec. 1979.

———. 1987c. *Truth and Beauty: Aesthetics and Motivations in Science*. Chicago: University of Chicago Press.

———. 1988. Foreword to Arthur Stanley Eddington, *The Internal Constitution of the Stars*, rev. ed., pp. vii–xii. Cambridge, Eng.: Cambridge University Press.

———. 1992. "The Series Paintings of Claude Monet and the Landscape of General Relativity." Lecture, Inter-University Centre for Astronomy and Astrophysics, Pune, India, 28 Dec. 1992. Reprinted in S. Chandrasekhar, *Selected Papers*, vol. 7, pp. 135–67. Chicago: University of Chicago Press, 1997. Page numbers in notes refer to the reprint.

———. 1995. *Newton's* Principia *for the Common Reader*. Oxford: Oxford University Press.

———. n.d.1. "Science in Pre- and Post-Independent India." CA, Addenda, Box 23, Folder 1.

———. n.d.2. Fragment beginning "I shall strictly adhere." CA, Addenda, Box 70, Folder 6.

———. n.d.3. "How I Came to Writing My Book on the *Principia*." CA, Addenda, Box 70, Folder 7.

———. n.d.4. "Historical Notes on Some Astrophysical Problems." CA, Addenda, Box 70, Folder 6.

———. n.d.5. Handwritten reminiscence, CA, Box 2, Folder 1.

Chandrasekhar, Subrahmanyan, and Christian Møller. 1935. "Relativistic Degeneracy." *Monthly Notices of the Royal Astronomical Society* 95: 673–76.

Chandrasekhar, Subrahmanyan, with George Gamow and Merle A. Tuve. 1938. "The Problem of Stellar Energy." *Nature* 141: 982.

Chandrasekhar, Subrahmanyan, and Mario Schönberg. 1942. "On the Evolution of the Main-Sequence Stars." *Astrophysical Journal* 96: 161–72.

Chandrasekhar, Subrahmanyan, and Enrico Fermi. 1953a. "Magnetic Fields in Spiral Arms." *Astrophysical Journal* 118: 113–15.

———. 1953b. "Problems of Gravitational Stability in the Presence of a Magnetic Field." *Astrophysical Journal* 118: 116–41.

Chandrasekhar, Subrahmanyan, and Valeria Ferrari. 1984. "On the Nutku-Halil Solution for Colliding Impulsive Gravitational Waves." *Proceedings of the Royal Society* 398: 55–74.

Christy, Robert F. 1994. Interview by Sandra Lippincott, 21 June 1994, OHP.

Colgate, Stirling. 1967. "Stellar Coalescence and the Multiple Supernova Interpretation of Quasi-Stellar Sources." *Astrophysical Journal* 150: 163–92.

Colgate, Stirling, and Montgomery H. Johnson. 1960. "Hydrodynamic Origin of Cosmic Rays." *Physical Review Letters* 5: 235–38.

Colgate, Stirling, and Richard White. 1966. "The Hydrodynamic Behavior of Supernovae Explosions." *Astrophysical Journal* 143: 626–81.

Compton, Herbert. 1904. *Indian Life in Town and Country.* New York: G. P. Putnam's.

Cowling, T. G. 1966. "The Development of the Theory of Stellar Structure." *Quarterly Journal of the Royal Astronomical Society* 7: 121–37.

———. 1978. Interview by David DeVorkin, 22 Mar. 1978, AIP.

Cronin, James W. 1998. "Subrahmanyan Chandrasekhar." *The Observatory* 118: 24.

Dalitz, Richard H. 1997. "Some Recollections of S. Chandrasekhar." In Wali 1997, pp. 142–55.

Danin, D. S. 1989. "The Passionate Sobriety of Youth." In I. M. Khalatnikov, ed., *Landau: The Physicist and the Man*, pp. 78–83. Oxford: Pergamon Press.

DeVorkin, David H. 2000. *Henry Norris Russell. Dean of American Astronomers.* Princeton: Princeton University Press.

Dimopoulos, Savas, and Greg Landsberg. 2001. "Black Holes at the Large Hadron Collider." *Physical Review Letters* 87: 161602–604.

Dirac, P. A. M. 1937. "The Cosmological Constants." *Nature* 139: 323–24.

Dirac, P. A. M., R. Peierls, and M. H. L. Pryce. 1942. "On Lorentz Invariance in the Quantum Theory." *Proceedings of the Cambridge Philosophical Society* 38: 193–200.

Douglas, Allie Vibert. 1956. *The Life of Arthur Stanley Eddington.* London: Thomas Nelson & Sons.

Eddington, Arthur Stanley. 1917. "The Radiation of the Stars." *Nature* 99: 445.

———. 1918. *Report on the Relativity Theory of Gravitation.* London: Physical Society of London.

———. 1919. "Globular Clusters, Cepheid Variables, and Radiation." *Nature* 103: 25–27.

———. 1920. "The Internal Constitution of the Stars." *Nature* 106: 14–20.

———. 1923. *Mathematical Theory of Relativity.* Cambridge: Cambridge University Press.

———. 1924a. "The Relation Between the Masses and Luminosities of the Stars." *Nature* 113: 786–88.

———. 1924b. "A Comparison of Whitehead's and Einstein's Formulas." *Nature* 113: 192.

————. 1926. *The Internal Constitution of the Stars.* Cambridge.: Cambridge University Press; 2nd ed., 1930; reissued 1988 with a foreword by Chandra.

————. 1927. *Stars and Atoms.* New Haven: Yale University Press.

————. 1929a. "The Masses, Luminosities and Effective Temperatures of the Stars." *The Observatory* 52: 349.

————. 1929b. *Seen and Unseen World.* London: Allen & Unwin.

————. 1930a. "The Effect of Stellar Boundary Conditions: A Reply." *Monthly Notices of the Royal Astronomical Society* 90: 808–9.

————. 1930b. "The Problem of Stellar Luminosity." *Nature* 125: 489.

————. 1930c. "The Connection of Mass with Luminosity of Stars." *The Observatory* 53: 208–11.

————. 1931a. "The Analysis of Stellar Structure." *The Observatory* 54: 34–36.

————. 1931b. "Upper Limits to the Central Temperature and Density of a Star." *The Observatory* 54: 97–99.

————. 1931c. "Upper Limits to the Central Temperature and Density of a Star." *Monthly Notices of the Royal Astronomical Society* 91: 444–46.

————. 1932. "The Hydrogen Content of Stars." *Monthly Notices of the Royal Astronomical Society* 92: 471–81.

————. 1935a. "Relativistic Degeneracy." *The Observatory* 58: 37–39.

————. 1935b. "On 'Relativistic Degeneracy.'" *Monthly Notices of the Royal Astronomical Society* 95: 194–206.

————. 1935c. "Note on 'Relativistic Degeneracy.'" *Monthly Notices of the Royal Astronomical Society* 96: 20–21.

————. 1935d. "The Pressure of a Degenerate Electron Gas and Related Problems." *Proceedings of the Royal Society* A 152: 253–72.

————. 1935e. "The External Radii of Stars of the Generalised Standard Model." *The Observatory* 58: 176.

————. 1936a. *Relativity Theory of Protons and Electrons.* Cambridge: Cambridge University Press.

————. 1936b. "Constitution of the Stars." *Scientific Monthly,* Nov., pp. 385–95.

————. 1939a. "The Hydrogen Content of White Dwarf Stars in Relation to Stellar Evolution." *Monthly Notices of the Royal Astronomical Society* 99: 595–606.

————. 1939b. "The Hydrogen Content of White Dwarf Stars in Relation to Stellar Evolution." *The Observatory* 62: 171–74.

————. 1942. "On Lorentz Invariance in the Quantum Theory, II." *Proceedings of the Cambridge Philosophical Society* 38: 201–9.

————. 1943. Review of Chandrasekhar, *The Principles of Stellar Dynamics. Nature* 151: 91.

————. 1946. *Fundamental Theory.* Cambridge: Cambridge University Press. Edited by Edmund T. Whittaker, longtime colleague of Eddington's and professor of mathematics at Edinburgh University. Whittaker chose the title.

————. 1958a. *The Nature of the Physical World.* Reprint, Ann Arbor: University of Michigan Press. Original publication 1928, Cambridge: Cambridge University Press. Based on Eddington's Gifford Lectures, University of Edinburgh, Jan.–Mar. 1927.

————. 1958b. *The Expanding Universe.* Reprint, Ann Arbor: University of Michigan Press. Original publication 1933, Cambridge: Cambridge University

Press. From a series of lectures presented at the International Astronomical Union, Cambridge, Mass., in 1932.

———. 1958c. *The Philosophy of Physical Science.* Reprint, Ann Arbor: University of Michigan Press. Original publication 1939, Cambridge: Cambridge University Press. Based on Tanner Lectures, Trinity College, Cambridge, Easter term, 1938.

———. 1959. *New Pathways in Science.* Reprint. Original publication 1935, Cambridge: Cambridge University Press.

Einstein, Albert. 1923. "Fundamental Ideas and Problems of the Theory of Relativity." In *Nobel Lectures: Physics 1901–1921,* pp. 482–90. Amsterdam: Elsevier, 1967.

———. 1939. "On a Stationary System with Spherical Symmetry Consisting of Many Gravitating Masses." *Annals of Mathematics* 40: 922–36.

———. 1949. "Reply to Criticisms." in P. A. Schilpp, ed., *Albert Einstein: Philosopher-Scientist.* New York: Open Court.

Elbert, Donna. 1997. "On Working with Chandra." In Wali 1997, pp. 41–45.

Enz, Charles P. 2002. *No Time to Be Brief: A Scientific Autobiography of Wolfgang Pauli.* Oxford: Oxford University Press.

Fowler, Ralph H. 1926. "On Dense Matter." *Monthly Notices of the Royal Astronomical Society* 87: 114–22.

———. 1936. *Statistical Mechanics,* 2nd rev. ed. Original publication 1929, Cambridge: Cambridge University Press.

Fowler, William A. 1966. "The Stability of Supermassive Stars." *Astrophysical Journal* 144: 180–200.

———. 1984. Interview by John Greenberg, 31 May 1984, OHP.

Fowler, William A., and Fred Hoyle. 1964. "Neutrino Processes and Pair Formation in Massive Stars and Supernovae." *Astrophysical Journal* suppl. 9: 201–318.

Frisch, Otto. 1979. *What Little I Remember.* Cambridge: Cambridge University Press.

Galison, Peter, and Barton Bernstein. 1989. "In Any Light: Scientists and the Decision to Build the Superbomb, 1942–1954." *Historical Studies in the Physical Sciences* 19, pt. 2, 267–349.

Gamow, George. 1939. "Physical Possibilities of Stellar Evolution." *Physical Review* 55: 718–25.

———. 1968. Interview by Charles Weiner, 25 Apr. 1968, AIP.

Gandhi, Mohandas K. 1983. *Autobiography: The Story of My Experiments with Truth.* Reprint, New York: Dover Publications. Original publication 1948, London: Public Affairs Press.

Giddings, Steven, and Scott Thomas. 2002. "High Energy Colliders As Black Hole Factories: The End of Short Distance Physics." *Physical Review* D 65: 056010-1–056010-12.

Gleick, James. 2003. *Isaac Newton.* New York: HarperCollins.

Gott, J. Richard. 2002. *Time Travel in Einstein's Universe: The Physical Possibilities of Travel Through Time.* London: Phoenix.

Greene, Brian. 2000. *The Elegant Universe: Superstrings, Hidden Dimensions, and the Quest for the Ultimate Theory.* London: Vintage.

Greenstein, Jesse. 1974. Interview by Paul Wright, 31 July, AIP.

————. 1982. Interview by Rachel Prud'homme, 16 Mar., OHP.

————. 1985. "On the Redshift of Sirius B." *Quarterly Journal of the Royal Astronomical Society* 26: 279–88.

Greenstein, Jesse, and Thomas A. Matthews. 1963. "Red-shift of the Unusual Radio Source: 3C 48." *Nature* 197: 1041–42.

Greenstein, Jesse, and Maarten Schmidt. 1964. "The Quasi-Stellar Radio Sources 3C 48 and 3C 273." *Astrophysical Journal* 140: 1–34.

Greenstein, Jesse, John Beverley Oke, and Harry Shipman. 1971. "Effective Temperature, Radius, and Gravitational Redshift of Sirius B." *Astrophysical Journal* 169: 563–66.

Hamada, T., and Edwin E. Salpeter. 1961. "Models for Zero-Temperature Stars." *Astrophysical Journal* 134: 683–98.

Hardy, G. H. 1931. "The Analysis of Stellar Structure." *The Observatory* 54: 40.

————. 1940. *Ramanujan*. Cambridge: Cambridge University Press.

————. 1941. *A Mathematician's Apology*. Cambridge: Cambridge University Press.

Harrison, B. Kent, Kip S. Thorn, Masamo Wakano, and John A. Wheeler. 1965. *Gravitational Theory and Gravitational Collapse*. Chicago: University of Chicago Press.

Hawking, Stephen, and Werner Israel, eds. 1987. *300 Years of Gravitation*. Cambridge: Cambridge University Press.

Heisenberg, Werner. 1926. "Mehrkörperprobleme und Resonanz in der Quantenmechanik. II." *Zeitschrift für Physik* 41: 239–67.

————. 1962. Interview by Thomas S. Kuhn, 30 Nov., AIP.

————. 1971. *Physics and Beyond: Encounters and Conversations*. Trans. P. Heath. New York: Harper.

————. 1974. *Across the Frontiers*. Trans. P. Heath. New York: Harper Torchbooks.

Herant, Marc, Stirling Colgate, and Chris Fryer. 1997. "Neutrinos and Supernovae." *Los Alamos Science* no. 25: 164–79.

Herken, Gregg. 2002. *Brotherhood of the Bomb: The Tangled Lives and Loyalties of Robert Oppenheimer, Ernest Lawrence, and Edward Teller*. New York: Henry Holt.

Holloway, David. 1994. "How the Bomb Saved Soviet Physics." *Bulletin of the Atomic Scientists* 50, no.6. www.thebulletin.org/issues/1994/nd94/nd94 Holloway.html.

Hoyle, Fred, and William A. Fowler. 1960. "Nucleosynthesis in Supernovae." *Astrophysical Journal* 132: 565–90.

————. 1963. "On the Nature of Strong Radio Sources." *Monthly Notices of the Royal Astronomical Society* 125: 169–76.

Jeans, James. 1931. "The Analysis of Stellar Structure." *The Observatory* 54: 36–38.

Johansson, Sven. 1993. "The Nobel Prize for Physics." In Tore Frängsmyr, ed., *Nobel Lectures in Physics: 1981–1990*, pp. 135–37. Singapore: World Scientific.

Kanigel, Robert. 1991. *The Man Who Knew Infinity: A Life of the Genius Ramanujan*. New York: Charles Scribner's Sons.

Kerr, Roy. 1963. "Gravitational Field of a Spinning Mass as an Example of Algebraically Special Metrics." *Physical Review Letters* 11: 237–38.

Khalatnikov, I. M. 1989. "Landau, Bohr, and Kapitza: Letters 1936–41." Trans. J. B. Sykes. In I. M. Khalatnikov, ed., *Landau: The Physicist and the Man*, pp. 308–19. Oxford: Pergamon Press.

Khan, K. A., and Roger Penrose. 1971. "Scattering of Two Impulsive Gravitational Plane Waves." *Nature* 229: 185–86.

Khariton, Yuli, and Yuri Smirnov. 1993. "The Khariton Version." *Bulletin of the Atomic Scientists* 49, no. 3. www.thebulletin.org/issues/1993/may93/may93Khariton.html.

Khariton, Yuli, Kitor Adamskii, and Yuri Smirnov. 1998. "The Way It Was." *Bulletin of the Atomic Scientists* 54, no. 6. www.bullatomsci.org/issues/1996/nd96/nd96khariton.html.

Khriplovich, Iosif B. 1992. "The Eventful Life of Fritz Houtermans." *Physics Today* 45: 29–37.

Kilmister, Clive W. 1994. *Eddington's Search for a Fundamental Theory: A Key to the Universe.* Cambridge: Cambridge University Press.

Kragh, Helge. 1990. *Dirac: A Scientific Biography.* Cambridge: Cambridge University Press.

Kuiper, Gerard P. 1941. "White Dwarfs: Discovery, Observations, Surface Conditions." In Lundmark et al. 1941, pp. 3–39; discussion session, pp. 64–69.

Landau, Lev Davidovich. 1932. "On the Theory of Stars." *Physikalische Zeitschrift der Sowjetunion* 1: 285–88.

———. 1938. "Origin of Stellar Energy." *Nature* 141: 334–35.

Lang, Kenneth, and Owen Gingerich. 1979. *A Source Book in Astronomy and Astrophysics, 1900–1975.* Cambridge: Harvard University Press.

Lee, T. D. 1950. "Hydrogen Content and Energy-Productive Mechanism of White Dwarfs." *Astrophysical Journal* 111: 625–40.

Littlewood, J. E. 1953. *Littlewood's Miscellany,* ed. and with a foreword by Béla Bollobás. Cambridge: Cambridge University Press.

Lundmark, Knut, Serge Gaposchkin, Bengt Edlén, C. P. Gaposchkin, F. J. M. Stratton, and C. S. Beals, eds. 1941. *Novae and White Dwarfs.* Colloque International d'Astrophysique, 17–23 July 1939, Paris, under the presidency of Henry Norris Russell. Paris: Hermann.

McCrea, William H. 1935. "The International Astronomical Union Meeting in Paris 1935." *The Observatory* 58: 256–65.

———. 1951. "Edward Arthur Milne: Obituary Notice." *Monthly Notices of the Royal Astronomical Society* 111: 161–70.

———. 1996. "Subrahmanyan Chandrasekhar." *The Observatory* 116: 121–24.

———. 1998. "Sir William McCrea Replies." *The Observatory* 118: 24.

McVittie, George C. 1978. Interview by David DeVorkin, 21 Mar. 1978, AIP.

Marshak, Robert Eugene. 1940. "The Internal Temperature of White Dwarf Stars." *Astrophysical Journal* 92: 321–46.

Merrill, Paul W. 1938. *The Nature of Variable Stars.* New York: Macmillan.

Miller, Arthur I. 1972. "The Myth of Gauss' Experiment on the Euclidean Nature of Physical Space." *ISIS* 63: 345–48.

———. 1994. *Early Quantum Electrodynamics: A Source Book.* Cambridge: Cambridge University Press.

———. 1998. *Albert Einstein's Special Theory of Relativity: Emergence (1905) and Early Interpretation (1905–1911).* Reprint, London: Springer-Verlag; original publication 1981, Reading, Mass.: Addison-Wesley.

———. 2000. *Insights of Genius: Imagery and Creativity in Science and Art.* Reprint,

Cambridge, Mass.: MIT Press; original publication 1996, New York: Springer-Verlag.

———. 2002. "Erotica, Aesthetics and Schrödinger's Wave Equation." In Graham Farmeloe, ed., *It Must Be Beautiful: Great Equations in Modern Science*, pp. 80–101. London: Granta.

Milne, Edward Arthur. 1923. "Those in Authority: R. H. Fowler, M.A. Trinity, Senior Lecturer." *Granta* 32: 469.

———. 1930a. "The Analysis of Stellar Structure." *The Observatory* 53: 330–34.

———. 1930b. "The Analysis of Stellar Structure." *Monthly Notices of the Royal Astronomical Society* 91: 4–55.

———. 1931. "The Analysis of Stellar Structure." *The Observatory* 54: 34.

———. 1932. "The Theory of Stellar Structure II. Energy-Generation." *Zeitschrift für Astrophysik* 5: 337–47.

———. 1935a. "The Configuration of Stellar Masses." *The Observatory* 58: 52.

———. 1935b. "The External Radii of Stars of the Generalised Standard Model." *The Observatory* 58: 174–76.

———. 1945a. "R. H. Fowler." *Obituary Notices of Fellows of the Royal Society* 5: 61–78.

———. 1945b. "Ralph H. Fowler." *Monthly Notices of the Royal Astronomical Society* 105: 80–87.

———. 1947. "Sir James Jeans." *Obituary Notices of Fellows of the Royal Society* 5: 573–89.

Misner, Charles, Kip S. Thorne, and John A. Wheeler. 1970. *Gravitation*. New York: Freeman.

Moore, Walter. 1991. *Schrödinger: Life and Thought*. Cambridge: Cambridge University Press.

NASA. 1972. *Apollo 15 Preliminary Science Report*. Special Publication SP-289. Washington, D.C.: Science and Technical Information Office.

Oppenheimer, J. Robert. 1963. Interview by Thomas S. Kuhn, 18 and 20 Nov. 1963, AIP.

Oppenheimer, J. Robert, and Robert Serber. 1938. "On the Stability of Stellar Neutron Cores." *Physical Review* 54: 504.

Oppenheimer, J. Robert, and Hartland Snyder. 1939. "On Continued Gravitational Contraction." *Physical Review* 56: 455–59.

Oppenheimer, J. Robert, and Georg M. Volkoff. 1939. "On Massive Neutron Cores." *Physical Review* 55: 374–81.

Osterbrock, Donald E. 1996. "Walter Baade, Observational Astrophysicist, 2: Mount Wilson 1931–1937." *Journal of the History of Astronomy* 27: 301–48.

———. 1997. *Yerkes Observatory, 1892–1950: The Birth, Near Death, and Resurrection of a Scientific Research Institution*. Chicago: University of Chicago Press.

———. 1999. "Chandra and His Students at Yerkes Observatory." In G. Srinivasan, ed., *From White Dwarfs to Black Holes: The Legacy of S. Chandrasekhar*, pp. 199–237. Chicago: University of Chicago Press.

———. 2001. "Who Really Coined the Word Supernova/Who First Predicted Neutron Stars?" *Bulletin of the American Astronomical Society* 33: 1330–31.

Pais, Abraham. 1986. *Inward Bound: On Matter and Forces in the Physical World*. Oxford: Oxford University Press.

———. 2000. *The Genius of Science: A Portrait Gallery of Twentieth-Century Physicists.* Oxford: Oxford University Press.

Pauli, Wolfgang. 1955. "The Influence of Archetypal Ideas on the Scientific Theories of Kepler." In C. G. Jung and W. Pauli, *The Interpretation of Nature and the Psyche,* pp. 219–79. New York: Pantheon.

———. 1979. *Wissenschaftlicher Briefwechsel mit Bohr, Einstein, Heisenberg, u.a. I: 1919–1929.* A. V. Hermann, K. v. Meyenn, and V. F. Weiskopf, eds. Berlin: Springer-Verlag.

Payne, C. 1925. "The Relative Abundances of the Elements." In *Stellar Atmospheres,* Harvard Observatory Monograph no. 1. Cambridge: Harvard University Press. Excerpt in Lang and Gingerich 1979, pp. 245–48.

Payne-Gaposchkin, C. 1968. Interview by Owen Gingerich, 5 Mar. 1968, AIP.

Payne-Gaposchkin, C. and Katherine Haramundanis. 1970. *Introduction to Astronomy.* Englewood Cliffs, N.J.: Prentice-Hall.

Peierls, Rudolf. 1936. "Note on the Derivation of the Equation of State for a Degenerate Relativistic Gas." *Monthly Notices of the Royal Astronomical Society* 96: 780–84.

Pellam, J. R. 1989. "Lev Davidovich Landau." In I. M. Khalatnikov, *Landau: The Physicist and the Man,* pp. 198–204. New York: Pergamon Press.

Penrose, Roger. 1965. "Gravitational Collapse and Space-Time Singularities." *Physical Review Letters* 14: 57–59.

———. 1969. "Gravitational Collapse: The Role of General Relativity." *Revista del Nuovo Cimento* numero speciale 1: 252–76.

Plaskett, Henry H. 1951. "Edward Arthur Milne." *Monthly Notices of the Royal Astronomical Society* 111: 170–72.

Plummer, H. C. 1948. "Arthur Stanley Eddington." *Obituary Notices of the Royal Society* 5: 113–25.

Raman, C. V. 1930. "The Molecular Scattering of Light." Nobel lecture, 11 Dec. 1930, Stockholm. In *Nobel Lectures in Physics, 1922–1941,* pp. 267–75. New York: Elsevier, 1965.

Ramaseshan, S. 1996. "S. Chandrasekhar and C. V. Raman — Some Letters." *Current Science* 70: 104–7.

———. 1997. "Chandrasekhar — Some Reminiscences." In Wali 1997, pp. 101–12.

Rees, Martin. 1997. "Reminiscing about Chandra's Research." In Wali 1997, pp. 183–86.

———. 1998. "Astrophysical Evidence for Black Holes." In R. W. Wald, ed., *Black Holes and Relativistic Stars,* pp. 79–101. Chicago: University of Chicago Press.

Rhodes, Richard. 1986. *The Making of the Atomic Bomb.* New York: Simon & Schuster.

———. 1995. *Dark Sun: The Making of the Hydrogen Bomb.* New York: Simon & Schuster.

Rosenfeld, Léon. 1966. "Nuclear Physics, Past and Future." in M. Nève de Mévergnies, P. Van Assche, and J. Vervier. eds., *Nuclear Structure Study with Neutrons.* Amsterdam: North Holland.

———. 1967. "Niels Bohr in the Thirties." In S. Rozenthal. ed., *Niels Bohr: His Life and Work as Seen by His Friends and Colleagues,* pp. 114–36. New York: Wiley.

Rüger, Alexander. 1988. "Atomism from Cosmology: Erwin Schrödinger's Work on Wave Mechanics and Space-Time Structure." *Historical Studies in the Physical Sciences* 18: 377–401.

Russell, Henry Norris. 1913. "Giant and Dwarf Stars." *The Observatory* 36: 325.

———. 1914. "Relations Between Spectra and Other Characteristics of Stars." *Popular Astronomy* 22: 275–94.

———. 1941. "Address." In Lundmark et al. 1941, pp. 1–6.

———. 1945. "Arthur Stanley Eddington, 1882–1944." *Astrophysical Journal* 101: 133–35.

Saha, Meghnad. 1920. "Ionisation in the Solar Chromosphere." *Philosophical Magazine* 40: 479–88.

Sakharov, Andrei. 1990. *Memoirs.* New York: Knopf.

Salpeter, Edwin E. 1961. "Energy and Pressure of a Zero-Temperature Plasma." *Astrophysical Journal* 134: 669–82.

———. 1999. "Neutron Stars Before 1967 and My Debt to Chandra." In G. Srinivasan. ed., *From White Dwarfs to Black Holes: The Legacy of S. Chandrasekhar,* pp. 27–29. Chicago: University of Chicago Press.

Sandage, Allan R., and Martin Schwarzschild. 1952. "Inhomogeneous Stellar Models II: Models With Exhausted Cores in Gravitational Contraction." *Astrophysical Journal* 116: 463–76.

Schmidt, Maarten. 1963. "A Star-like Object with Large Red-Shift." *Nature* 197: 1040.

Schödel, R., et. al. 2002. "A Star in a 15.2-Year Orbit Around the Supermassive Black Hole at the Centre of the Milky Way." *Nature* 419: 694–96.

Scott, David R., Alexei Leonov, and Christine Toomey. 2004. *Two Sides of the Moon: Our Story of the Cold War Space Race.* New York: Simon & Schuster.

Serber, Robert. 1992. *The Los Alamos Primer.* Berkeley, Calif.: University of California Press. Based on five lectures given by Serber in Apr. 1943 as an indoctrination course for the start of the Los Alamos Project. The first five chapters are Serber's lectures, annotated by him. Introduction by Richard Rhodes, pp. ix–xxi; preface by Serber, pp. xxii–xxxiii.

Singh, Amar Kuman. 1963. *Indian Students in Britain: A Survey of Their Adjustment and Attitudes.* London: Asia Publishing House.

Singh, Simon. 2004. *Big Bang: The Most Important Scientific Discovery of All Time and Why You Need to Know About It.* London: Fourth Estate.

Slater, Gilbert. 1936. *Southern India: Its Political and Economic Problems.* London: George Allen & Unwin.

Smart, W. G. M. 1931. "The Analysis of Stellar Structure." *The Observatory* 54: 38.

Sommerfeld, Arnold. 1923. *Atomic Structure and Spectral Lines.* Trans. Henry L. Brose from 3rd German edition. London: Methuen.

———. 1928a. "Zur Elektronentheorie der Metalle auf Grund der Fermischen Statistik." *Zeitschrift für Physik* 47: 1–32.

———. 1928b. "Zur Elektronentheorie der Metalle auf Grund der Fermischen Statistik, insbesondre über den Volta-Effekt." *Zeitschrift für Physik* 47: 38–60.

Stoner, Edmund C. 1929. "The Limiting Density in White Dwarf Stars." *Philosophical Magazine* 7: 63–70.

———. 1930. "The Equilibrium of Dense Stars." *Philosophical Magazine* 9: 944–63.

―――. 1932a. "The Minimum Pressure of a Degenerate Gas." *Monthly Notices of the Royal Astronomical Society* 92: 651–61.

―――. 1932b. "Upper Limits for Densities and Temperatures in Stars." *Monthly Notices of the Royal Astronomical Society* 92: 662–76.

Strömgren, Bengt. 1932. "The Opacity of Stellar Matter and the Hydrogen Content of the Stars." *Zeitschrift für Astrophysik* 4: 118–52.

―――. 1976. Interview by Lillian Hoddeson and Gordon Baym, 6 and 13 May 1976, AIP.

Struve, Otto, and Velta Zebergs. 1962. *Astronomy of the Twentieth Century.* New York: Macmillan.

Stuewer, Roger H. 1986. "Gamow's Theory of Alpha Decay." In Edna Ullman-Margalit, ed., *The Kaleidoscope of Science,* pp. 147–86. Englewood Cliffs, N.J.: Humanities Press.

Swerdlow, Noel M. 1997. "Chandrasekhar's Research on Newton's *Principia.*" In Wali 1997, pp. 201–5.

Telegdi, Valentin. 1997. "Recollections about Chandra." In Wali 1997, pp. 206–9.

Teukolsky, Saul A. 1997. "Chandra at Caltech." In Wali 1997, pp. 76–79.

Thorne, Kip. 1994. *Black Holes and Time Warps: Einstein's Outrageous Legacy.* New York: Norton.

―――. 1998. "Probing Black Holes and Relativistic Stars with Gravitational Waves." In R. M. Wald, ed., *Black Holes and Relativistic Stars,* pp. 41–78. Chicago: University of Chicago Press.

Ulam, Stanislaw. 1976. *Adventures of a Mathematician.* New York: Scribner's.

United States Atomic Energy Commission. 1954. *In the Matter of J. Robert Oppenheimer.* Cambridge: MIT Press.

Vandervoort, Peter O. n.d. "S. Chandrasekhar: Incidental Lessons." Unpublished manuscript in the possession of Professor Vandervoort, University of Chicago.

Van Maanen, Adriaan. 1917. "Two Faint Stars with Large Proper Motion," *Publications of the Astronomical Society of the Pacific* 29: 258–59.

Venkataraman, Janaki, Sridevi Rao, and Nirupama Nityanandan. 1983. "Our Nobel Family." *Aside,* Dec., pp. 12–18.

Wali, Kameshwar C. 1991. *Chandra: A Biography of S. Chandrasekhar.* Chicago: University of Chicago Press.

―――. ed.. 1997. *S. Chandrasekhar: The Man Behind the Legend.* London: Imperial College Press.

Warwick, Andrew. 2003. *Masters of Theory: Cambridge and the Rise of Mathematical Physics.* Chicago: University of Chicago Press.

Weinberg, Steven. 1972. *Gravitation and Cosmology: Principles and Applications of the General Theory of Relativity.* New York: John Wiley & Sons.

―――. 1977. *The First Three Minutes: A Modern View of the Origin of the Universe.* New York: Basic Books.

Westfall, Robert S. 1996. "Technical Newton." *Isis* 87: 701–6.

Weston Smith, Meg. 1990. "E. A. Milne and the Creation of Air Defence: Some Letters from an Unprincipled Brigand, 1916–1919." *Notes and Records of the Royal Society* 44: 241–55.

Wheeler, John A., Kent Harrison, and J. A. Wakano. 1958. "Matter-Energy at High Density: End Point of Thermonuclear Evolution." In *Solvay — Onzième*

Conseil de Physique — La Structure et l'évolution de l'univers, pp. 125–41; "Discussion of Wheeler's Report," pp. 147–48. Brussels: R. Stoops.

Wheeler, John A., and Ken Ford. 1998. *Geons, Black Holes and Quantum Foam: A Life in Physics.* New York: Norton.

Wolpert, Stanley. 1997. *A New History of India,* 5th ed. Oxford: Oxford University Press; 1st ed. 1977.

Woolsey, Stan E., A. Heger, and T. A. Weaver. 2002. "The Evolution and Explosion of Massive Stars." *Reviews of Modern Physics* 74: 1015–71.

Xanthopoulos, Basilis C. 1991. Foreword. In S. Chandrasekhar, *Selected Papers,* vol. 6: *The Mathematical Theory of Black Holes and of Colliding Plane Waves.* Chicago: University of Chicago Press.

Zwicky, Fritz. 1939. "On the Theory and Observation of Highly Collapsed Stars." *Physical Review* 55: 726–43.

Biographical Sketches

Adams, Walter Sydney (1876–1956) One of the most important American observational astronomers of his day, who made groundbreaking measurements of the white dwarf star Sirius B. Eddington dismissed his findings as "absurd."

Ambartsumian, Viktor Amazaspovich (1908–1996) Armenian astrophysicist whom Chandra met during a visit to Leningrad in 1934.

Anderson, Carl David (1905–1991) American physicist who worked under Robert Millikan at Caltech studying cosmic rays. In 1932 Anderson discovered the positron, the antiparticle of the electron, which Dirac had originally predicted in his theory.

Anderson, Wilhelm Robert Karl (1880–1940) Russian-born physicist of German extraction who realized that a full understanding of white dwarf stars required the theory of relativity.

Artin, Emil (1898–1962) Viennese-born mathematician whose specialty was the theory of numbers.

Aston, Francis William (1877–1945) English physicist who was awarded the 1922 Nobel Prize in chemistry for his precise measurements of the masses of isotopes.

Atkinson, Robert d'Escourt (1893–1981) Welsh-born astrophysicist who, with Fritz Houtermans, worked on early theories of thermonuclear reactions in stars, following a suggestion by George Gamow.

Ayyar, Chandrasekhara Subramanya (1885–1960) Chandra's father, often referred to affectionately by his family as Babuji.

Baade, Walter (1893–1960) German-born astronomer who, with Fritz Zwicky, came up with the concept of the neutron star in 1933.

Balakrishnan (1914–1998) One of Chandra's younger brothers, to whom Chandra was particularly close. He became an outstanding pediatrician and author, whose writings in both Tamil and English were read throughout southern India.

Balakrishna, Sitalakshmi (1891–1931) Chandra's mother.

Bethe, Hans Albrecht (b. 1906) German-born American physicist who discovered the thermonuclear reactions that produce energy in stars. He played an important role in the development of the atomic and hydrogen bombs and later returned to astrophysics.

Bohr, Niels Henrik David (1885–1962) Danish physicist who formulated the first quantum theory of the atom.

Born, Max (1882–1970) German physicist who coined the term "quantum mechanics" and did important work in that field.

Chandrasekhar, Lalitha (b. 1910) Chandra's wife and soulmate for fifty-nine years.

Chandrasekhar, Ramanathan (1837–1906) Chandra's grandfather, a professor of mathematics and author of several mathematics books.

Chandrasekhar, Subrahmanyan ("Chandra") (1910–1995) Indian-born American astrophysicist who discovered the maximum mass for a stable white dwarf, the Chandrasekhar limit.

Chowla, Sarvadaman (1907–1995) Indian mathematician who was Chandra's close friend at Trinity College, Cambridge, during 1930 and 1931.

Colgate, Stirling (b. 1925) American physicist who formulated the first realistic models of supernovae, based on his work developing hydrogen bombs. He is still working at Los Alamos.

Compton, Arthur Holly (1892–1962) American physicist who discovered how x-rays change their wavelength when they are deflected by electrons, known as the Compton effect, which provided proof of the particle nature of radiation.

Cowling, Thomas George (1906–1990) English astrophysicist who was Milne's student and a friend of Chandra's.

Critchfield, Charles Louis (1910–1994) American nuclear physicist whose early results on how stars such as the Sun shine influenced Bethe. After working on the Manhattan Project, Critchfield taught intermittently in universities but could not resist the lure of Los Alamos and eventually moved there.

Cronin, James (b. 1931) Nobel prize–winning American physicist who was a close friend of Chandra's at the University of Chicago.

Dingle, Herbert (1890–1978) English physicist, editor of the *Monthly Notices of the Royal Astronomical Society*, with whom Milne corresponded at the height of his dispute with Eddington. Dingle is best remembered for his controversial criticism of Einstein's special theory of relativity.

Dirac, Paul Adrien Maurice (1902–1984) English physicist celebrated for his eccentricities and best known for the Dirac equation, which accurately describes electrons. At Trinity College, Cambridge, Chandra idolized him and often sought his advice.

Douglas, Alice ("Allie") Vibert (1894–1988) Canadian astronomer who studied with Eddington in 1921 and was later his biographer. Vibert completed her Ph.D. in astrophysics at McGill University, making her one of the first women to be awarded a Ph.D. in science in North

America. She was professor of astronomy at Queens University from 1946 to 1964 and was made a Member of the Order of the British Empire.

Eddington, Arthur Stanley (1882–1944) The world's greatest astrophysicist of his day, and Chandra's archrival and nemesis.

Eddington, Winifred (1878–1954) Eddington's sister.

Einstein, Albert (1879–1955) Indisputably the greatest physicist of the twentieth century. Einstein discovered, among much else, the special and general theories of relativity and the iconic equation $E = mc^2$. Yet he refused to accept the most spectacular consequence of his life's work: black holes.

Fermi, Enrico (1901–1954) Italian-born American physicist whose many discoveries included Fermi-Dirac statistics. He was a close colleague and friend of Chandra's at the University of Chicago.

Feynman, Richard Phillips (1918–1988) American physicist legendary for his antics and brilliance, who worked on the Manhattan Project.

Finlay-Freundlich, Erwin (1885–1964) German astrophysicist, a colleague of Einstein's at the University of Berlin. He was director of the Astrophysical Observatory in Potsdam and editor of the *Zeitschrift für Astrophysik* when Chandra visited in 1932.

Fowler, Ralph Howard (1889–1944) English physicist who first applied quantum mechanics to astrophysics. Dirac and Chandra were among his most distinguished Ph.D. students.

Fowler, William Alfred (1911–1995) American astrophysicist who studied how supernovae produced the chemical elements in the universe.

Fuchs, Klaus (1911–1988) German-born nuclear physicist who fled Germany for England in 1933. A Communist, he worked on the British atomic bomb project and then the Manhattan Project at Los Alamos while passing key information on atomic and hydrogen bomb development to the Soviets. He was arrested in 1950, when he was the head of the theoretical physics division at the Atomic Energy Establishment at Harwell, England.

Gamow, George (1904–1968) Flamboyant Russian-born American nuclear physicist who initiated modern research on the thermonuclear processes that power stars.

Ganesan, Angarai Seshia (1900–86) The husband of Chandra's sister Rajam. Ganesan completed his Ph.D. in physics in 1926 at Imperial College, London, and went on to do research with C. V. Raman, leading to the discovery of the Raman effect.

Greenstein, Jesse Leonard (1909–2002) American astrophysicist who was a colleague and friend of Chandra's at the University of Chicago before moving to Caltech, where he did important work on quasars.

Hardy, Godfrey Harold (1877–1947) English mathematician and colleague of Chandra's at Trinity College, Cambridge. He was responsible for discovering the enigmatic Indian genius Srinivasa Ramanujan, bringing him to Cambridge, and nurturing his career.

Hawking, Stephen William (b. 1942) The world's best-known astrophysicist today, Hawking has played a key role in exploring the properties of black holes. Among his many major advances is his discovery that black holes can evaporate by emitting radiation, known as Hawking radiation.

Heisenberg, Werner Karl (1901–1976) German physicist whose many major discoveries included quantum mechanics and the uncertainty principle. He visited Madras when Chandra was a student there and greatly encouraged him.

Hill, Archibald Vivian (1886–1977) English biologist who won a Nobel Prize in 1922 for his adroit use of physics to discover how muscles function. He played an important role in defense research during World War I and influenced Fowler and Milne.

Houtermans, Friedrich ("Fritz") Georg (1903–1966) German physicist who did pioneering work with Gamow and Atkinson in the late 1920s on the thermonuclear reactions that power stars.

Hoyle, Fred (1915–2001) English astrophysicist who advanced our knowledge of how heavier elements are generated from hydrogen and helium.

Hubble, Edwin Powell (1889–1953) American astronomer who discovered that galaxies are receding at speeds related to their distances from Earth.

Hutchins, Robert Maynard (1899–1977) As president of the University of Chicago he played a key role in hiring Chandra and keeping him there.

Jeans, James Hopwood (1877–1946) English astrophysicist, also famous as a mathematician, theoretical physicist and popularizer of science. He was an archrival of Eddington's.

Kerr, Roy (b. 1934) New Zealand physicist who in 1963 discovered a solution to Einstein's general theory of relativity that turned out to describe the properties of rotating black holes. At the time, he later recalled, he did not realize the significance of his result.

Kuiper, Gerard Peter (1905–1973) Dutch-born American astronomer whose early observations of white dwarfs supported Chandra's theory. The two were colleagues at Yerkes Observatory.

Landau, Lev Davidovich (1908–1968) Russian physicist who discovered the maximum mass of a stable white dwarf in 1932, independently of Chandra. The two met in the Soviet Union in 1934.

Lee, Tsung-Dao (b. 1926) Chinese-born American physicist who carried out research on the constitution of white dwarfs. He discussed the subject with Chandra when he worked at Yerkes Observatory.

Lewis, Gilbert Newton (1875–1946) American physical chemist who coined the term "photon" for the smallest particle of light.

Littlewood, John Edensor (1885–1977) English mathematician who, along with G .H. Hardy, made important advances in the theory of numbers. He spent his entire academic career at Cambridge and lived in Neville's Court, Trinity College, for fifty-one years.

McCrea, William Hunter (1904–1999) English astrophysicist, a friend of Chandra's in the 1930s.

Marshak, Robert Eugene (1916–1992) American physicist whose Ph.D. thesis included Bethe's new theory of the thermonuclear processes that make stars shine.

Millikan, Robert Andrews (1868–1953) American physicist known for his talent as a researcher and as a creator of physics departments at the University of Chicago and Caltech.

Milne, Edward Arthur (1896–1950) English astrophysicist whose theory of the constitution of stars clashed with Eddington's and then with Chandra's

Møller, Christian (1904–1980) Danish physicist who wrote a paper with Chandra in support of his theory of white dwarfs and against Eddington's.

Oppenheimer, Julius Robert (1904–1967) American physicist who co-authored critical papers on neutron star formation and collapsing stars. He directed the Manhattan Project.

Osterbrock, Donald E. (b. 1924) American astrophysicist, a student of Chandra's, who made important contributions to our understanding of nebulae and quasars. He is also an eminent historian of astrophysics.

Pauli, Wolfgang (1900–1958) Famously ironic Viennese physicist who discovered the Pauli exclusion principle and postulated the existence of the neutrino.

Payne-Gaposchkin, Cecilia Helena (1900–1979) A student of Eddington's at Trinity College, Cambridge, she obtained a Ph.D. in astrophysics at Radcliffe College and went on to become the first female professor at Harvard.

Peierls, Rudolf Ernest (1907–1995) German-born English physicist who wrote papers in support of Chandra's theory of white dwarfs.

Penrose, Roger (b. 1931) English mathematician who applied topology to black holes.

Plaskett, Harry Hemley (1893–1980) Canadian-born pioneer in observational astrophysics and a close friend of Milne's.

Pryce, Maurice Henry Lecorney (1913–2003) English physicist who, along with Dirac and Peierls, wrote a paper in support of Chandra's theory of white dwarfs.

Raman, Chandrasekhara Venkata (1888–1970) India's most illustrious physicist, who had a rocky relationship with his famous nephew Chandra.

Ramanujan, Srinivasa (1887–1920) Indian mathematician whose brilliance in number theory and Cinderella-like career were an inspiration to Chandra's generation.

Rees, Martin John (b. 1942) English astrophysicist and Astronomer Royal, who made essential contributions to the study of the formation of black holes.

Robertson, Howard Percy (1903–1961) American researcher on the general theory of relativity who frequently discussed his work with Einstein, his colleague at Princeton. At Caltech he gave valuable advice to Oppenheimer and Snyder in their research on collapsing stars.

Rosenfeld, Léon (1904–1974) Belgian physicist who was a confidant of Chandra's during the 1930s.

Russell, Henry Norris (1877–1957) Dean of American astronomers, a powerhouse in American science during the first half of the twentieth century.

Rutherford, Ernest (1871–1937) New Zealand–born English physicist who made many important discoveries relating to radioactivity and in atomic and nuclear physics. He was a major force at the Cavendish Laboratory in Cambridge.

Saha, Meghnad (1894–1956) Indian astrophysicist who used Bohr's atomic theory to relate a star's spectral lines to the chemical constitution of its outer layers.

Sakharov, Andrei Dmitrievich (1921–1989) Russian physicist who did significant work on the Soviet hydrogen bomb along with Zel'dovich.

Schmidt, Maarten (b. 1929) Dutch-born American astronomer who deciphered the puzzle of quasars.

Schrödinger, Erwin (1887–1961) Viennese-born physicist whose wide and diligently pursued interests included biology, literature, and women. He is best known for the equation and paradox named after him.

Schwarzschild, Karl (1873–1916) German astronomer and physicist who provided the first exact solution to the equations of Einstein's theory of general relativity in 1916. What the scientific community long considered to be a flaw in his solution turned out to be a black hole.

Scott, David Randolph (b. 1932) American astronaut who flew three space missions, the last one as commander of *Apollo 15,* which landed on the Moon and was the first expedition into outer space with astronauts specifically trained for scientific exploration. He writes and broadcasts on space travel, and advised on the film *Apollo 13.*

Serber, Robert (1909–1997) American nuclear physicist, coauthor with Oppenheimer of an early paper on neutron cores inside stars. He went on to play an important role in the Manhattan Project.

Shapley, Harlow (1885–1972) Director of the Harvard Observatory who offered Chandra a research position there.

Snyder, Hartland (1913–1962) American physicist who, with Oppenheimer, wrote the first paper to explore what happens when a collapsing star becomes smaller than its Schwarzschild radius. During World

War II he worked on the development of radar and in the 1950s made valuable contributions to the design of the most powerful particle accelerator of its day, the Cosmotron at the Brookhaven National Laboratory in New York State.

Sommerfeld, Arnold Johannes Wilhelm (1868–1951) German physicist renowned for his research and teaching.

Steigman, Gary (b. 1941) American astrophysicist at Ohio State University who has made important contributions to our understanding of the early universe.

Stoner, Edmund Clifton (1899–1968) English physicist who, in 1924, came close to discovering the Pauli exclusion principle. He then took a break from atomic physics and looked into the nature of white dwarf stars. He and Chandra arrived at similar results simultaneously and independently.

Stratton, Frederick John Marrian (1881–1960) English astronomer known for his work in spectroscopy. He spent his professional career at Cambridge University and played an active role in the Royal Astronomical Society, of which he was president from 1933 to 1935.

Strömgren, Bengt Georg Daniel (1908–1987) Danish astrophysicist who was a friend of Chandra's in the 1930s and his colleague at the University of Chicago. In 1957 he became the first professor of astrophysics at the Institute for Advanced Study in Princeton. He returned to Copenhagen in 1967 to assume the prestigious honorary position first held by Niels Bohr.

Strutt, John William (Lord Rayleigh) (1842–1919) English scientist who researched in both experimental and theoretical physics. Chandra admired his mathematical rigor.

Struve, Otto Ludwig (1897–1963) Russian-born American astronomer who put Yerkes Observatory on the map by hiring dynamic young researchers such as Chandra.

Teller, Edward (1908–2003) Hungarian born American physicist whom Chandra met in the 1930s. They became colleagues at the University of Chicago but fell out over politics.

Thomson, William (Lord Kelvin) (1824–1907) Irish-born Victorian-era scientist whose research spanned electricity, magnetism, hydrodynamics, mathematical physics, geology, and the theory of heat (thermodynamics). Many of his discoveries were highly practical as well as deeply theoretical. He played a key role in establishing the first successful transatlantic telegraph cable. Thomson formulated a temperature scale relating to the laws of thermodynamics, known as the kelvin scale.

Thorne, Kip Stephen (b. 1940) American physicist known for his work on black holes, gravitational wave detection, and time machines. Chandra spent part of 1971 with his group at Caltech.

Trimble, Charles John Agnew (1883–1958) Classmate of Eddington's at Trinity College, Cambridge, and his lifelong friend and companion.

Trimble taught mathematics, and his students remembered him as "a great trainer of minds."

Ulam, Stanislaw (1909–1984) Polish-born American mathematician who contributed important ideas to the development of the hydrogen bomb.

Volkoff, George Michael (1914–2000) Canadian physicist best known for the paper he wrote with Oppenheimer on neutron stars.

Von Neumann, John (1903–1957) Hungarian-born American polymath who discussed astrophysics with Chandra at Trinity College, Cambridge, during 1934 and 1935.

Waerden, Bartel Leendert van der (1903–1996) Dutch scientist who did important work in mathematics, physics, and the history of mathematics. In 1931 he became a colleague of Heisenberg's at the University of Leipzig.

Wali, Kameshwar C. (b. 1935) Indian-born American theoretical physicist who wrote the first biography of Chandra.

Wentzel, Gregor (1898–1978) German-born physicist who was a close friend of Chandra's at the University of Chicago.

Wheeler, John Archibald (b. 1911) American physicist who made important contributions to nuclear physics, the development of the atomic and hydrogen bombs, and general relativity, particularly black hole physics.

White, Richard Harold (b. 1934) American physicist at Livermore National Laboratory whose expertise with computers was essential to early research into supernovae with Stirling Colgate.

Whittaker, Edmund Taylor (1873–1956) English physicist who arranged for the posthumous publication of Eddington's book on his fundamental theory.

Woosley, Stanford ("Stan") Earl (b. 1945) American astrophysicist who studies how massive stars evolve from placid beginnings to violent deaths as supernovae. He carries out his extensive research at the University of California at Santa Cruz.

Yang, Chen Ning (b. 1922) Chinese-born American physicist who befriended Chandra as a young instructor in the physics department at the University of Chicago.

Zel'dovich, Yakov Borisovich (1914–1987) Russian scientist who did critical work in the Soviet hydrogen bomb program and made important contributions to astrophysics.

Zwicky, Fritz (1888–1974) Quirky Swiss-born American astronomer. One of his flashes of brilliance was to link supernovae with neutron star formation. He published this work with Walter Baade.

Glossary

accretion disk A broad, flat disk of gas rotating around a black hole. The inner parts of the disk rotate faster than the outer ones, and as the inner particles rub against their slower neighbors they heat up from the friction and give off x-rays. The disk is continually being dragged into the black hole like water down a drain, from the inside outward.

alpha particle The nucleus of a helium atom, made up of two neutrons and two protons.

annihilation The process by which a particle and an antiparticle come together and disappear in a blaze of light.

annihilation hypothesis The idea that the energy that powers stars is generated by the mutual annihilation of protons and electrons. This was proposed before the discovery of antimatter.

antimatter Matter made up of antiparticles.

antiparticle A particle of the same mass as its counterpart but having the opposite charge. For example, the positron is the antiparticle of the electron.

astronomer A scientist who uses telescopes to study cosmic objects.

astrophysicist A scientist who uses mathematical equations representing laws of nature to analyze and understand measurements made by astronomers.

atom The smallest part of an element that can characterize that element. An atom is made up of a nucleus with a positive charge and enough electrons outside the nucleus to give the whole a net electrical charge of zero.

atomic bomb (fission bomb) A bomb whose explosive energy comes from a chain reaction of nuclear fissions involving uranium or plutonium nuclei.

beta decay The transformation of a neutron into a proton, electron, and neutrino.

binary system Two objects that rotate around a common center, such as two stars, a star and a black hole, a star and a neutron star, a star and a white dwarf, two neutron stars, or two black holes.

black hole The well in spacetime into which a collapsing star falls when it becomes extremely dense and its gravity is so enormous that it pulls space and time in around itself. The gravity inside a black hole is so strong that not even light can escape from it.

Bohr's theory of the atom The earliest successful theory of the atom based on quantum theory as it was known before the discovery of quantum mechanics. It is based on a visualization of the atom as a minuscule solar system, made up of a central nucleus surrounded by electrons in specified orbits.

Bose-Einstein statistics A theory that calculates the probabilities for how values of position, velocity, and spin are distributed among elementary particles whose spin values can be only whole numbers, including zero; photons are an example of particles that obey Bose-Einstein statistics.

centrally condensed configurations Milne's term for all stars except white dwarfs.

chain reaction A sequence of nuclear fissions in which neutrons are produced and go on to trigger future fissions, releasing further neutrons, and so on.

Chandrasekhar limit The maximum mass of a stable white dwarf, about 1.4 times the mass of the Sun.

Chandrasekhar-Schönberg limit The maximum mass that a star's helium core can attain without collapsing during the process of hydrogen burning.

charge of the electron The electrical charge carried by a single electron. Until quarks were postulated in the 1960s, this was considered the basic unit of electrical charge. But quarks can have $1/3$ or $2/3$ of the charge of the electron.

classical physics (Newtonian physics) The physics of matter moving at speeds much lower than the speed of light, to which the special theory of relativity need not be applied.

classical statistics The manner in which values of position and velocity are distributed among particles in a perfect gas.

cold degenerate matter Highly compressed matter no longer radiating light or heat.

collapsed configurations Milne's term for white dwarfs. To avoid concluding that white dwarfs collapse completely, Milne made the ad hoc hypothesis that since all gas laws may be invalid deep in the interior of a white dwarf, it must have a core made up of some material that cannot be further compressed under any circumstances.

core bounce A recoil that occurs when the core of a massive collapsing star reaches its limit of compression. At this point the core stiffens and recoils like a spring.

cosmic rays High-energy particles raining down on Earth from outer space. They were discovered in 1911 and named by Robert Millikan in 1925.

cosmological redshift The lengthening of wavelengths of light from stars, caused by the expansion of the universe.

critical mass The minimum amount of uranium or plutonium needed to make an atomic bomb explode.

degeneracy The limit at which densely packed electrons cannot be compressed any further. They retaliate by generating an outward degeneracy pressure and becoming rock hard.

degeneracy pressure The pressure exerted by the electrons in a gas in a condition of degeneracy. Degeneracy pressure is what makes metals hard and maintains the stability of white dwarf stars whose mass is below the Chandrasekhar limit.

degenerate core The rock-hard core that results from the degeneracy pressure of densely packed electrons.

density, average The mass of an object divided by its volume.

density of particles The number of particles divided by the volume of the enclosure that contains them.

deuterium An isotope of hydrogen whose atoms consist of a single electron and a nucleus with a proton and a neutron. Deuterium, also known as "heavy hydrogen," has chemical properties identical to those of ordinary hydrogen.

deuteron A nucleus of deuterium.

dimensionless number A quantity that describes a physical system or process and is a pure number, with no associated unit. An example is the fine-structure constant.

Dirac equation The mathematical equation that describes the properties of an electron.

Doppler shift, optical The change in the wavelength of radiation caused by the relative motion of the source and an observer. When a star moves away from Earth, its light appears to have longer wavelengths; that is, it is redshifted. This is important for measuring the speed of a star relative to Earth.

dwarf star A star that, like the Sun, lies on the main sequence of the Hertzsprung-Russell diagram.

Eddington's paradox Although white dwarf stars continually lose heat and contract, they lack enough energy to cool down and expand to the density of terrestrial matter.

electromagnetic force The force that governs electricity and magnetism.

electromagnetic spectrum The gamut of wavelengths and frequencies of light, or radiation, extending from the most energetic (short-wavelength gamma rays) to the weakest (long-wavelength radio waves).

electron An elementary particle with one unit of negative electric charge and a spin of ½.

electroweak theory A unified theory of the electromagnetic force and the weak nuclear force.

element A substance that cannot be made into simpler substances by chemical means. Atoms with the same number of protons or electrons are atoms of the same element. Each element has a place in the periodic table. An atom's electrons determine its chemical properties, that is, how it combines with other atoms.

elementary particle A subatomic particle, such as an electron, a proton, a neutron, a neutrino, or a quark.

E-numbers Eddington's version of Dirac's spinors.

equation of state An equation that relates pressure to density.

event horizon The rim of a black hole. Anything falling past this horizon cannot escape.

Fermi-Dirac statistics A theory that calculates the probabilities for how values of position, velocity, and spin are distributed among electrons in a quantum gas or an atom.

fine-structure constant A pure number of great importance in atomic physics, related to the distance between closely spaced spectral lines, that is also a measure of the strength with which electrons interact with one another. It incorporates the Planck constant, the speed of light, and the charge of the electron.

fission bomb See atomic bomb.

fundamental theory Eddington's name for the theory he was working on that would unify the general and special theories of relativity with quantum theory.

fusion bomb See hydrogen bomb.

gas laws Mathematical descriptions of the properties of collections of particles. The applicable gas law depends on the density of particles in the gas. For very high densities, quantum mechanics must be taken into account.

gas pressure The net outward pressure exerted by the gas particles within a star as calculated from the gas law pertaining to that star, which may behave as a perfect gas or as a quantum gas.

general theory of relativity A widening of Einstein's special theory of relativity to include gravitational phenomena. It is essential to the understanding of black holes and neutron stars.

giant-dwarf theory of stars Russell's idea that dwarf stars do not obey the perfect gas law because of their high density, while low-density giants do. Eddington disproved this for stars on the main sequence of the Hertzsprung-Russell diagram.

giant star A very large, bright star lying above the main sequence of the Hertzsprung-Russell diagram.

gravitational pressure The net inward pressure caused by the gravitational attraction of the molecules of gas that make up a star.

gravitational redshift The shift of light toward longer wavelengths caused by increasingly strong gravitational fields.

gravitational waves Ripples in spacetime caused most dramatically by movements of objects with extremely intense gravitational fields such as black holes and neutron stars.

group theory The branch of mathematics that studies the symmetry properties of equations.

half-life The time it takes for half of a sample of radioactive nuclei to decay. The half-life of a nucleus is a property of the nucleus itself and does not depend on how many nuclei are packed together or on the temperature of the sample.

Hawking radiation Radiation emitted by a black hole through processes related to quantum physics.

helium The second most abundant element in the universe, with a nucleus containing two protons and two neutrons, surrounded by two electrons.

Hertzsprung-Russell (H-R) diagram A graph that plots the brightness of stars against their temperature.

Hubble constant The rate at which the speed of recession of a galaxy increases with its distance from Earth, following Hubble's law.

Hubble's law Galaxies recede from us at a speed proportional to their distance from Earth.

hydrogen The simplest and most abundant chemical element in the universe, with a single electron and a nucleus that is a single proton.

hydrogen bomb (fusion bomb, nuclear bomb) A bomb whose explosive energy comes from nuclear fusion, that is, by the combining of light nuclei such as hydrogen and deuterium. An atomic bomb supplies the heat necessary to trigger this thermonuclear reaction.

isotopes Atoms of the same chemical element that differ in the number of neutrons in their nuclei. Examples are deuterium and tritium, which are isotopes of hydrogen, and ^{238}U and ^{235}U, both isotopes of uranium. Since the nuclei of different isotopes have the same number of protons, the isotopes are virtually indistinguishable in chemical reactions.

jet A beam of fast-moving gas emitted from the vicinity of a black hole, for example.

Kerr black hole A spinning black hole defined only by its spin and mass.

law of conservation of energy The energy after a process occurs must be the same as the energy beforehand.

LIGO Laser Interferometer Gravitational Wave Observatory (LIGO), a device for detecting gravitational waves.

LISA Laser Interferometer Space Antenna, an orbiting detector for gravitational waves planned for launch in 2011.

main sequence The diagonal line on the Hertzsprung-Russell diagram on which most stars appear. These stars are at the most robust stage of their life.

Manhattan Project The massive U.S. effort during World War II to build an atomic bomb.

mass A measure of the amount of matter a body contains. A body's mass is the same wherever it is in the universe, whereas its weight depends on its gravitational environment. Since the masses of stars and other celestial bodies are so great, they are often expressed as multiples of the mass of the Sun.

mass-energy equivalence The equivalence of mass to energy and vice versa, as expressed in the iconic equation $E = mc^2$.

mass-luminosity relation The relationship between the mass and luminosity of stars on the main sequence of the Hertzsprung-Russell diagram.

mean molecular weight The average number of free electrons per nucleus after atoms are stripped of their electrons by the high-energy radiation inside stars. The exact figure depends on the mix of chemical elements in the star.

metric A measure of distance in spacetime.

neutrino An elementary particle with zero electrical charge. It was originally hypothesized to have no mass; today we know that it has a minute amount.

neutrino opacity The resistance of a star to the escape of neutrinos from its interior.

neutron An elementary particle with no electric charge and a mass about that of the proton.

neutron star A highly compact, dense star made up almost entirely of neutrons, with a mass about that of the Sun but only about 12 miles in diameter.

Newtonian physics See classical physics.

nonrelativistic Describing events in which the speeds of objects are well below that of light, so that the special theory of relativity does not need to be applied.

non-relativistic degeneracy A degenerate gas of electrons in which the pressure is too low for the electrons to move at speeds close to that of light.

nonrelativistic degeneracy pressure The pressure in a degenerate gas of electrons in which the electrons are moving at speeds much lower than that of light.

nova A star that undergoes a sudden increase in brightness to about a thousand times that of the Sun.

nuclear bomb See hydrogen bomb.

nuclear fission The splitting of a nucleus. A neutron hitting a uranium nucleus can split it, releasing an enormous amount of energy, as predicted by the mass-energy equivalence. Atomic bombs and nuclear reactors are powered in this way.

nuclear force The attractive force between the particles that make up the nucleus, such as neutrons and protons. Unlike the electric force, in which oppositely charged bodies attract, in the nuclear force there is attraction between a neutral neutron and a positively charged proton.

nuclear fusion The binding of light nuclei such as protons, deuterons, and alpha particles under the high temperatures and densities found in the interior of stars. These thermonuclear reactions generate the energy that makes stars shine and hydrogen bombs explode.

nucleus The massive central part of the atom, which has a positive charge. The nucleus contains protons and (except in the case of hydrogen) neutrons.

opacity A measure of the resistance that light encounters as it tries to escape from matter.

Pauli exclusion principle No two electrons in an atom or a dense gas can simultaneously have the same position, velocity, and spin.

perfect gas A hypothetical collection of particles whose density is low enough for its pressure, volume, and temperature to be predicted by the perfect gas law.

perfect gas law The equation of state for how the particles in a perfect gas behave. If the temperature of a perfect gas is held constant, an increase in pressure results in a decrease in volume, and vice versa; if pressure is held constant, an increase in volume results in an increase in temperature, and vice versa.

periodic table A table of the elements ordered by increasing number of electrons, an arrangement that allows comparison of their chemical properties.

photon (light quantum) The particle of light, corresponding to an electromagnetic wave, according to wave-particle duality.

Planck constant A universal constant of nature that is at the basis of quantum theory.

planetary nebula A luminous shell of gas ejected from a dying red giant star. What remains at the center is a white dwarf.

positron The antiparticle of the electron. Electrons and positrons differ only in that the positron has a positive charge.

proton An elementary particle with a positive electric charge. It is the nucleus of the hydrogen atom.

pulsar A rotating neutron star that emits radio signals at precise intervals.

quantum electrodynamics A theory that combines quantum mechanics and special relativity to study the interaction of light with electrons.

quantum gas A gas that has to be studied using quantum physics because the density of its electrons is high enough that their wave-particle duality has to be taken into account.

quantum mechanics The theory that describes how elementary particles such as electrons and protons interact.

quantum physics The study of light and matter under conditions in which phenomena such as wave-particle duality become important.

quantum statistics The manner in which values of position and velocity are distributed among particles in a quantum gas.

quarks Elementary particles currently considered to be the fundamental building blocks of matter; neutrons and protons are made up of quarks.

quasar The bright center of a young, distant galaxy, powered by a supermassive black hole.

radiation Energy traveling in the form of photons.

radiation pressure The outward pressure exerted by the light produced within a star.

radiative transfer The study of how radiation moves through matter.

radioactive dating A way to measure the time that has elapsed since a rock was last molten, and thus to ascertain the age of Earth, by estimating the original amount of uranium in the rock. Since the half-life of a radioactive substance is not affected by temperature or pressure, a sample of the substance will decay at a predictable rate.

radio galaxy A galaxy that emits strongly at radio wavelengths.

radio lobes Huge, radio-wave-emitting clouds of gas stirred up by jets of high-velocity electrons emitted by a quasar.

Raman effect A means to explore the structure of molecules by investigating the frequency of the light they scatter by the absorption and re-emission of photons.

red giant A bloated star ten to a hundred times the diameter of the Sun and in the twilight of its life.

redshift The shift of spectral lines to redder (longer wavelengths) in the spectra of astronomical objects apparently moving away from the Earth, such as quasars or galaxies.

relativistic Relating to objects moving at speeds close to the speed of light, as described in the special theory of relativity.

relativistic degeneracy The limit of compression for a gas of electrons of sufficiently high density that they are moving at speeds close to that of light.

relativistic degeneracy pressure The way in which a densely packed gas of electrons traveling at speeds close to that of light retaliates to being squeezed. The electrons generate an outward pressure that makes them rock-hard, which explains the stability of white dwarf stars whose mass is below the Chandrasekhar limit. (*See also* relativistic degeneracy.)

relativity See general theory of relativity, special theory of relativity.

Saha's equation An equation, based on the quantum theory of the atom, that relates the spectral lines from the light of a star to the elements that make up the star's surface layers.

Schwarzschild black hole A black hole that is not spinning.

Schwarzschild radius The size of a black hole, akin to its event horizon.

(The Schwarzschild radius is a distance, the event horizon is a surface.) Every object has a Schwarzschild radius.

Schwarzschild singularity At the Schwarzschild radius, Schwarzschild's solution to Einstein's general theory of relativity yields a singularity. The Schwarzschild radius is the size of the opening in spacetime (the event horizon) into which a collapsing star falls on the way to becoming a singularity. In other words, the Schwarzschild radius (event horizon) is not a singularity but a black hole. The star within the black hole becomes a singularity when it collapses to a point of infinite density and zero volume.

shock front The front of a shock wave, a narrow band of matter denser than the surrounding medium. A shock front can result from a violent pressure change, such as that caused by an explosion.

shock wave A wave made up of particles traveling at a speed greater than the speed of sound in the gas through which the wave is traveling. The sonic boom of a supersonic aircraft is produced by a shock wave.

singularity The point in a physical theory at which a quantity becomes infinite.

spacetime The three dimensions of space and one of time considered as a single four-dimensional entity.

special theory of relativity The theory used to study matter moving at speeds comparable to that of light.

spectral lines Lines, each corresponding to a certain frequency, that emerge when light of different frequencies is passed through a spectrograph.

spectrograph A device used to break up light into its component frequencies.

speed of light According to Einstein's special theory of relativity, the speed of light is constant no matter what the motion of the source of the light is; in empty space it is 186,000 miles per second. Thus the speed of light is the cosmic speed limit, and time is a relative quantity. Since light takes time to travel across space, we observe the universe as it was in the past; the farther away we look, the more distant the time we see.

spin A property of elementary particles that emerged from the analysis of spectral lines. Its mathematical representation has a form analogous to what one would expect for a spinning top; hence the term "spin."

spinors Mathematical quantities, intermediate between vectors and tensors, used in the Dirac equation.

standard model Eddington's theory of the constitution of the stars.

Stoner-Anderson equation Equation of state for a relativistically degenerate gas very similar to that deduced by Chandra. Chandra's later theory of white dwarfs was more detailed and complete than Stoner's.

supergiant A rare type of star, typically a thousand times the diameter of the Sun and more than ten times the diameter of a red giant.

supermassive black hole A black hole millions of times the mass of the Sun.

supernova The explosion of a massive star powered by the gravitational energy released by the collapse of the star's core, a billion times as luminous as the Sun.

supernova relic The neutron star left behind after a supernova explosion.

supernova remnant The expelled remains of a supernova, visible as a small nebula.

thermonuclear reaction Nuclear fusion reaction that occurs in the dense interior of a star and that also provides the power for a hydrogen bomb.

topology The branch of mathematics that studies the properties of surfaces when they are distorted.

tritium An isotope of hydrogen, each of whose atoms consists of a single electron and a nucleus with a proton and two neutrons. It is an even heavier hydrogen isotope than deuterium.

Type Ia supernova The explosion of a white dwarf in a binary system that has accreted enough mass from its companion to exceed the Chandrasekhar limit.

Type II supernova The death throes of an old, massive star undergoing gravitational collapse.

uncertainty principle In quantum theory, the principle that the position and velocity of an elementary particle cannot be measured simultaneously with the same accuracy; the more accurately the particle's position is measured, the less accurately its velocity is known.

universal constant of nature One of the numbers at the foundation of scientific theories, including the charge of the electron, the Planck constant, and the speed of light.

wave-particle duality The ability of an elementary particle to behave as either a wave or a particle.

weak nuclear force The force that governs radioactive processes such as beta decay.

white dwarf An old star that has burned up all of its nuclear fuel and contracted to about the size of Earth, with a mass less than the Chandrasekhar limit.

Index